ENVIRONMENTAL MICROBIOLOGY

Environmental Microbiology

P D Sharma

Alpha Science International Ltd.
Harrow, U.K.

P D Sharma
Formerly, Professor of Botany
University of Delhi
Delhi, India

Alpha Science International Ltd.
Hygeia Building, 66 College Road
Harrow, Middlesex HA1 1BE, U.K.

ISBN 1-84265-276-1

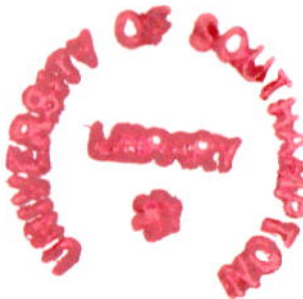

Printed in India

PREFACE

As biotic component of the ecosystem, the role of microorganisms expanded during recent past, and interest of microbiologists shifted to discovery and application of new microbes and their products to environment, human health and welfare. Several textbooks have been written on microbial ecology, which is essentially the study of interactions between microorganisms themselves as well as between them and their environments. However, there have been only few attempts to include the subject matter on expanding applications of microbial activities to environment in the form of a complete textbook on environmental microbiology. Activities of microorganisms can be regulated and manipulated to benefit humans. Such applications can impact humans directly or indirectly through effects on animals, plants or ecosystem health in general. New or tailored strains of microbes have been discovered and successfully used in maintenance of the quality of environment through minimizing the use of conventional chemicals in municipal waste disposal, public health and agriculture.

Unlike any traditional textbook of microbial ecology, the subject matter of the present textbook includes sequentially arranged areas accommodated in 20 chapters. These subject areas presented in the text include: (i) introductory chapter on the discipline followed by chapters on diversity of microbes and the environments in which they grow under natural conditions; (ii) chapters on devices for sampling, processing, enumeration and activity of microorganisms; (iii) chapters on impact of microbial activity on environment in terms of element cycling, fate of organic and metal pollutants, and role of microorganisms in bioremediation of polluted sites of the environment; (iv) chapter on role of microbes in agriculture in terms of nitrogen fixation and biological control of pathogens, pests and weeds; (v) chapters on environmental transmission and control of pathogens, including domestic waste treatment; and (vi) chapter on laboratory experiments designed to detect and quantify the microorganisms, and understand their activites in different kinds of environments.

The present textbook is designed primarily to fulfill the requirements of undergraduate and postgraduate students in the disciplines of environmental microbiology, microbiology, microbial ecology, biotechnology and botany. The book may also serve as a reference for scientists, researchers or professionals interested in exploring the role of microorganisms in public health, waste disposal and bioremediation. The overall objectives of the text are to identify and highlight the microorganisms involved in environmental microbiology, the nature of different environments where they grow, the methodologies used to monitor the microbes and their activities, and to evaluate the success achieved in the applications of microbes in solving environmental problems through regulation and manipulation of their activities. Suitable strains of microorganisms have been successfully used for treatment and disposal of municipal and other wastes, bioremediation of polluted sites and biological control of pathogens, pests and weeds. Available information on these and related environmental issues has been included in the text of appropriate chapters of the book.

I am thankful to Mr. N. K. Mehra for undertaking the task of publication of the book and Mr. S. Mehra for technical assistance in reproduction of Figs. 7.6 and 7.7.

P D Sharma

CONTENTS

1

INTRODUCTION

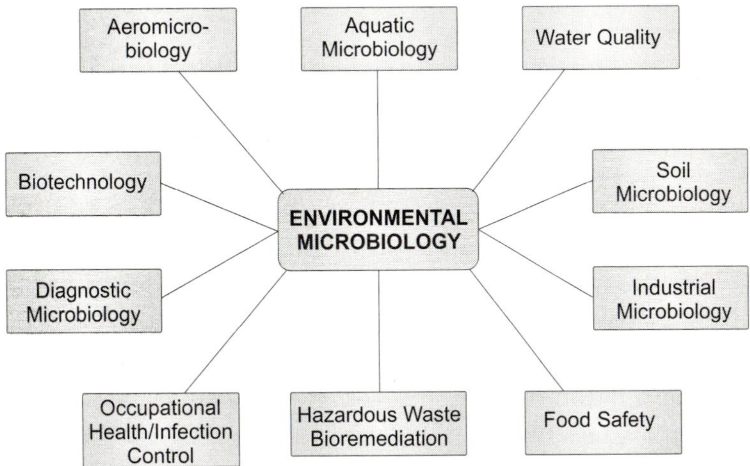

Chapter Outline

- *History of Environmental Microbiology*
- *Interrelations with Other Fields of Microbiology*
- *Modern Environmental Microbiology*

Environmental microbiology is usually defined as "the study of the applied effects of microorganisms on the environment and on human activity, health, and welfare". These effects can impact humans directly, as in the case of microbial disease, or indirectly through actions on animals, plants, or ecosystem health in general. The actions can be beneficial, as nitrogen fixation by *Rhizobium* spp., or detrimental as transmission of bacterial or viral pathogens of humans.

HISTORY OF ENVIRONMENTAL MICROBIOLOGY

It was in the 1970s that a new area of microbiology emerged and developed into the field of environmental microbiology. The roots of this field can be traced back to microbial ecology, which comprises the study of the interactions of microorganisms within an environment—be it air, water, or soil. The primary difference between these two fields is that environmental microbiology is an applied field in which the driving question is **how can we use our understanding of microorganisms in the environment to benefit society?** Thus, the two fields are related but not synonymous because they emphasise different viewpoints and address different problems.

In fact several events occurred simultaneously during the 1970s that highlighted the need for a better understanding of environmental microorganisms leading to development of this new field of microbiology. Two major groups of events occurred during this period. One was the emergence of a series of **new waterborne and food borne pathogens** that posed a threat to both, human and animal health. Second, in the same time frame was **frequent contamination of both surfacewater and groundwater supplies with organic and inorganic chemicals as a result of past waste disposal practices.**

The initial focus of environmental microbiology was on water quality and the fate of pathogens in the environment in the context of public health safety. Until the 1960s it was thought that threats from waterborne diseases had been eliminated. However, later several other agents such as viruses and protozoa were discovered which were more resistant to disinfection than enteric bacteria. Waterborne outbreaks caused by the protozoan parasite, *Giardia* is such an example. Pathogens in our food supply are the second area of immense concern. It is very likely that imported foods and vegetables carry human pathogens. The protozoan pathogen, *Cyclospora* is such an example. Table 1.1 lists many of the important pathogens that have emerged during the past 30 years or so.

The developing field of environmental microbiology also expanded to several other areas of applied research. These include microbial interactions with chemical pollutants in the environment, and the use of microorganisms for resource

TABLE 1.1 Some recently discovered pathogenic microbes having significant impact on human health

Pathogen	*Mode of transmission*	*Disease/Symptoms*
Rotavirus	Waterborne	Diarrhea
Legionella	Waterborne	Legionnaire's disease
E. coli 0157: H7	Waterborne Foodborne	Enterohemorrhagic fever, kidney failure
Hepatitis E Virus	Waterborne	Hepatitis
Cryptosporidium	Waterborne Foodborne	Diarrhea
Calicivirus	Waterborne Foodborne	Diarrhea
Helicobacter pylori	Foodborne Waterborne	Stomach ulcers
Cyclospora	Foodborne Waterborne	Diarrhea

production and resource recovery. Chemical pollutants in soil and ground water have pronounced effects on human population, in terms of both: (i) potential diseases caused due to intake of these chemicals, and (ii) the economic impact of cleaning up the contaminated environment. Some of the chemical pollutants routinely found in our soil and ground water are shown in Table 1.2. The cost of cleanup or remediation of the contaminated sites may be too high. It has thus been recognised that biological cleanup alternatives, known as **bioremediation** may have profound economic

TABLE 1.2 Some common organic and inorganic pollutants found in the environment and the potential for cleanup using bioremediation

Chemical class	*Frequency of occurrence*	*Status of bioremediation technologies*
Gasoline, fuel oil	Very frequent	Established
Polycyclic aromatic hydrocarbons	Common	Emerging
Creosote	Infrequent	Emerging
Alcohols, ketones, esters	Common	Established
Ethers	Common	Emerging
Chlorinated organics	Very frequent	Emerging
PCBs	Infrequent	Emerging
Nitroaromatics (TNT)	Common	Emerging
Metals (Cd, Cr, Cu, Hg, Ni, Pb, Zn)	Common	Possible
Nitrates	Common	Emerging

advantages over traditional physical and chemical remediation techniques. Bioremediation through use of microorganisms has now become an effective and established alternative for cleanup of contaminated sites. The acceptance of bioremediation as a feasible cleanup alternative has opened the door for environmental microbiologists.

INTERRELATIONS WITH OTHER FIELDS OF MICROBIOLOGY

Since environmental microorganisms affect so many aspects of life and are easily transported between environments, the field of environmental microbiology interfaces with a number of subspecialities, including soil, aquatic and aeromicrobiology, as well as bioremediation, water quality, occupational health and infection control, food safety, and industrial microbiology (Fig. 1.1).

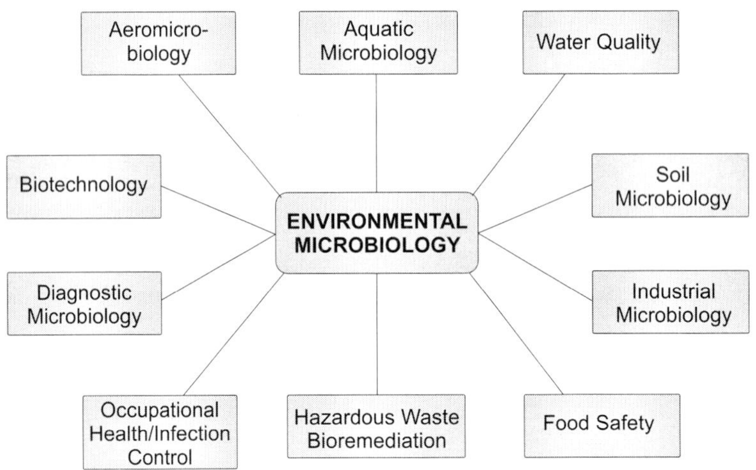

FIG. 1.1

Interrelations of environmental microbiology with other fields of microbiology

MODERN ENVIRONMENTAL MICROBIOLOGY

Modern environmental microbiology has a much wider scope. It is more than the study of pathogens and bioremediation. There are many different areas of this field. These areas infact are the recently recognised problems in different fields of microbiology addressed by enviromental microbiologists. These include the discovery and identification of new microbes and their products that may have practical application for protection of environment; protection of human health; and commercial application. Different areas and microbial issues of modern environmental microbiology are as follows:

1. **Agriculture and Soil Microbiology.** It includes the issues such as biological control, including introduced microbial biocontrol agents to reduce disease; nitrogen fixation, including introduced rhizobia; and nutrient cycling, including introduced mycorrhizal fungi.
2. **Biogeochemistry.** It includes carbon and mineral cycling; control of acid mine drainage; and control of loss of fixed nitrogen.
3. **Aeromicrobiology.** It concerns with collection and detection of pathogens or other microbes in aerosols and microbial movement in aerosols.
4. **Food Quality.** It includes detection of pathogens in foods and their elimination.
5. **Water Quality.** It includes removal of organic and inorganic contaminants, detection of pathogens and their elimination.
6. **Wastewater (Sewage) Treatment.** It includes biodegradation of waste and reduction of pathogens.
7. **Bioremediation.** It aims at biodegradation of organic contaminants; immobilisation or removal of inorganic contaminants present in contaminated soil and water environments; and enhanced remediation of organic-and/or metal-contaminated sites by using introduced bioremediative agents, usually bacteria.
8. **Biotechnology.** It includes detection of pathogens or other microbes in the environment, detection of microbial activity in the environment, and genetic engineering.
9. **Resource Production.** It includes production of alcohol and single-cell protein from cheap or waste materials.
10. **Resource Recovery.** It includes recovery of oil and metals mediated by microorganisms.

The abovesaid issues will be elaborated at appropriate places in the concerned chapters of the book.

2

MICROBIAL DIVERSITY IN THE ENVIRONMENT

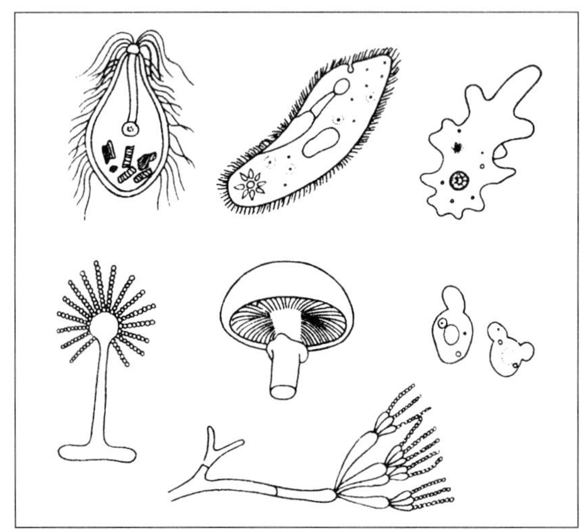

Chapter Outline

- *Viruses*
- *Bacteria*
- *Fungi*
- *Algae*
- *Protozoa*

There exists a great diversity and complexity among microorganisms in the environment (Fig. 2.1). In environmental microbiology, we categorise microbes as prokaryotes or eukaryotes, both of which profoundly affect human health and welfare and are essential for maintaining life as we know it (Fig. 2.2). Viruses are unique as they are acellular and consist solely of nucleic acids and proteins. The smallest cellular microbes are bacteria which are prokaryotic. Larger and more complex are eukaryotic fungi, algae and protozoa.

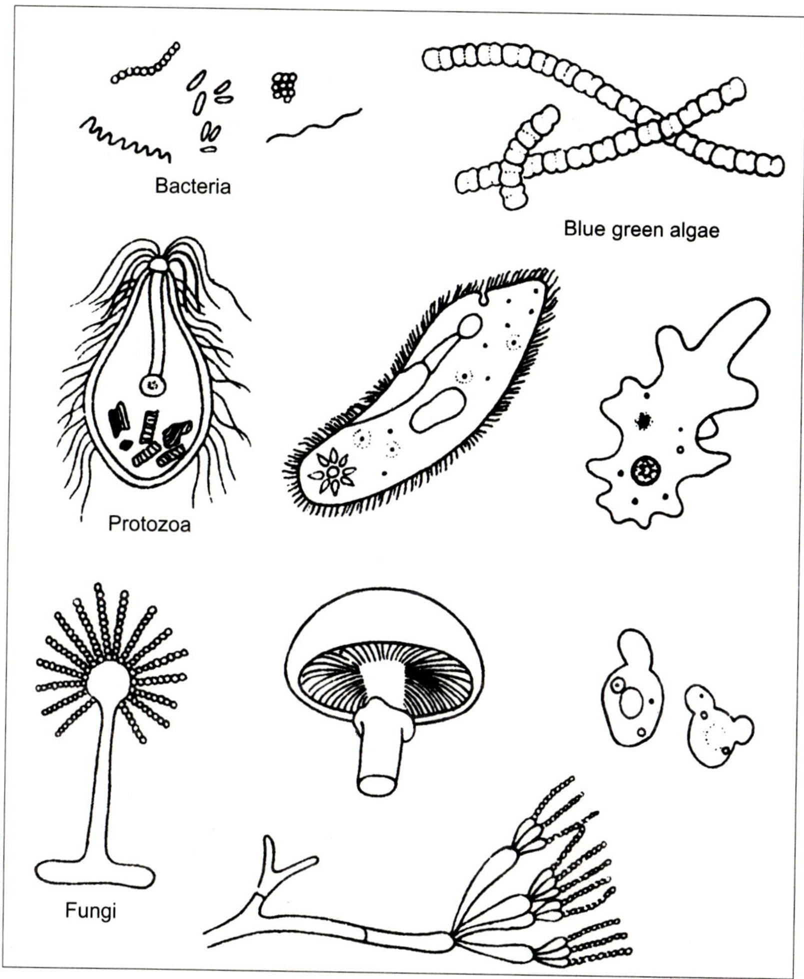

FIG. 2.1

The diversity and complexity among environmental microbes

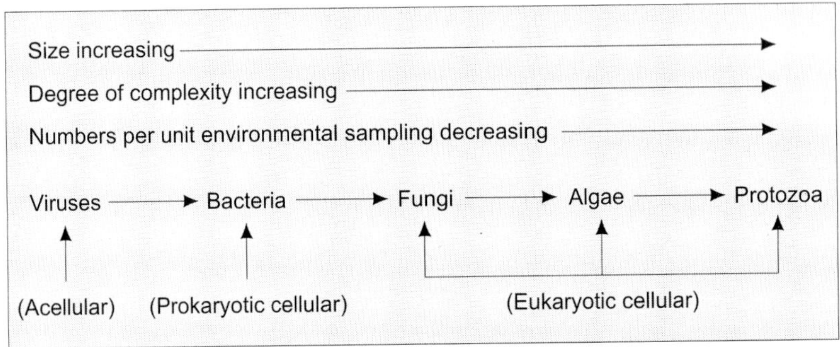

FIG. 2.2

Scope and diversity of microbes included in environmental microbiology

Viruses are important because as pathogens they have the ability to infect other live organisms, including man, and cause diseases. **Bacteria** can also cause infections but are very important in other areas of environmental microbiology including biochemical transformations that affect nutrient cycling, bioremediation, waste disposal and plant growth. **Fungi** are equally important, and are involved in these activities, particularly in surface environments. **Algae** affect surface water quality and can also produce toxins, but their overall impact in environmental microbiology arena is not as significant as that of bacteria and fungi. Finally, the **protozoa** are recognised pathogens that affect human health directly and may result into outbreak of some diseases. We shall now briefly consider only the salient features of these main groups of microorganisms because of their importance and relevance to environmental issues. In this book I will not include the details of their structure, metabolism and reproduction which are available in textbooks of general microbiology.

VIRUSES

They are acellular, infectious agents. Since they have no associated metabolism one can argue whether they are organisms at all. They have no such activity except replication, that too only in cytoplasm of live cells. They consist of a fragment of nucleic acid (DNA or RNA, never both) surrounded by a layer of protein. They cause important diseases.

Viroids are minute infectious agents of plants which are composed of single naked RNA molecules. The first report of these pathogenic RNA molecules in the early 1970s aroused much interest, since it was previously thought that infectious nucleic acids could survive outside a cell only if they are encapsulated in a protein capsid. Viroid RNAs are very unusual: they are circular, uniquely folded and so small that the largest one so far described (CEV: citrus exocortis viroid) is only 371 nucleotides long-about 1/10th of the size of the smallest RNA virus. Viroids so far are known to cause diseases in plants only.

Prions (proteinaceous infective particles) or slow viruses are a group of recently discovered infectious particles. They are associated with certain degenerative diseases of the central nevous system such as **kuru**, a disease restricted to a few cannibalistic tribes of New Guinea and transmitted by eating uncooked brain. and other diseases such as scrapie of sheep and goats and **Crutzfeldt-Jacob** disease of humans and animals. Prions appear to be proteins or glycoproteins with no associated nucleic acid of any kind. Since replication of the prion-protein presumably occurs by normal process, this implies that these proteins may be able to bring about the production of their own requisite mRNA. If this were so, prions would have revolutionary impact on molecular biology since they would constitute the only example where genetic information passes from protein to nucleic acid and not the other way around. However, recent evidence shows that scrapie glycoprotein may be a product of a normal host gene, indicating that these infectious agents may be more conventional than thought previously.

BACTERIA

This group includes the true bacteria (eubacteria) and related organisms i.e. archaebacteria, blue-green bacteria (cyanobacteria or oxyphotobacteria), mycoplasmas, rickettsiae and chlamydiae.

They are the true prokaryotes. Reproduction is by simple fission. Nutritional processes are very diverse. Many can thrive in both oxygen-rich and oxygen-free environments. Growth occurs over a wide range of temperatures, usually at a neutral pH. They occur in almost every terrestrial and aquatic environments.

Archaebacteria have following features (in common) whereby they differ from other bactetia:

(1) Their 16s rRNA molecules are similar to each other, but differ greatly from those of other bacteria and from eukaryotes.

(2) Their walls do not contain peptidoglycan, but a range of other unique polysaccharides.

(3) Their cell membrane is of a single layer of glycero-hydrocarbon-glycerol chains instead of a bi-layer of phospholipids arranged tail to tail.

(4) Their ribosomes are insensitive to chloramphenicol.

(5) They inhabit extreme environments.

They include the methanogens (the methane-generating bacteria of anaerobic muds); the extreme red halophiles; the salt-loving bacteria of saturated brine and salted fish; and the thermoacidophiles, found in hot sulphur spring or smouldering coal wastes. It is suggested that they represent a very ancient lineage which diverged from the eubacteria very early in the evolutionary process, and have survived only in these specialised ecological niches.

The rickettsiae, chlamydiae and mycoplasmas also are prokaryotes. The rickettsiae and chlamydiae are intracellular parasites of eukaryotes. The rickettsiae are tiny rods that are transmitted by arthropods and that multiply only in living cells. The chlamydiae are among the smallest bacteria which were formerly considered viruses. The mycoplasmas are the smallest organisms that can be cultivated outside living tissues. Some are involved in lung disorders.

There are some intracellular parasites of prokaryotes. This is a group of small, highly motile bacteria, that adhere to the wall of the host, penetrate through to the periplasm where they replicate, causing eventual lysis of cell. Organisms of this type are called **bdellovibrios**, and perhaps occur in soil.

The blue-green bacteria are typical prokaryotes. They evolve oxygen during photosynthesis and have chlorophyll **a**, a pigment also found in all algae. They are very ancient. They are perhaps major primary producers in the world's oceans, and many are ecologically important as fixers of atmospheric nitrogen.

Myxobacteria (also called fruiting bacteria) form fruit bodies, specialised multicellular structures like slime molds. Some cells at tips of fruit body produce cysts. They are most complex in behaviour and life cycle patterns found in prokaryotes.

Actinomycetes are Gram-positive prokaryotes. Majority of them are mycelial and a large group of filamentous bacteria which characteristically show branching patterns just like those of fungi, to give rise to a spreading mycelium. Also, like fungi, they often readily produce spores when grown in culture. They are easily isolated from soil. Most of these, particularly *Streptomyces,* produce antibiotics, and so are of great value to the pharmaceutical industry. Their main characterstics are:

(1) Mycelial (majority), Gram-positive prokaryotes.

(2) Some (euactinomycetes) develop only in mycelial state reproducing by unicellular specialised spores (e.g. *Streptomyces, Micromonospora, Actinoplanes, Streptosporangium, Thermoactinomyces*); in others (proactinomycetes) mycelial development is transitory and often limited, specialised spores not produced and reproduce primarily by fragmentation into short rod-shaped cells (e.g. *Bifidobacterium, Actinomyces, Mycobacterium, Nocardia, Geodermatophilus*); and the rest are certain Gram-positive, unicellular bacteria which do not form endospores and which are distinguished from the coryneform group by their

regular cell shape, and include the lactic acid bacteria (*Streptococcus, Leuconostoc, Pediococcus, Lactobacillus*) and micrococci (*Staphylococcus, Micrococcus, Sarcina)*.

(3) Cell wall has some peptidoglycan as in nearly all Gram-negative prokaryotes. However, a very large number (almost 60) of other peptidoglycan types have been found in these organisms. The chief variations are the nature of the diaminoacid in position 3; the presence, number and nature of additional amino acids which form interpeptide bridges; and the position of the cross link between peptide chains.

FUNGI

Some of the chief characteristics are as follows:

(1) Heterotrophic eukaryotic microbes obtaining their food in a soluble form by uptake through plasma membrane in a manner similar to prokaryotes.

(2) Have a thick cell wall usually made of polysaccharides, nearly always with chitin microfibrils.

(3) Motile stages absent, never form flagella. Those with motile spores, included in kingdoms Protozoa and Chromista are not true fungi.

(4) Have a typically branched growth or **mycelium** made up of individual filaments, the **hyphae**. Some as yeasts are unicellular i.e. non-mycelial.

(5) Mycelia may be coenocytic, or septate.

(6) Asexual reproduction by a variety of spores.

(7) Sexual spores are also produced. In some large fruit bodies are produced. Life cycles may be simple to complex.

ALGAE

They are pigmented eukaryotes, ranging in size from microscopic forms to giant kelps of marine waters. The chief characteristics are as follows:

(1) Source of energy is an oxygen-producing photosynthesis occurring in chloroplasts. Within the chloroplast, there are found a variety of chlorophylls and carotenoids. Chlorophyll **a** always present.

(2) Exhibit a wide range of morphological types.

(3) Many are motile, usually by flagella.

(4) Cell wall characteristically of polysaccharide made up of components as pectin, cellulose or xylan, that is sometimes calcified with calcium carbonate. Wall of some red seaweeds contain the agar gel.

(5) Can produce sexually or asexually and can show very complicated life cycles.

PROTOZOA

They are commonly defined as unicellular, eukaryotic animals. To a microbiologist, however, it is not a very helpful definition. It would perhaps be more appropriate to say that they are a group of unicellular non-photosynthetic, eukaryotic microorganisms which normally obtain their food by phagocytosis and which possess no true cell wall. They are involved in many blood and tissue diseases. The movement is an important aspect in life cycle. A few of their characteristics are as follows:

(1) On the basis of the type of **movement**, protozoa are divided into three major groups:

   (a) Amoeboid motion as in *Amoeba* and similar forms.

   (b) Flagellar movement as in flagellate protozoa, some of which are the colourless counterparts of particular algae. Some as trypanosomes (cause of sleeping sickness) have simple flagellum, while in others as *Trichonympha* (an organism inhabiting the guts of termites where it is responsible for the wood cellulose eaten by insect) there is very complex flagellar arrangement.

   (c) Ciliary movement in ciliates, as in *Paramecium,* whose surface covered with cilia, that are shorter than flagella and have a co-ordinated motion so that waves of contraction pass over them.

(2) Food, taken by phagocytosis of solid particles.

(3) They have generally no cell wall.

(4) Life cycles less complicated than algae or fungi, exceptions being the forms like *Plasmodium.*

(5) Degree of specialisation in single cells, as in *Paramecium.*

Some authors also study **slime molds** and **water molds** alongwith protozoa, thus prefer to treat them protists rather than fungi. They have been traditionally studied by mycologists. However, evidence is accumulating to show that phylogenetically they are protists. The **slime molds**, like fungi form macroscopic fruit bodies. But their feeding phase is amoeboid. They live on surface of decaying vegetation. There are two main groups: (a) the cellular slime molds, whose vegetative stage is single amoeboid cells that may aggregate to form a **pseudoplasmodium** as in *Dictyostelium,* and (b) acellular slime molds, where a single amoeba produces a multinucleate **plasmodium** of indefinite size and shape which moves over the surface of the substratum engulfing food particles as it goes.

Majority of the **water molds** are the oomycetes originally included among the fungal kingdom. They are shown phylogenetically related to Chromista most obviously because they form flagellated motile cells. However, their nutritional processes and vegetative states appear more like fungi than protists.

3

AQUATIC ENVIRONMENTS

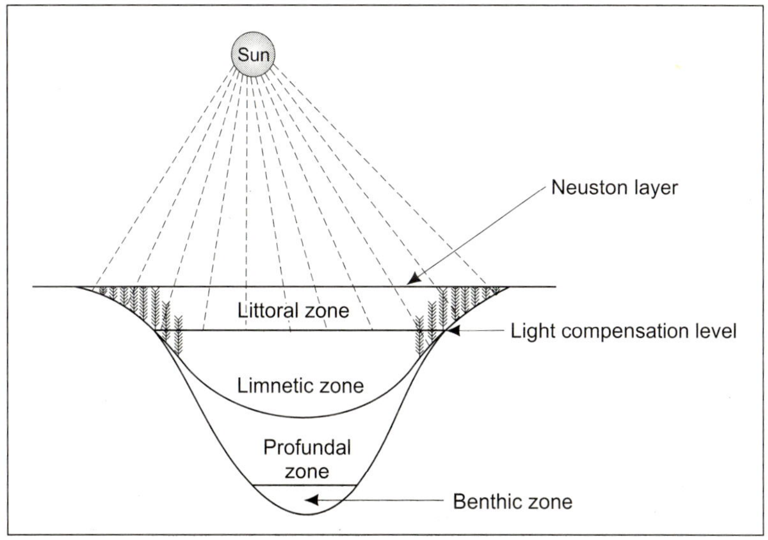

Chapter Outline

- *How Do Microbes Grow in Aquatic Environments?*
- *Types of Aquatic Environments*

The study of microbes and microbial communities in water environments is broadly known as **aquatic microbiology**. Aquatic environments occupy roughly more than 70% of earth's surface. Although most of this area is occupied by oceans, there is also a broad spectrum of other aquatic environments including estuaries, harbors, river system, lakes, wetlands, springs and aquifers. Microbial communities of aquatic environments include viruses, bacteria, fungi, algae and other microbes.

HOW DO MICROBES GROW IN AQUATIC ENVIRONMENTS?

Microorganisms grow in aquatic environments in different ways. Accordingly, they exhibit different life-forms or life habits, each with its own characteristic pattern of structure in terms of organisation and composition. These patterns may also be referred to as their structural habitats which are as follows.

Planktons

They refer to the microbial communities suspended in the water column, also called floating forms. Forms swimming freely in water are called **nektons**, whereas those swimming or resting on water surface as **neustons**. Microbes attached to plants in water are called **periphyton**. Photoautotrophic microbes within planktonic community which are chiefly algae and cyanobacteria, are collectively referred to as **phytoplankton**. Suspended heterotrophic bacterial populations are called **bacterioplankton**, and protozoan populations make up the **zooplankton**. Together these three groups make up the microbial planktonic community. The relationship and interdependence of the various microbial components within a general planktonic food web is shown in Fig. 3.1. Phytoplanktons are the primary producers. Open oceans have relatively low primary production due to low levels of essential nutrients, nitrogen and phosphorus. Coastal areas are productive because of the introduction of dissolved and particulate organic material from river outflows and surface run off from the terrestrial environment. Freshwater lakes, like the open seas are often low-productive, particularly those with large, deep, nutrient-poor (oligotrophic) bodies of water. In contrast, smaller and shallower, freshwater bodies tend to be nutrient rich (eutrophic). Nutrient loading, which cause eutrophication of lakes result from terrestrial runoff, rivers feeding into lake, and plant debris. Human activities, like disposal of municipal wastewater and run off of fertilisers from agricultural fields also cause nutrient loading.

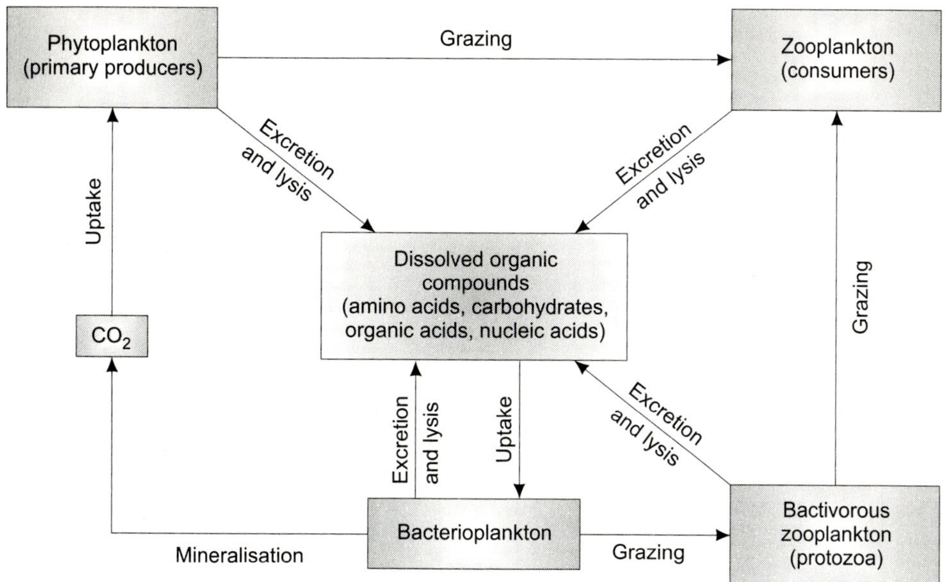

FIG. 3.1

The microbial loop in a planktonic food web

Benthos

This is a transition zone between the water column and the mineral subsurface. This interface collects the organic matter that settles from the water column or that is deposited from the terrestrial environment. The interface is a diffuse and non-compacted mixture of organic matter, mineral particulate matter, and water. This zone is characterised by a dramatic increase in the concentration of microorganisms (may be five times more) compared with the planktonic environment. The benthic habitat is an important feature of the aquatic environment. This environment supports and favours the formation of conjoint aerobic and anaerobic microenvironments.

The sediment zone supports a physiologically diverse aquatic microbial community. Fermentative bacteria metabolise dissolved organic materials into organic acids, such as acetic acid, and CO_2. These acids act as electron donors for strictly anaerobic bacteria which utilise CO_2 as the final electron acceptor in anaerobic respiration, thus generating CH_4. This methanogenic activity in turn supports the activity of methane-oxidising bacteria, which under aerobic conditions, can utilise methane and other one-carbon compounds as energy source, generating CO_2.

Microbial Mats

In benthic habitat the sediment-planktonic interface results into a microenvironment in which the combined aerobic and anaerobic activity supports a diversity of microbial populations. This microbial diversity tends to increase further in microbial mats, which are an extreme example of an interface aquatic habitat in which many microbial groups are laterally compressed into a thin mat of biological activity. These groups interact with each other in close spatial and temporal physiological relationships. The mats are from several millimeters to one centimeter thick and are vertically stratified into distinct layers. Microbial mats have been found associated with environments like surface-planktonic interface of hot springs, deep sea vents, hypersaline lakes, and marine estuaries. They are considered self-sufficient. An example of a laminated form of a microbial mat is shown schematically in Fig. 3.2.

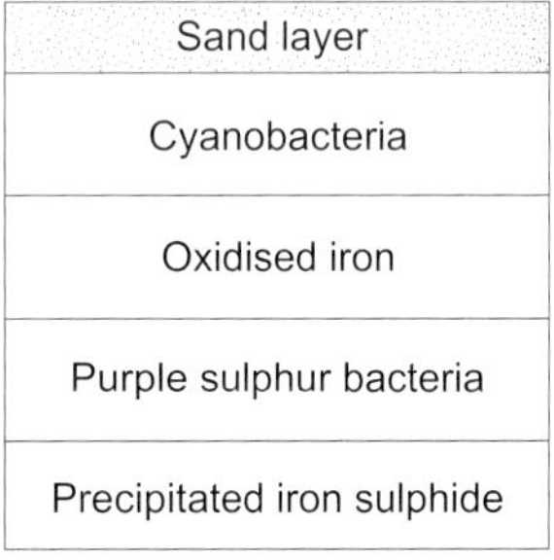

FIG. 3.2

Schematic representation of a part of vertical transverse section of a microbial mat

Cyanobacteria occupy the upper zone of the mat, where they have access to sunlight. A layer of sediment rich in oxidised iron may form directly beneath this zone. The origin of this layer is not known, but it appears to form a barrier between aerobic and anaerobic phototrophic activities. Purple sulphur bacteria can form a distinct layer just below the aerobic-anaerobic interface. They are also photosynthetic but not capable of photolysis of water. They utilise reduced sulphur compounds as electron donors, which are provided by sulphate-reducing bacteria. Below the purple sulphur bacteria is an extensive black layer enriched by the precipitation of FeS.

Microbial mats are unique communities and often found in extreme environments or in the environments with fluctuating conditions. The cyanobacteria can tolerate extreme conditions as high temperatures, high saline waters. Microbial mats are of evolutionary significance. Fossilised microbial mats, known as **stromatolites**, dating back 3.5 billion years were among the first indications of life on earth. At that time, under poor oxygen conditions, stromatolites with anoxygenic phototrophic purple and green sulphur bacteria were perhaps formed on earth.

Biofilms

A biofilm is a layer of organic matter and microorganism formed by the attachment and proliferation of bacteria on the surface of an object. In most cases this surface is submerged in nonsterile water or surrounded by a moist environment. Solid surfaces suitable for bacterial colonisation include inert surfaces such as rocks and hulls of the ships and even living surfaces such as submerged portions of aquatic plants. Biofilms are characterised by the presence of bacterial extracellular polymers **glyocalyx** that create a visible slimy layer on solid surface. Glyocalyx provides a matrix for attachment of bacterial cells and forms an internal architecture of the biofilm community. Biofilms have been extensively studied for their role in nutrient cycling and pollution control in aquatic environments, as well as for their role in public health.

Biofilm development is complex and usually initiated by the attachment of bacteria to a solid surface (Fig. 3.3). Dissolved organic molecules of a hydrophobic nature accumulate at the solid-water interface and form a conditioning film. Bacteria approach the solid surface because of water flow and/or active motility. Permanent attachment of bacteria requires two stages—(i) reversible attachment (phase I), which is a transitory physico-chemical reaction (with time, it can become permanent), and (ii) irreversible attachment (phase II), which is biologically mediated stabilisation reaction. The initial adhesion (phase I) is controlled by various

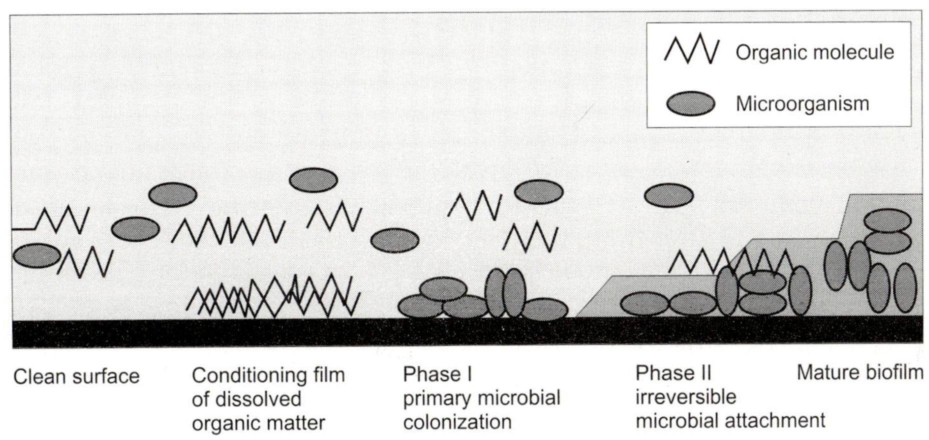

FIG. 3.3

Steps in biofilm formation

attractive or repulsive physico-chemical forces leading to passive, reversible attachment to the surface. An irreversible attachment is a biological, time-dependent process related to the proliferation of bacterial exopolymers forming a chemical bridge to the solid surface (phase II). By a combination of colonization and bacterial growth, the mature biofilm is formed. It is characterized by cell clusters surrounded by water-filled voids.

Biofilm formation is a good strategy of survival in nutrient-poor environments like rock surfaces. Bacteria predominantly found attached to rock surfaces take full advantage of continuous renewal of nutrients provided by water flow and thus act as biological filter to remove dissolved organic materials from flowing waters. This filtration process from the water by these attached communities represents a water purification system in natural environments. This system has long been exploited for use in water purification from municipal sewage or industrial sources. In trickling filters and fluidised bed reactors biofilm biomass is maximised by providing porous network of solid supports for bacterial attachment. In this way microbial colonisation at a solid-liquid interface can be beneficial, as in nutrient cycling and water purification systems. However, in some cases as in industries using water pipelines in cooling towers or heat exchange structures, the biofilm formation is detrimental as it lowers the flow capacity of pipes and decreases heat-exchange efficiency. Biofilms due to presence of extracellular matrix are more resistant to antibiotics and disinfectants. Higher concentrations of these chemicals can affect public and cause environmental health problems. Therefore, alternative approaches like impregnation of materials such as plastics, with biocides or antibiotics have been developed.

TYPES OF AQUATIC ENVIRONMENTS

Chiefly on the basis of physico-chemical conditions and general microbial characteristics four main types of aquatic environments are recognised: (i) inland surface waters, also referred to as freshwater environments (lakes, rivers, streams), (ii) estuarine environments or brackish waters, (iii) marine environments (seas, oceans, harbours), and (iv) ground waters.

Freshwater Environments

These are inland bodies of waters like springs, rivers and streams, and lakes that are not directly influenced by marine waters. The science focusing on the study of all aspects (physical, chemical, geological, and biological) of freshwater habitats is called **limnology**, whereas the study of freshwater microbes as **microlimnology**. As compared to marine and terrestrial habitats, freshwater habitats occupy a relatively small portion of earth's surface. But these habitats are of much importance to mankind. There are two general types of freshwater environments: (i) **lentic** or standing waters such as pond, lakes, swamps and bogs, and (ii) **lotic** or running waters, such as springs, stream and rivers. These freshwater environments have very different physico-chemical make-ups and correspondingly characteristic populations and communities of microorganisms.

Springs

Springs form wherever subterranean water reaches the earth's surface. There are many types of springs. For instance, melted snow and ice in mountains feed **cold springs**. **Thermal springs** also called warm or hot springs originate from volcanic areas or great depths. Geysers are an example. On the basis of distinct mineral or chemical composition, there may be sulphur and magnesium springs, acid springs and radioactive springs. Bacteria and algae are the chief microorganisms found in springs. These are photosynthetic and the primary producers. The largest portion of dissolved organic matter originates from surrounding terrestrial systems that includes plant exudates, dead plants, animals and microbes transported as terrestrial runoff, seepage and wind.

Rivers and streams

Springs, as they flow away from their subsurface source, merge with other water sources to form streams and rivers that eventually flow into other water bodies as lakes or seas. An interface where the subsurface water table meets and interacts with the streambed is called a **hyporheos**. As a stream progresses and becomes larger, it tends to accumulate organic matter and heterotrophic populations, inoculated from surrounding terrestrial environment. Thus a profile of microbes in river resembles very often with terrestrial microbial communities. Most physico-chemical characteristics of rivers and streams (temperature, volume, velocity and chemical make up) are determined by the geography and climate of the area through which they flow. Mountain streams have fast currents and low temperature, whereas rivers flowing through plains have higher temperatures. River are usually not very deep, though larger ones may have pools. There are seasonal fluctuations in rivers and streams. Photosynthetic microbes, including phytoplankton are the primary producers.

As streams develop into rivers they tend to acquire more dissolved organic matter, which limits the light penetration and hence the photoautotrophs. There is an increase in heterotrophic populations. Due to their flow patterns, streams and rivers are well aerated for most of their routes. Hence heterotrophs are predominantly aerobic or facultative anaerobic. Sewage outflows are especially such areas, where heterotrophs tend to increase in numbers downstream.

Lakes

As compared to ponds, lakes have exceedingly larger surface area, up to 100,000 km^2 and are very deep. Though regarded lentic, lakes have inflows and outflows, wind-generated turbulence and temperature-generated mixtures, all of which collectively create a dynamic environment. Some salt lakes have a high salt content and may represent an extreme environment. Some lakes are characterised by their chemical makeup as bitter lakes (rich in $MgSO_4$), borax lakes (rich in $Na_2B_4O_7$) and soda lakes (rich in $NaHCO_3$). Lakes are most complex of the freshwater environments and so also their microbial communities and their interactions. Based on the morphometric (depth, dimension, geology of shores, currents etc.) and physico-chemical (temperature, p^H, oxygen level) parameters, a lake can be subdivided into different zones or subsections (Fig. 3.4). These are: (i) **littoral zone**, which is the edge of the lake where sunlight can penetrate effectively to its bottom. The air-water interface including the upper few millimeters of the water column is known as the **neuston layer**. The general structure of the neuston is shown in Fig. 3.5. This upper layer can range from 1 to 10 μm in depth. Most scientists consider it an extreme environment due to intense solar radiation, large temperature fluctuations and accumulation of toxic substances including heavy metals. The upper layer that interacts with the atmosphere consists of a water-lipid mixture having increased surface tension. Below this is a layer of organic matter that accumulates from organic matter rising up the water column. (ii) **limnetic zone**, which refers to the surface layer of open water away from the littoral zone, where light readily penetrates.

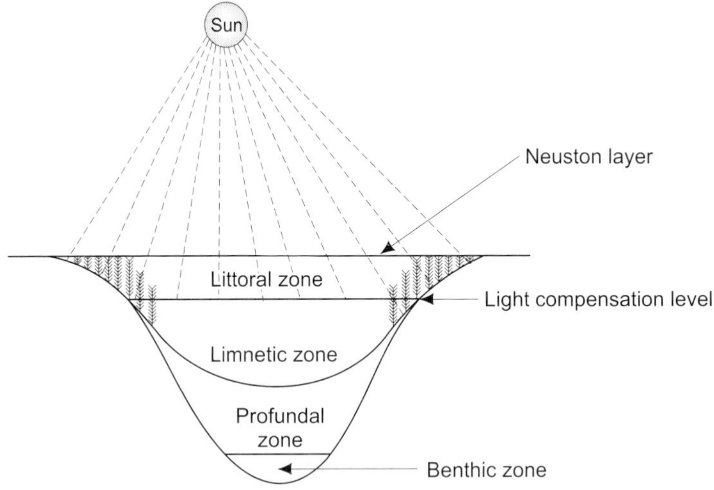

FIG. 3.4

Diagrammatic sketch showing the three major zones of a freshwater body as a lake

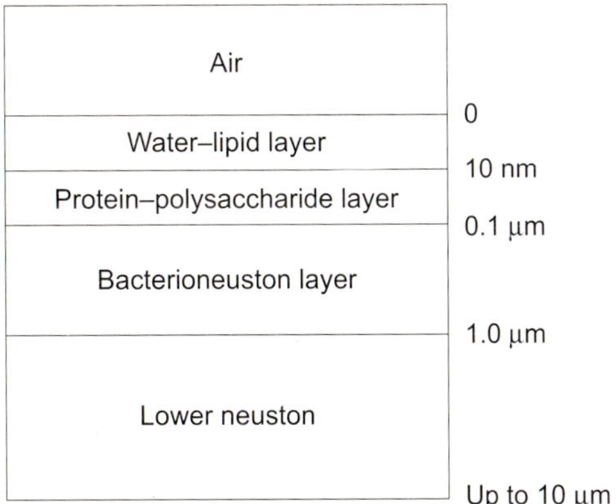

FIG. 3.5

Schematic representation of the neuston

(iii) **profundal zone** is the area below limnetic zone where light intensity is less than 1% of the sunlight (the light compensation point), and (iv) **benthic zone**, which consists of the bottom of the lake and associated sediments.

Temperature in lakes is an important factor and during summer these exhibit **thermal stratification**. During summer the temperature of the upper layers of water may rise to 22-24°C whereas that of bottom layers as low as only 4°C. Between the range of 4°C to 22°C there are generally differentiated three different zones (regions). These are, (i) **epilimnion**, the upper zone showing vertical gradient of only gradually decreasing temperature from the surface, followed by (ii) **thermocline** or **metalimnion**, a middle zone with a rapid change in temperature, which falls rapidly, and (iii) **hypolimnion**, the lower bottom cold zone where no temperature gradient is evident. Due to intense primary productivity in presence of sunlight and oxygen-rich conditions nutrients become depleted and limiting in epilimnion. The characteristics of epilimnion are reversed in hypolimnion which has low temperature and oxygen levels, lack of light penetration and a high mineral content.

Microbial communities in lakes are the most extensively studied aquatic environments and in many cases are microbially most complex. They contain extensive primary and secondary productive microbes interacting dynamically. In the littoral zone the phytoplanktons are predominantly algae followed by cyanobacteria. Filamentons and epiphytic algae occur as periphytons. Limnetic zone is also dominated by phytoplanktons forming distinct community gradients based upon the wavelength and the light intensity penetrating to a given depth (Fig. 3.6). Different photosynthetic phytoplanktons have their characteristic light absorption spectra.

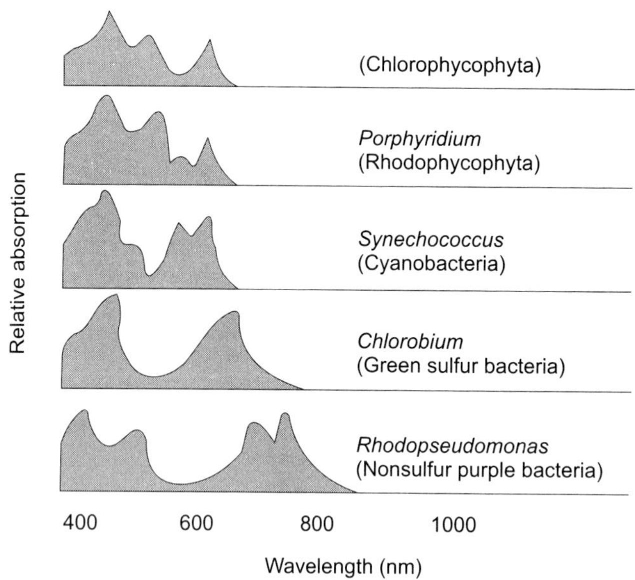

FIG. 3.6

Graph showing the light absorbance spectrum of common phytoplanktonic algae and photosynthetic bacteria. It is evident that each of these groups has a different profile which enables them to take advantage of their niche. In general organisms capable of utilizing longer wavelengths are found deeper in the water column avoiding competition with organisms higher in the water column that absorb the shorter wavelength. (Adapted from Atlas and Bartha, 1993)

Besides photoautotrophs lakes have extensive heterotrophic communities also. Their concentrations vary with depth and secondary production is 20 to 30% of primary production. There are three zones that generally have elevated numbers of heterotrophs. The first is the neuston layer rich in proteins and fatty acids that create localised eutrophic conditions. The second zone where heterotrophs are markedly higher is thermocline, where organic debris tends to settle and accumulate. The third zone characterised by higher heterotrophs is the upper layer of the benthos. Viruses in freshwater environments can be very abundant and can utilise bacteria, cyanobacteria and microalgae. Protozoa are also important predators of aquatic microbes, particularly algae and bacteria.

Estuarine Environments (Brackish Waters)

Brackish water is a wider term used to describe water that is more saline than fresh water but less saline than true marine waters. Often these are transitional areas between fresh and marine waters. The best known example of brackish water is an estuary. An **estuary** is a semi-enclosed coastal body of water (as a part of a river) that meets the sea. It has a free connection with the open sea, thus strongly affected by tidal action and within it sea water is mixed with fresh water from land drainage. Examples of estuaries are river mouths, coastal bays, tidal marshes and water bodies behind barrier beaches. Thus estuaries are transitional zones or ecotones between the freshwater and marine habitats.

Estuaries are highly variable environments because salinity can change drastically over a relatively short distance. Dramatic change may also occur at a given point in the estuary as a function of the time of the day or season of the year. Seasonal increases in fresh water due to rainfall or snowmelt will decrease the salinity at a given point in the estuary. High tide will increase the salt content at a given spot. The variation in salinity can range from 10% to 32% with the average salinity of fresh water being 0.5% (cf. salinity of marine waters 33 to 37% with an average of 35%). Microbes and plants are adapted to fluctuation in salinity in the estuaries. Despite this estuaries are very productive environments. Mangrove swamps are highly productive. These swamps are an important transition community because they help filter contaminants and nutrients from water, they stabilise sediments, and protect the shoreline from erosion. In general estuarine primary production (10 to 45 mg c/m^3/ day) is not always enough to support the secondary populations. Estuaries tend to be turbid and thus light penetration is poor. Primary producers numbers range from 10^0 to 10^7 organisms/ml water. Despite low primary production, heterotrophic activity is high (150 to 230 mg c/m^3/day). Local runoff and organic carbon are brought abundantly by the rivers flowing into the estuaries.

Marine Environments

Like those of lakes, marine water environments are highly diverse. The study of the sea in all of its aspects i.e. physical, chemical geological and biological is termed **oceanography**.

The chief ecological features of marine environment are: (i) the sea is big, covering about 70% of the earth's surface. (ii) the sea is deep and continuous, not separated as are land and fresh water. All the oceans are connected. Temperature, salinity and depth are the chief barriers to free movement of marine organisms. (iii) it is in continuous circulation due to wind stress set up by air temperature differences between poles and equator. (iv) it is dominated by waves of many kinds and tides produced by the pull of moon and sun. (v) the sea is salty, with an average salinity of 35 parts of salt (weight basis) per 1000 parts of water, or 3.5 per cent, that is usually written as 35% i.e. parts per 1000 (cf. salinity of fresh water that is less than 0.5%). The chief salts are chlorides, sulphates, bicarbonates, carbonates and bromides of sodium, magnesium, calcium and potassium, of which sodium chloride is present in maximum amount. (vi) dissolved nutrients are in a low concentration, that is an important limiting factor in determining size of marine populations.

As in ponds and lakes, seas also exhibit a distinct **zonation**. The various zones of sea are shown in Fig. 3.7. Generally, there is a continental shelf extending for a distance offshore, beyond which the bottom drops off steeply as the continental slope then levels off somewhat (the continental rise) before dropping down to a deeper, but more level, plain. The shallow-water zone on the continental shelf is the **neritic** (near shore) **zone.** The zone between high and low tides (also called the littoral zone) is known as the **intertidal zone.** The region of the open sea beyond the continental shelf is called **oceanic region,** which comprises the region of the continental slope and rise–the **bathyal zone**; area of the ocean 'deeps'–**abyssal region;** and light compensation zone separating an upper thin **euphotic** or **photic zone** from a vastly thicker **aphotic zone.** Light can penetrate through photic zone to a depth of about 200 m depending on water turbidity. Within these primary zones (based chiefly on physical factors), there may occur distinct secondary zones, horizontal as well as vertical, in such waters.

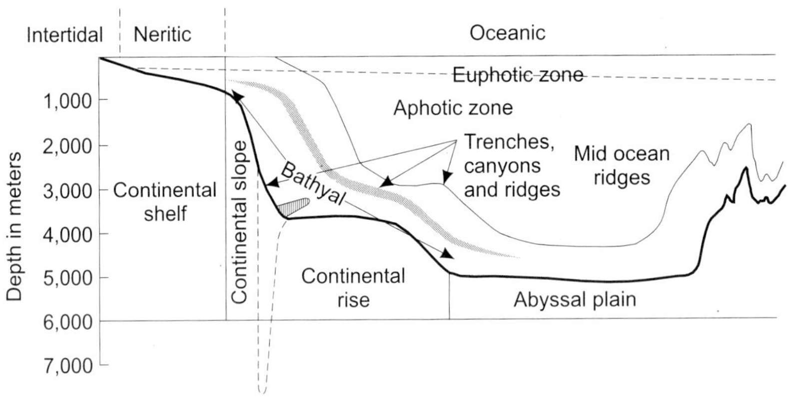

FIG. 3.7

Diagrammatic sketch showing horizontal and vertical zonation in the sea

Different zones in a sea can also be differentiated on the basis of habitats. From a microbiological view point **four** major habitats are recognised. These are, (i) **neuston,** the surfae of the sea (air-water interface), (ii) **pelagic zone,** a broad term used for the water column or planktonic habitat, which may be subdivided on the basis of precise depth in the water column. The upper 100 m of water column is called **epipelagic zone,** which is the habitat of large proportion of photosynthetic microbes. Further depths are designated as mesopelagic, bathypelagic and abyssopelagic habitats and the final sea-sediment interface is **benthopelagic zone** (benthos), (iii) **epibiotic** habitat refers to surfaces on which attached communities occur, and the (iv) **endobiotic** habitat pertains to organisms found within the tissues of other organisms such as fish. *Epulopiscium fishelsoni* is the example of such endobiotic bacterium.

Marine waters contain diverse microbial habitats. Different microbial communities predominate at different locations in the ocean. Usually in deep waters, microbial concentrations are highest in the neuston droping markedly below this region. Immediately below neuston, the numbers are on average close to 10^7 organisms/ml and decrease by more than a log at a depth of 100 m. Vertically, like lakes, numbers increase at the thermocline that exists below the zone of primary production (50 m). At greater depths, the numbers of heterotrophs quickly diminish, being very low at a depth of 200 m. However, thermocline-induced stratification is not as dramatic in coastal waters because of the mixing of water by winds, currents, and temperature. Due to this number of bacteria is more or less uniform at all depths except when the weather is calm.

An interesting phenomena in phytoplankton of marine environments are **algal blooms**, which occur in eutrophic water body under usually warm, sunny and calm conditions. Under these nutrient-rich and favourable environmental conditions some algae and cyanobacteria grow rapidly resulting in blooms. Under very high eutrophic conditions algal blooms adversely affect the water quality. Such water tends to be scummy with foul smell making it unpleasant for recreation and even dangerous for swimming, fishing, or boating.

Algae in such blooms may produce potent toxins. Immense algal bloom composed of red-pigmented dinoflagellates are called **red tides.** These algae produce toxins that are taken up by fish causing extensive deaths. These toxins can also affect marine birds, mammals and even humans consuming fish caught in such waters. Fungi, protozoa and viruses in their role as parasites and predators are also important components of marine communities. Protozoa act as bacterial predators and bacteriophages are also prevalent on bacterial hosts. Marine fungi feed on both plants and animals in addition to bacteria and algae.

Groundwater Environments

Groundwater environment is found inland in the subsurface zone and includes shallow and deep aquifers. Microorganisms are the sole inhabitants of these environments and bacteria are predominant. Unlike other aquatic environments which harbour substantial populations of planktons, most of the bacteria are either attached or transiently suspended in these environments. In general, there are low levels of microbial activity due to poor nutrient levels.

Saturated zones of soil found beneath the vadose zone are commonly called **aquifers** and are composed of porous parental materials saturated with water. Underground aquifers serve as a major source of potable water. There are several types of aquifers, including shallow table aquifers and intermediate and deep aquifers that are separated from shall aquifers by confining layers. The confining layers are composed of materials like clay having very low porosity.

Shallow aquifers are most closely connected to the earth's surface. They receive water from rainfall providing recharge to adjacent streams and rivers. Groundwater flow in these aquifers is rapid that makes them aerobic. Confined aquifers within 300 m of the surface soil are termed **intermediate aquifers.** These have much slower flow rates. It is this aquifer system that supplies a major portion of drinking and irrigation water. **Deep aquifers,** more than 300 m deep are characterised by extremely slow flow rates, and hence anaerobic. They are not directly recharged or affected by surface rainfalls.

Since subsurface microbiology is still a developing field, information is limited in comparison with that for surface microorganisms. Shallow subsurface zones with rapid rate of water recharge have high numbers of microorganisms. Majority of these are bacteria, though fungi and protozoa are also present. Numbers of bacteria remain more or less the same throughout the system profile between 10^5 and 10^7 cells/g of sediment. This is due to low inorganic nutrients. Eukaryotic counts are also low as compared to surface soils.

Thus in shallow aquifers dominant population type is aerobic, heterotrophic bacteria, though small populations of eukaryotes, anaerobic and autotrophs may also be present. Organisms are not as diverse as in surface soils.

In deep aquifers microbial communities are different from those in shallow aquifers. There is wide diversity with total counts ranging from 10^6 to 10^7 cells/g of sediment. The type of microbes include aerobic and facultatively anaerobic chemoheterotrophs; denitrifiers; methanogens; sulphate-reducers; sulphur-oxidisers; nitrifiers and N_2-fixing bacteria. Low numbers of unicellular cyanobacteria, fungi, and protozoa may also be present in some samples. Some specific types of microbes have also been detected in some samples through the analysis of 16s ribosomal RNA gene sequences. These include alpha-proteobacteria (*Agrobacterium, Blastobacter*), beta-proteobacteria (*Alcaligenes, Zoolgloea*), gamma-proteo-bacteria (*Acinetobacter, Pseudomonas*) and high – G + C gram-positive bacteria (*Arthrobacter, Micrococcus, Terrebacter*).

4

TERRESTRIAL ENVIRONMENTS

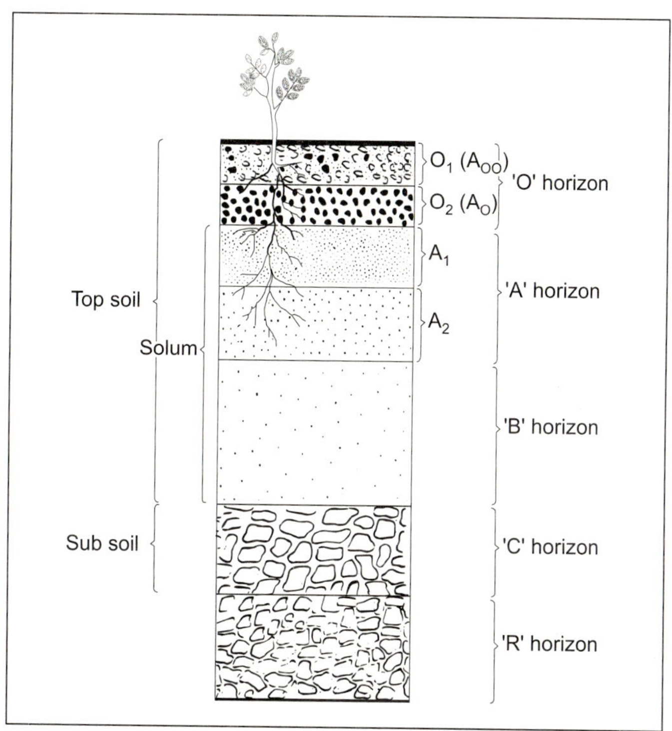

Chapter Outline

- *The Solid Phase*
- *The Liquid Phase or Soil Solution*
- *The Gas Phase or Soil Atmosphere*
- *Environment and Microorganisms in Surface Soil and Subsurface Habitats*

Among all the microbial environments, terrestrial ones are the richest and most complex. **Soil** constitutes the major habitat of terrestrial microorganisms and is very favourable for the growth and activity of microbes. Though terrestrial environments may vary from a rain forest to desert, all soils contain a rich diversity of active microorganisms.

An examination of a cross section of the earth (Fig. 4.1) reveals the **three** main regions (zones)—**surface soils,** the **unsaturated** or **vadose zone** and the **saturated zone,** also referred to as the underground **aquifers**. Each of these regions comprises a combination of mineral or solid, solution or liquid, and atmosphere or gas phases that is more commonly called a **porous medium**. Thus all porous media are three-phase systems consisting of (i) a solid or mineral inorganic phase that is often associated with organic matter, (ii) a liquid or solution phase, and (iii) a gas phase or atmosphere. All phases interact together when the system is perturbed and then move toward an equilibrium state to creat an unique environment. If left undisturbed, this environment remains fairly constant. The unique properties of any porous medium depend on the specific composition of each of these phases.

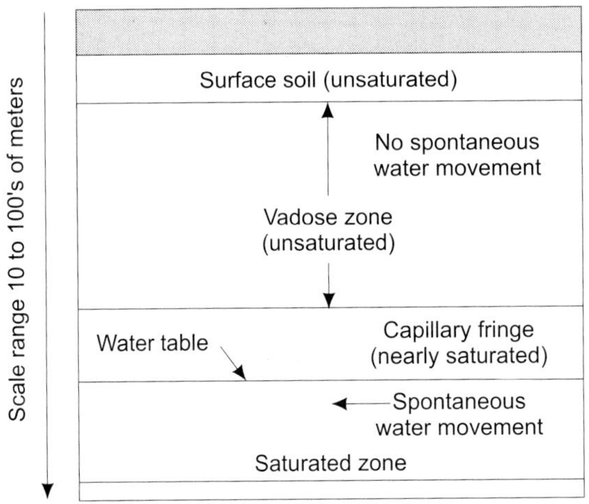

FIG. 4.1

A part of the cross section of the subsurface showing surface soil, vadose zone and saturated zone

THE SOLID PHASE

Typically, a porous medium contains 45 to 50% solids on a volume basis. Of this solid fraction, 95-99% is the **mineral fraction** in which silicon and oxygen predominate. Associated with minerals is also small amount of organic mater.

As a result of varying degrees of weathering of parent material-rock, the mineral particles of different sizes are formed. In a given sample of soil, there may be present, different-sized particles in different proportions. Depending upon their size (diameter basis), the International Society of Soil Science has given different names to these mineral particles, which are as follows:

Name of the particles	*Diameter range (mm)*
Clay	Less than 0.002
Silt	0.002-0.02
Fine sand	0.02-0.20
Coarse sand	0.20-2.0
Stones and Gravel	Above 2.0

Soil texture is determined by the relative proportion of mineral particles of different sizes present in the soil. The distribution (on a percent by weight basis) of chiefly sand, silt and clay within a porous medium defines its texture. Stones and gravels are excluded from the textural classes. To separate the particles of different sizes, the organic matter of soil is oxidised and inorganic cementation is removed to breakdown structural aggregates. Fine fractions, as clay and silt are removed from coarser material by sedimentation technique (Day, 1965) and the coarse fraction is dried and graded by sieving. Stones and gravels are retained by a 20 mm perforated plate sieve, the coarse sand retained on a 70 mesh wire sieve (0.23 mm per opening) while the fine sand passes through the 70 mesh. Silt and clay fractions are graded according to sedimentation velocity in a liquid column.

On the basis of the proportion of different-sized particles, soils are classified into different textural groups. Figure 4.2 shows an aribitrary textural classification of natural soils which may be used with laboratory particle size analyses, explained above or crudely with 'finger tests' in the field. As shown in Fig. 4.2, some of the various groups of soil may be as follows:

Textural group	*Relative proportion of different-sized mineral particles*
Sandy soil	85% Sand + 15% Clay or Silt or both
Loamy sand	70% Sand + 30% Clay or Silt or both
Loam soil	50% Sand + 50% Clay or Silt or both
Silt	90% Silt + 10% Sand

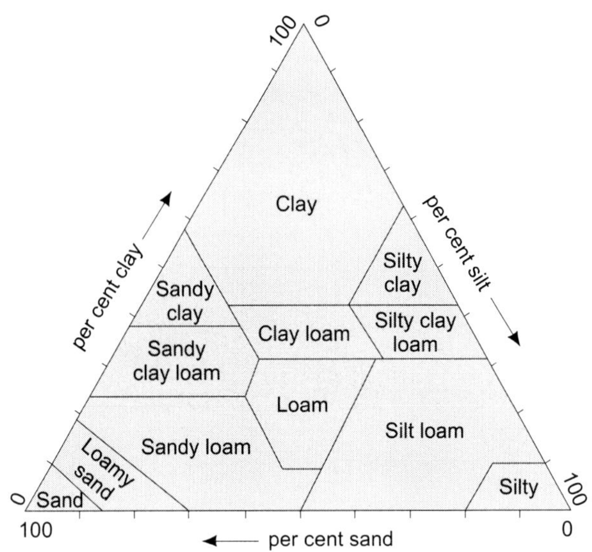

FIG. 4.2
Diagram to show classification of soil texture i.e. different textural groups of soil

Soil texture directly influences soil-water relationships, aeration and root penetration through its relationship with interparticle pore space. Indirectly it also affects the nutritional status of soil, the clay fraction being main source of many plant nutrients and of cation exchange activity. Sandy soils are nutrient-deficient due to high porosity. The relative proportion of the mineral and organic solids to the liquid or gas-filled pore space are determined by the particle size distribution of the mineral matter (soil texture) and the binding of these fundamental particles into larger units or aggregates.

Soils in nature are generally found in the form of aggregates. In fact the groups of mineral matrix (as shown in Fig. 4.3) are aggregated together to form larger units or aggregates (crumbs). The aggregated condition, depending upon the mode of arrangement of the groups of mineral matrix, gives a definite pattern to the soil, referred to as **soil structure.** The aggregates of soil are more or less well-defined structural units which are generally much larger than the textural particles, from a few millimeters to several centimeters in diameter.

On the basis of the final shape of these aggregates, following common types of soil structures are generally recognised in nature:

(a) **Prism-like.** Where the vertical axis is more pronounced, with its flattened sides, appearing like a pillar. The top of the pillar may be either round (**columnar**), or flat (**prismatic**).

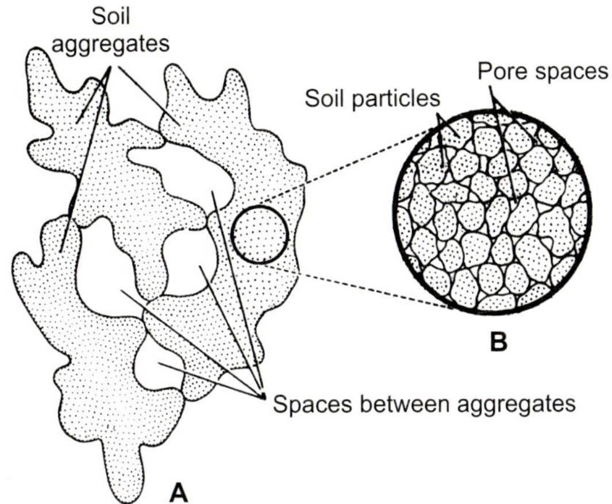

FIG. 4.3

Diagrammatic sketch to show soil structure. A—Soil aggregates and spaces between them. B—A part of soil aggregate to show soil particles and pore spaces between them

(b) **Platy.** Where the horizontal dimensions are more pronounced than the vertical ones. These appear as flattened, compressed pad-like structures, which may be thick (platy) or thin (laminar).

(c) **Block-like.** Where all the three dimensions are of more or less similar size and equally developed, thus making the aggregates as cubical structure, with round or flat faces.

(d) **Spheroidal.** Where all the axes of the aggregates are of more or less same length with their faces curved or irregular. These are further distinguished as **granular**, where the aggregates are less porous, and **crumby**, where the granules are more porous. Spheroidal soils are supposed best for plants' growth.

There are spaces between the mineral particles (Fig. 4.3B) as well as between the larger aggregates (Fig. 4.3A). The spaces, however, between the aggregates are much larger than those between the textural particles. Sandy soils have larger pore spaces (but total space area being lesser) than those in between textural particles of a clay or silt soil.

Detailed structure of soil is investigated by (i) separating and size grading stable aggregates by passing the sample through a nest of suitable-sized sieves with a standardized washing procedure (Kemper and Chepil, 1965), or (ii) by measuring pore space of undisturbed sample (Vomocil, 1965). Earthworms and similar animals are important in creating pore space and in stability of aggregates as they affect aeration of soil. Thus structure of a soil is of considerable importance in relation to root penetration.

Organic matter, the second component of the solid phase of porous media is defined as a combination of (i) live biomass, including animals, microbes, and plant roots; (ii) recognisable dead and decaying biological matter; and (iii) humic substances, the heterogenous polymers formed during decay and degradation of plant, animal and microbial biomass.

Organic matter is the chief source of mineral's return to soil. It has also a much water holding capacity. It can hold minerals as ions by absorption and binds clay particles into aggregates. It makes clay soils porous thus increasing aeration of soil and also causes easy percolation of water in soil.

Soil organic matter (humus) represents the equilibrium between input, originating from primary photosynthetic production, and the degradative and resynthetic processes associated with soil organisms. The input is largely of carbohydrates, lignin compounds, fats and proteins with smaller quantities of free amino acids, alkanes, terpenoides, carotenoids, flavonoids, alkaloids, polyphenols, resins etc. Most of these, as carbohydrates, cellulose etc. are easily decomposed and have short residence time in soil. Soluble sugars, lipids, proteins, amino acids, organic acids etc. are minor constituents of organic matter. However, the major constituents of humus are humic complexes, which make the greatest proportion of it in most soils. They comprise a mixture of diverse phenolic polymers like lignin, resins etc.

Organic matter to soil is contributed by surface-living plants and animals (dead leaves, twigs, branches of plants and bodies of animals) as well as by micro- and macro-organisms living in the soil (a variety of microorganisms and animals and dead roots of plants etc.). In thick forests, in each season there accumulates thousands of tons of litter on the soil surface. All dead, fresh organic matter fallen recently to the ground is called **litter**. In deciduous forests, where all leaves fall at one time, large amounts of litter are produced as a crop each year, whereas in evergreen forests the leaf fall is gradual with simultaneous decomposition. Thus litter here is less pronounced. In grassland, however, there is little accumulation of litter as compared with the forests. This is due to marked differences in the types of vegetations in two habitats. Just beneath the fresh litter there may be present material derived from the litter of previous season in which decomposition might have progressed. This partially decomposed litter is called **duff.** Thus initial input of organic matter is in the form of surface litter and underground dead root matter. But as a result of various soil processes, chiefly activities of micro-organisms in the soil, this organic matter (litter) is grossly modified, which is now known as **humus,** or **soil organic matter** in current practice. Humus is dark, finely divided, amorphous organic matter.

In humus formation the first step is the comminution and transport of the original plant fragments by the soil fauna. After this initial breakdown, commenced by the soil fauna, attack of microorganisms (bacteria, fungi, actinomycetes) increases rapidly and the original material is soon cleared of its labile components like soluble sugars, polysaccharides, proteins and fats, leaving a residue of more resistant lignin and lignin-derivatives. Thus in many soils, a large proportion of the humus is derived from the joint activity of the soil fauna and microorganisms.

The nature of the organic matter is governed both by input and by "soil metabolism", consequently vegetation, climate, parent material and topography all have a strong influence. The composition of the organic matter of soil is thus determined not only by litter chemistry but also influenced by the nature of microbiological resynthesis. The rate of humus formation and its accumulation are thus governed by nature of microorganisms in decomposition, temperature, moisture, aeration and pH of soil.

In temperate climates the range of organic matter types reflects the gross differences between soil groups. Depending upon the nature of parent matter, vegetation and climatic condition, there develop two common types of humus: **mor** humus and **mull** humus. These two terms were originally used in the late 1800s by Muller (see Howard, 1969). These are as follows:

(a) **Mor humus.** Also known as raw humus, mor is characteristic of true podsol (a type of soil) and formed at a low pH (below 3.8–4.0). Moderate rains alongwith a nutrient-deficient, well drained parent matter due to podsolisation produce a mor humus. Earthworms are absent. Thus litter does not become mixed with mineral soil. It decays slowly *in situ* to form O_2 layer of true mor. This is a compact, rather tough, black or brown peaty material in which the individual fragments are easily recognised microscopically as plant material. The most prominent feature of the podsol is the intense eluviation of the surface horizons with mobilisation of organic matter and its redeposition at lower levels. The fully developed podsol is usually very acid and its cation exchange complex is strongly desaturated. Thus, organic matter in acidic conditions does not decompose rapidly due to inhibition of microbial activities. Organic matter thus accumulates as peat or muck.

(b) **Mull humus.** In contrast with mor humus, mull is characteristic of brown forest soils and some other mollisols of temperate climates. Fine-textured, nutrient-rich rock with broad-leaved forest cover develop **mull** humus forms. It is mirobiologically rich than mor, supporting a greater bacterial and fungal population. It is associated with soil pH values above 5.0, an abundance of divalent cations and well-developed earthworm fauna. As a grey, brown-grey or blackish material, it is diffusely incorporated amongst the soil mineral particles by biological mixing. Since microbial decomposition is rapid, no plant or microbial remains are recognisable (*cf* mor) in mull. It is amorphous and colloidal in nature.

It is the humic fraction of soil organic matter that provides the stable, long-term nutrient base to microorganisms in a porous medium. Humic substances extremely have extremely complex structures. During degradation of organic matter in soil, a diversity of chemical components such as sugars, amino acids, lipids and phenolics are released. Some are degraded easily, whereas others like lignin of plant residue resist degradation or degraded slowly. Humic substances range in molecular weight from 700 to 300,000 and can be divided into three fractions based on their chemical structure: fulvic acid, humic acid and humin. Overall humus has a three-dimensional, sponge like structure that contains hydrophobic and hydrophilic areas. Being less polar than water, it provides favourable environment for solutes.

The zone of soil around the roots of plants characterised by intense microbial activity is known as **rhizosphere**, whereas the surface of root itself as **rhizoplane**. Large amounts of readily available materials are exuded from roots in the rhizosphere. These exudates are a mixture of sugars, organic acids and other soluble plant components diffusing out of roots in the rhizosphere. These provide nutrients to rhizosphere microorganisms.

THE LIQUID PHASE OR SOIL SOLUTION

From the point of view of plant nutrients in soil, there are two important chemical components—organic matter and the chemicals in solution. The soil solution is constantly changing. There exists a weak solution of various salts, alongwith other liquids and gases in the soil mass. This soil solution contains almost all the essential minerals. Complex mixtures of minerals as carbonates, sulphates, nitrates, chlorides, and organic salts of Ca, Mg, Na, K etc., are found as dissolved in water. The chemical nature of soil solution depends on the nature of the parent matter and the climatic and other factors (including chemical nature of organic matter) involved in pedogenesis. Thus, soil solutions of rivers, lakes, ponds, forests, and grassland differ widely from each other. These are of much ecological significance.

The **soil solution** is the primary source of inorganic nutrients for plant roots and, in the case of the more important cations, the concentration of this solution represents an equilibrium with the cations absorbed on the exchange complex. The cation exchanging property of soil is mainly due to its clay fraction and this property is caused by the presence of unsatisfied negative charges at the surface of clay and organic particles; metal cations and hydrogen ions are bound at these sites, sufficiently loosely to show exchange with ions in the bathing solution. This porperty of exchanging cation is called **cation exchange capacity** (CEC) of soils. The CEC of sandy soils is associated mainly with their organic content but that of finer-textured soils with their clay content. The presence of ion exchangers in soil is of much importance both in pedogenesis and in the soil-plant nutritional relationship. Most metallic elements which are taken up by plants are absorbed as cations but they exist in three forms in the soil (i) as poorly soluble components of mineral or organic matter, (ii) absorbed onto the cation exchange complex, and (iii) in small quantities in soil solution. Soils with more or less optimal concentrations of various nutrient solutes in them are called **eutrophic,** whereas those with suboptimal concentration of these nutrient salts as **oligotrophic.**

Soil acidity is associated with the presence of hydrogen and aluminium ions on the exchange complex and the existence of an equilibrium solution of hydrogen ions in the interstitial water of soil. It is expressed in terms of pH, which is defined as "the negative longarithm of hydrogen ion activity" where activity is understood to mean effective concentration. pH values of soil show much correlations with the soil type, vegetation type, profile horizon, thus affecting plant's growth, lime requirement and mineral nutrition.

The pH values of natural soils show much variation, ranging from 3.0 to 8.4. However, extreme values also do occur. In some unusually alkaline soils with high sodium carbonate content, pH values reach up to 10.0-10.5, while drained gleys have 2.0 or even less.

Soil pH strongly affects the microbial activities, as below pH 5.0 bacterial as well as fungal activities are reduced. Some soil-borne diseases, like root-rot of cotton, root-rot of tobacco, potato scab etc., may be controlled by lowering the pH of the soil, whereas others like club-root of crucifers and *Rhizcotonia* root rot by increasing the soil pH.

The amount of **soil water** present in the pore space of a medium is an important parameter in understanding the level and type of microbial activity in an environment. Like soil aeration, the water content of soil is also much related to its texture and structure, and also with the magnitude of such physical forces as capillary attraction, cohesion and adhesion.

Thus size of the mineral particles and their shape and number of the pore spaces are important in the amount of water retained by the soil. Sand being coarse-textured with larger particles can hold water only loosely due to bigger pore spaces. The water in such soils generally runs down rapidly reaching to deeper layers. Thus sandy soils are well drained. Clay soils, with much proportion of colloidal fractions as humus etc., can retain much water. Silts, like clays also are able to retain much water. Loams, which are a mixture of sand and silt and/or clay, are supposed as best soils for plants' growth, as such soils are very fertile, being rich in nutrients, have proper aeration, and hold fairly large amount of water.

Soil aeration and soil water are thus greatly influenced by the texture and structure of soils. As indicated above, sandy soils due to bigger pore spaces are unable to hold much water and are thus well drained. Such soils are also called **physically dry soils**. In some soils as those of halophytic conditions, although water is present in abundance, but due to high degree of salinity (much concentration of salts), it is not available to plants. Such soils are referred to as **physiologically dry soils**.

All the water present in the soil is not available to plants. The problem of physical description of the status and movement of water in the non-homogeneous, three-phase matrix of a water-unsaturated soil has exercised soil physicists for many years. Briggs (1897) visualised the retention of water in terms of the forces provided by the curved capillary menisci between soil particles, and proposed the classification of soil water as **hygroscopic, capillary** and **gravitational waters.** As shown in Figure 4.4A, B the water is present in the space (between the soil particles) as well as around the soil particles. In addition to these, there are also some other forms of water retained in the soil, such as **combined water** and the **water vapour**. Thus water in soil is generally present in the following five forms/states (Fig. 4.5):

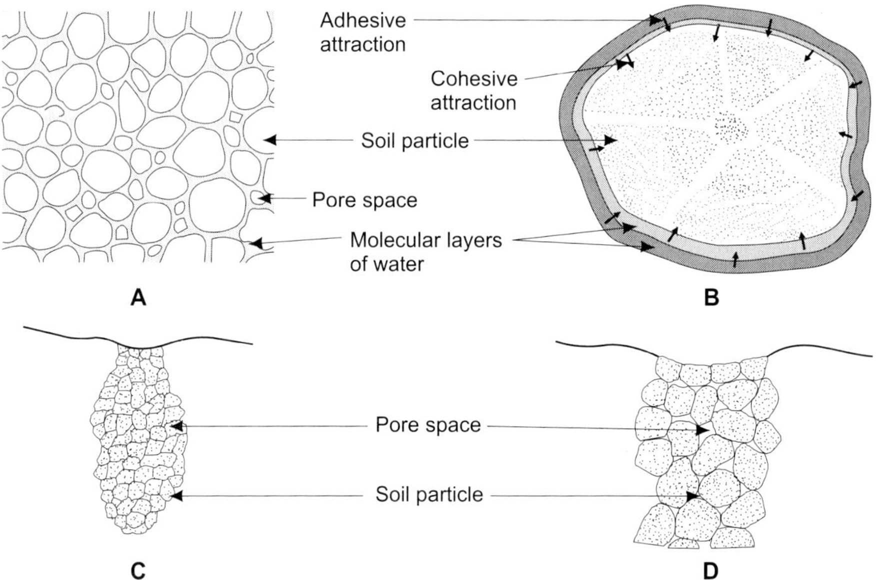

FIG. 4.4

Hypothetical diagrams to show the mode of the presence of water in the soil (A, B) and the relationship of soil texture with size of pore space (C, D). Note that fine-textured soil (C) has smaller pore spaces than coarse-textured soil (D). A—showing the water present in between and at surfaces of the soil particles. B—a soil particle enlarged to show the cohesive and adhesive forces that keep molecular layers of water held up around it

(a) **Hygroscopic water.** This is the water which is held by the surface forces of the soil particles. As shown in Fig. 4.4B soil particle is surrounded by a film of molecular layers of water. The water is held tightly around the particle as a result of cohesive and adhesive forces and it can not be easily removed by the plants.

(b) **Capillary water.** This water in soil is mainly held between spaces (pore spaces) of the soil particles and angles between them, forming a system of capillaries (Fig. 4.4A). This water, held by the interparticle meniscus effects and removed by the gravity is known us **capillary water**. This is the most important form of water to plants.

(c) **Gravitational water.** This is the water that moves downwards through a moist soil in response to gravity. It is thus removed from the soil by gravity alone. In fact gravitational water is surplus to the water retaining capacity (due to surface forces and interparticle meniscus effects) of soil and thus drains from it reaching to deep saturated zone of the earth - **ground water**, the upper surfaces of which is **water table**.

(d) **Combined water.** This is the form of water present as hydrated oxides of aluminium, iron, silicon etc., in the soil.

The above said four forms of water together with their possible relationships with the soil and plant water status, are shown in Fig. 4.5.

(e) **Water vapour.** Some water in soil is also present in its vapour state in the pore spaces between soil particles.

After rainfall, the water enters the soil and it fills the space between soil particles and the volume that can be filled is known as the **pore space** (Fig. 4.4C, D), which varies with the texture and structure of a soil. As stated earlier,

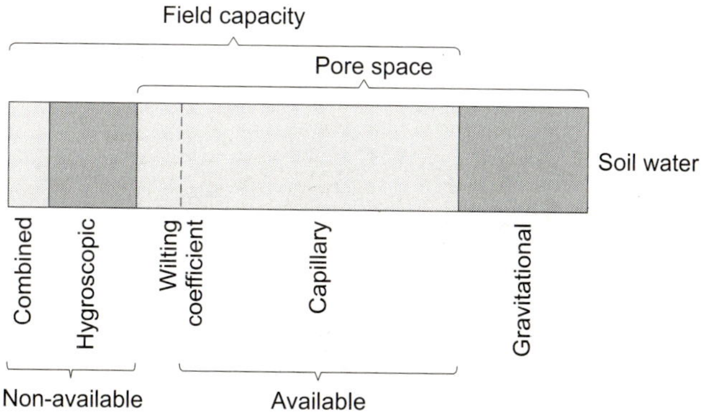

FIG. 4.5

Generalised diagram showing different forms of soil water and their possible relationship with soil and plant water status

all the water present in a soil is not available to plants. The total amount of water present in the soil is called **holard.** Of this, the amount of water that can be absorbed from soil by plants is known as **chresard** or available water and the rest, which is not absorbed by plants is called **echard** or non-available. Thus the amount of water between the full field-capacity as a maximum and the wilting coefficient as a minimum represents the available water (Fig. 4.5).

Water content of the soil alone does not determine the microbial level and their activity in the soil habitat. This is infact the **water movement**, which is generally the major mechanism responsible for the transport of chemicals and microorganisms. Water movement in a porous medium depends on the **soil water potential**, which is the work per unit quantity necessary to transfer an infinitesimal amount of water from a specified elevation and pressure to another point somewhere else in the porous medium. The soil water potential is a function of several forces acting on water, including matric and gravitational forces. Soil water potential is usually expressed in units of pressure (pascals, atmospheres or bars). In saturated zone, due to the regional hydraulic gradient there occurs general horizontal flow of water. In contrast, flow is generally downward in the unsaturated zone. In unsaturated zone, where the soil pores are not completely filled with water, there are several incremental forces that affect water movement. These forces are related to the amount of water present (Fig. 4.6). In very dry soils, there is an increment of adsorbed water that exists as an extremely thin film (order of Å in width). This thin film is very tightly held to particle surfaces by **surface forces** with soil water potentials ranging from –31 to –10,000 atm. This water is, therefore, essentially immobile. As water is added to this soil, a second increment of water forms as a result of **matric** or **capillary forces.** This

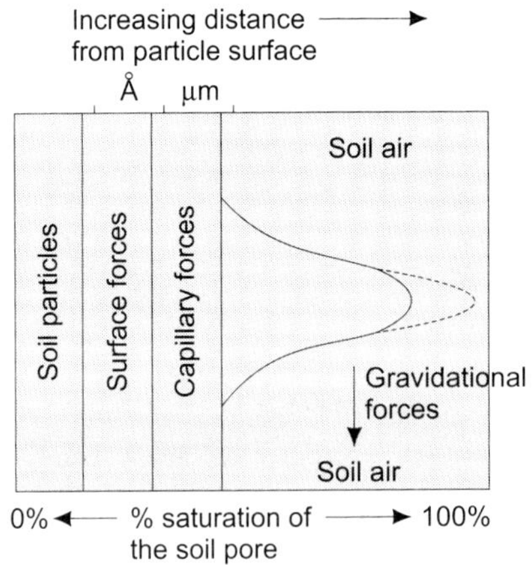

FIG. 4.6

The continuum of soil water

water exists as bridges between particle surfaces in close proximity and can actually fill small soil pores. This results in soil water potentials ranging from –0.1 to –31 atm. This water moves slowly from larger to smaller pores in any direction and is held against gravitational forces. The next increment of water added, free water, can be removed by **gravitational forces**. The soil water potential for free water ranges from 0 to – 0.5 atm. Although, there may be recognised different categories of water, in porous media they actually occur as a continuum rather than with sharply defined boundaries.

THE GAS PHASE OR SOIL ATMOSPHERE

As noted above, there are spaces in between mineral particles (pore spaces) as well as between aggregates of the soil. These spaces are occupied, in additon to various liquids, by the gases, which make the atmosphere of soil. Size and number of such spaces differ in different textural groups of soil. Thus coarse-textured soils like sand have larger pore spaces (but less in number) than the fine-textured soils like clay or silt. Soil aeration is thus influenced by pore spaces. Gases present in the soil atmosphere are more or less similar to those found in the external atmosphere. However, oxygen concentration is less and that of CO_2 higher in soil than those in the atmosphere. For aerobic soils, the nitrogen + argon content remains close to that of external atmosphere (79% v/v) while O_2 and CO_2 vary in complementary proportions to make up the remaining 21%. Anaerobic soils such as waterlogged soils are highly deficient in oxygen. By using suction and diffusion techniques for sampling the soil atmosphere in the fields, it has been shown that appreciable oxygen contents are found in the majority of soils unless they are almost entirely water saturated. From above towards the water table, oxygen content gradually falls and the CO_2 content rises proportionately. Normal, dry soil rarely has a CO_2 content of more than 0.5-1.0% or an O_2 content of less than c. 20%. Between about 15 to 30 cm above the water table, according to soil texture and structure, the CO_2 content reaches c. 5%. With very wet soils the CO_2 and O_2 concentrations may become equal at about 10%. These effects are closely dependent on soil texture and structure. Coarse-textured or well-structured soils have higher gaseous diffusive transfer rates than fine-textured or poorly structured soils under **wet conditions**. This is because they contain many large pores which remain gas-filled. The water-filled pores of the fine-textured soils form potent barriers to gaseous diffusion; O_2 diffuses ten thousand times more slowly in water than in gas. In **dry soil**, the situation is very often reversed, as the fine-textured soils have a greater total pore space and provide a large gas-filled cross-sectional area for diffusion. The major part of the gaseous movement in the soil is diffusive, though may also occur by mass flow. In this way, soil aeration regimes are very important for microbial activities in any porous medium.

ENVIRONMENT AND MICROORGANISMS IN SURFACE SOIL AND SUBSURFACE HABITATS

Both, surface soils and subsurface environments are three-phase porous medium systems composed of solid, liquid and gaseous phases. Differences between these environments are site dependent, based on a combination of the mineral and biological components of the porous medium These components in turn depend on the geology, topography, and climate of the site. Even at a given site, there may be differences in the porous media making up the surface soils and subsurface environments (both, vadose and saturated zones). Such differences bring out variations in the number and activity of microbes in these environments. In the following pages, we would discuss briefly the characteristics of the environment and microbial communities of **surface soils**, the **vadose zone** and the **saturated zone**.

Surface Soils

Surface soils are the weathered end product of the **five soil forming factors.** These factors are, **parent material, topography, time, climate** (rainfall, temperature, humidity/evaporation, wind) and **live organisms,** also called the **biosphere** (plants/ phytosphere, animals/zoosphere, microorganisms). Thus soil formation at a given site involves action of climate and live organisms on a specific soil parent material (which may be igneous, sedimentary or metamorphic rock), with a given topography over a given time period.

During soil formation, distinct layers or horizons often form. These horizons make up the **soil profile** which can be viewed by excavating a section and observing the vertical layering of the same. A **soil horizon** is defined as "a layer which is approximately parallel to the soil surface and that has properties produced by soil forming processes but that are unlike those of adjoining layers."

Although, profiles of different types of soil differ markedly in respect of their physico-chemical and biological properties, Figure 4.7 shows a hypothetical soil profile with its principal horizons. It is not always true that all these horizons are always present in each profile. The different horizons of a soil profile historically in Russian terminology, were classified into ABC terminology. However, according to the present one in general use, the soil profile (Fig. 4.7) consists of the following **five** main horizons:

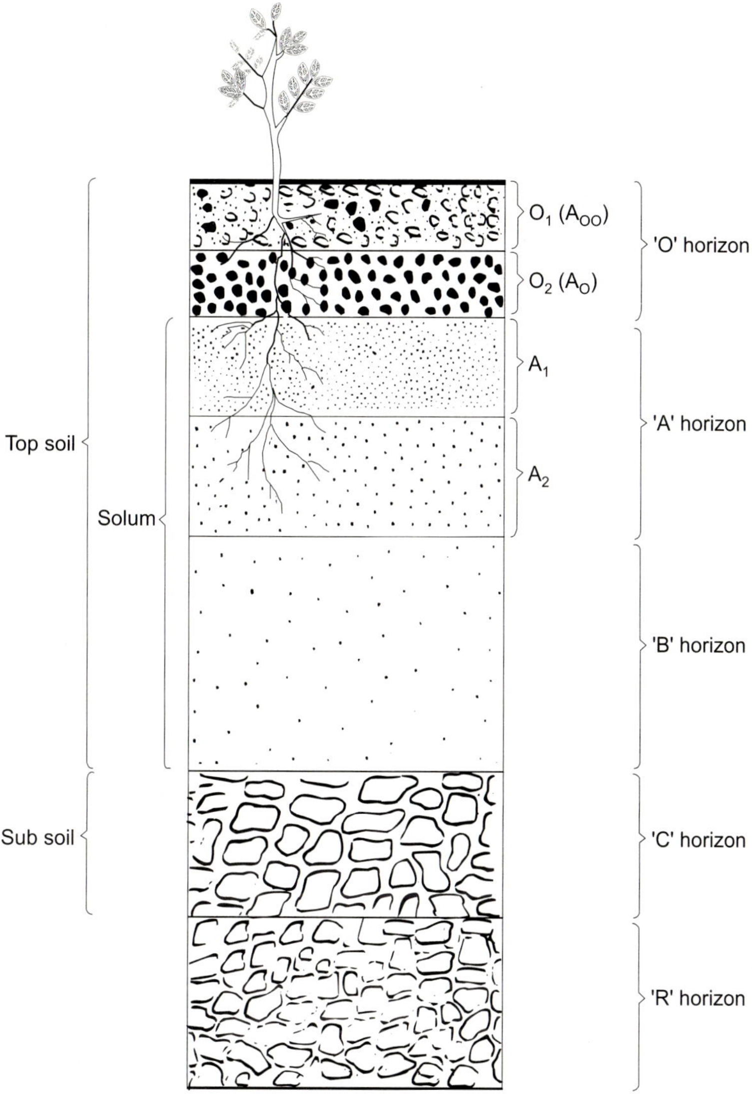

FIG. 4.7

Hypothetical diagram of the soil profile to show principal horizons

The 'O' horizons

These are the organic horizons forming above the surface of the mineral matrix, mainly composed of fresh or partially decomposed organic matter. This horizon is well developed in forests and may be completely absent in grasslands. This horizon is divided into following two sub-layers:

1. **O_1 (A_{00}) region.** This is the uppermost layer consisting of freshly fallen dead organic matter as dead leaves, branches, flowers and fruits, dead parts of animals etc. These do not show evident breakdown.
2. **O_2 (A_0) region.** It is just below the O_1 region in which decomposition has begun. Thus organic matter is found under different stages of decomposition and microorganisms like bacteria, fungi, actinomycetes are frequently found. Upper layers contain detritus in initial stage of decomposition, in which material can be faintly recognised, whereas the lower layers contain fairly decomposed matter, the **duff.**

The 'A' horizons

These are the mineral horizons formed either at or adjacent to the surface. These are rich in organic matter and/or show downward loss (eluviation) of soluble salts, clay, iron or aluminium, being consequently rich in silica or other resistant minerals. This is thus also known as zone of **eluviation** - downward loss or leaching.

This horizon is divided into following two sub-layers:

1. **A_1 region.** It is dark and rich in organic matter. The amorphous, finely divided organic matter here becomes mixed with the mineral matter, which is now known as humus, which is dark brown or black coloured. This region having a mixture of finely divided organic matter and the mineral matter is also called **humic** or **melanized region.** In forest soils this region is less deeper than those of the grasslands.
2. **A_2 region.** This region is of light colour in which the mineral particles of large size as sand are more, with little amount of organic matter. Chiefly in areas with heavy rainfall, the mineral elements and organic chemicals are rapidly lost downwards in this region, making it light-coloured. This is thus also known as **podsolic** or **eluvial zone** or zone of leaching.

The 'B' horizons

These are the mineral horizons forming below the surface in which one or more of the following features can be present (i) enrichment with inwashed clay (lessivation), iron, aluminium, manganese or organic matter. (ii) residual enrichment with sesquioxides or silicate clays which has occurred other than by the removal of carbonates or readily soluble salts. (iii) sesquioxide coatings of mineral grains sufficient to give a more intense colour than horizons above or below. (iv) alteration of the original rock material to give silicate clays or oxides in conditions where (i), (ii) and (iii) do not apply.

This is just below the 'A' horizon, and can also be divided into B_1 (A_3), B_2 and B_3 regions, depending upon the stages of soil development in the area. B_1 (A_3), if present is in the process of successional development, the presence of which depends upon the extent of development of horizons, above and below it.

The 'B' horizon is dark-coloured and coarse-textured due to the presence of silica-rich clay organic compounds, hydrated oxides of' aluminium, iron etc. Since, the chemicals leached from A_2 region, become collected in this horizon, it is also known as zone of **illuviation** or **illuvial zone.** This zone is poorly developed in dry area.

A_1, A_2 and B collectively are also known as mineral soil or **solum**.

The 'C' horizons

These are the **mineral horizons** below the 'B' but excluding true bedrock and without any characteristics of 'A' or 'B' horizons. It consists of incompletely weathered, large mases of rocks.

The 'R' horizons

This is the parent, unweathered, bedrock, upon which there is collected water.

A large proportion of the terrestrial environment contains surface soils that are unsaturated and aerobic except brief periods of rainfall and irrigation. A significant portion of these environments contains saturated or wetland environments, including swamps, marshes and bogs. These ecosystems are of increasing interest to environmental microbiologists for their potential to treat polluted waste streams such as sewage effluents and acid mine drainage.

For most of the time these areas remain saturated. Since wetlands are very productive (in terms of vegetative growth) and also inhibit aerobic activity of microbes they accumulate high proportions of organic matter (> 20% of a dry wt. basis). Particularly in temperate areas, these ecosystems are very important. Bogs are quite extensive world over. These are composed of deep layers of waterlogged peat and a surface layer of vegetation. Peat is composed of dead remains of plants accumulated since last more than thousands of years (since production rates of biomass exceed decomposition rates). Highly anaerobic conditions prevail and also these are highly acidic (p^H 3.2 to 4.2). Under such conditions microbial growth is suppressed. Canada has the most extensive bog systems in the world, with 129,500,000 ha. or 18.4% of the land area composed of bogs.

Distribution and Microbial Diversity in Surface Soils

There are diverse microbes in surface soils, and culturable microorganism concentration can reach 10^8/g of dry soil, though direct counts may be still larger. In general microbial colonies are found in a nonuniform patchlike distribution on surface of soil particles. Despite the large number of microbes, they make up only a small fraction of total organic carbon and a very small proportion of the soil volume (0.001%) in most soils.

The weathered component of minerals (a source of micronutrients) and organic matter (a carbon and nitrogen source) are two of the primary differences between surface soils and subsurface materials as environments for microorganisms. These differences in nutrient content are reflected in a higher and more uniform distribution of number of microorganism and their activity in surface soil environments. Microbial distribution is also dependent on soil texture and structure. Pore space is an important factor as it also controls water content to some extent. Most microbes are attached, sorbed to solid surfaces, and the remainder are free-living. Exopolysaccharides formed during microbial growth create a pseudoglue that helps in orienting adjacent clay particles and cementing them together to form a microaggregate.

Surface soils harbor indigenous populations of bacteria (including actinomycetes), fungi, algae and protozoa. Besides these, viruses pathogenic on these microbes may also be found. **Bacteria** are invariably the most abundant organisms found in surface soils in terms of numbers. There have been estimated over 10,000 species of bacteria. Culturable numbers vary with specific sites, and may be 10^7 to 10^8 cells/g of soil, whereas total populations (including viable and nonculturable) can exceed 10^{10} cells/g. Anaerobes increase with soil depth. **Actinomycetes** are important component, especially under high p^H, high temperature, or water stress. Their number is generally smaller than bacteria. They are unique to utilise chitin, cellulose and hemicellulose found in soil.

Indigeneous soil bacteria can be classified on the basis of their growth characteristics and affinity for carbon substrates into two broad categories: (i) **k-selected** or **autochthonous,** which metabolise slowly in soil, utilising slowly released soil organic matter as a substrate, and (ii) **r-selected** or **zymogenous,** that are adapted to intervals of dormancy and rapid growth, depending on substrate availability, following the addition of fresh substrate or amendment to the soil. Bacterial genera that dominate the typical surface soils and those critical to environmental microbiology have been listed in Tables 4.1, 4.2 and 4.3.

Fungi other than yeasts are aerobic and are abundant in most surface soils. Their numbers range from 10^5 to 10^6/g of soil. Despite their low numbers compared to bacteria, fungi usually contribute a higher proportion of the total soil microbial biomass (Table 4.4). This is due to their relatively large size (hyphae 2 to 10 µm in diam.) Yeasts growing anaerobically (fermentation) are less numerous (up to 10^3/g soil) than aerobic mycelial fungi. Fungi are most abundant in 'O' and A horizons. They are found associated with rhizosphere as well as nonrhizosphere soils.

TABLE 4.1 Characteristics of bacteria, actinomycetes, and fungi of surface soils

Characteristic	*Bacteria*	*Actinomycetes*	*Fungi*
Population	Most numerous	Intermediate	Least numerous
Biomass	Bacteria and actinomycetes	have similar biomass	Largest biomass
Degree of branching	Slight	Filamentous, but some fragment to individual cells	Extensive filamentous forms
Aerial mycelium	Absent	Present	Present
Growth in liquid culture	Turbid	Pellets	Pellets
Growth rate	Exponential	Cubic	Cubic
Cell wall	Murein, teichoic acid, and lipopolysaccharide	As in bacteria	Chitin or cellulose
Fruit bodies	Absent	Simple	Complex
Competitiveness for simple organics	Most competitive	Least competitive	Intermediate
Nitrogen fixation	Yes	Yes	No
Oxygen relations	Aerobic, anaerobic	Mostly aerobic	Aerobic except yeast
Moisture stress	Least tolerant	Intermediate	Most tolerant
Optimum pH	6–8	6–8	6–8
Competitive pH	6–8	> 8	< 5
Competitiveness in soil	All soils	Dominate dry, high-pH soils	Dominate low-pH soils

TABLE 4.2 Dominant culturable soil bacteria in surface soils

Organism	*Characteristics*	*Function*
Arthrobacter	Heterotrophic, aerobic, gram variable. Up to 40% of culturable soil bacteria.	Nutrient cycling and biodegradation.
Streptomyces	Gram-positive, heterotrophic, aerobic actinomycete. 5-20% of culturable bacteria.	Nutrient cycling and biodegradation, antibiotic production by *Streptomyces scabies.*
Pseudomonas	Gram-negative heterotroph, aerobic or facultatively anaerobic, possess wide array of enzyme systems, 10-20% of culturable bacteria.	Nutrient cycling and biodegradation, including recalcitrant organics, biocontrol agent.
Bacillus	Gram-positive, aerobic heterotroph, produce endospores, 2-10% of culturable soil bacteria.	Nutrient cycling and biodegradation, biocontrol agent, *(Bacillus thuringiensis).*

Fungi are important components of the soil with respect to nutrient cycling and especially decomposition of organic matter, both simple sugars and complex polymers like cellulose and lignin. Their role in decomposition is increasingly important as the soil p^H declines because fungi tend to be more acid tolerant than bacteria. They are important in development of soil structure also as their hyphae physically entrap soil particles. Several species of fungi are important root-parasites and some of them affect plant growth in mycorrhizal association with their roots.

TABLE 4.3 Examples of important heterotrophic surface soil bacteria

Organism	*Characteristics*	*Function*
Actinomycetes (*Streptomyces*)	Gram-positive, aerobic, filamentous	Produce geosmins "earthy odor," and antibiotics
Bacillus	Gram-positive, aerobic, spore former	Carbon cycling, production of insecticides and antibiotics
Clostridium	Gram-positive, anaerobic, spore former	Carbon cycling, (fermentation), toxin production
Methanotrophs (*Methylosinus*)	Aerobic	Methane oxidizers that can cometabolize trichloroethene (TCE) using methane monooxygenase
Alcaligenes eutrophus	Gram-negative, aerobic	2,4-D degradation via plasmid pJP4
Rhizobium	Gram-negative, aerobic	Symbiotic nitrogen fixation with legumes
Frankia	Gram-positive, aerobic	Symbiotic nitrogen fixation with nonlegumes
Agrobacterium	Gram-negative, aerobic	Important plant pathogen, cause crown gall disease

TABLE 4.4 Range of approximate biomass of each major component of the biota in a typical grassland soil (Killham, 1994)

Component of soil biota	*Biomass (tons/ha)*
Plant roots	Up to 90 but generally about 20
Bacteria	1–2
Actinomycetes	0–2
Fungi	2–5
Protozoa	0–0.5
Nematodes	0–0.2
Earthworms	0–2.5
Other soil animals	0–0.5
Viruses	Negligible

As stated above fungi alongwith other microbes, chiefly bacteria (including actinomycetes) play vital role in the decomposition of organic matter in soil, thus releasing the nutrients locked up in the dead organic matter of plant, animal and microbial matter and bringing about the recycling of nutrients in nature. In soil, microbes oxidise organic carbon to CO_2 and liberate bound materials. The decomposition process actually begins while the plants are still intact in their environments i.e. at the senescent stage and this continues till plant remains are totally decomposed. The microbes appearing at different stages differ in their species–composition and are infact nutritional and ecological groups of microbes. The sequence of the appearance of microbes on decaying plant remains, both on and in the soil has been worked out mostly for fungi. It has been shown that the primary colonisers on senescent plant parts are likely to be weak parasites and/or saprophytes of sugar fungi group (mucorales). On dead tissues, the ascomycetes and their imperfect forms (cellulose decomposers) are the secondary colonisers. At this stage, some mucorales may also appear as secondary saprophytic sugar fungi. Finally, the basidiomycetes (lignin decomposers) appear in the succession. Thus chemical make-up of the substrates determines the qualitative features of the fungal flora appearing at a particular stage of succession. This scheme is shown below:

Senescent tissue		**Dead tissue**	
Primary colonisers		**Secondary colonisers**	
Stage 1a	Stage 1	Stage 2	Stage 3
Weak parasites	Primary saprophytic sugar fungi, living on sugars and carbon compounds (mucoraceous phycomycetes)	Cellulose decomposers and associated secondary saprophytic sugar fungi, sharing products of cellulose decomposition (ascomycetes and some mucorales)	Lignin decomposers and associated fungi (basidiomycetes and others)

→ Fungal succession

Microbes, such as fungi and bacteria (including actinomycetes) play a vital role in the formation of brown or black organic complexes which remain in a dynamic stage, through the decomposition of organic materials and through the synthesis of substances that also become part of the humus. Humus is sparingly soluble in water but is rendered into solution by alkali. It contains amino acids, purines, pyrimidines, aromatics, uronic acids, amino sugars, pentose and hexose sugars, sugar alcohols, methyl sugars, aliphatic acids etc. Humus is the finally brokendown amorphous state of dead organic matter. In humus formation, microbes attack and degrade plant and animal remains. The plant residues comprise chiefly **cellulose, hemicellulose, pectic substances** and **lignin**.

Cellulose decomposing fungi include chiefly ascomycetes and deuteromycetes. These are chiefly species of the genera, *Aspergillus, Penicillium, Chaetomium, Trichoderma, Fusarium, Stachybotrys, Memnoniella, Humicola, Phoma, Cladosporium, Altemaria, Acremonium, Myrothecium, Thielavia* etc. They have a strong cellulose synthesising machinery, which breakdown cellulose to simple sugars. Three types of hemi-cellulases are involved in the degradation of **hemicellulose**. These are endo-enzymes, exo-enzymes and glycosidases. The fungi involved in their breakdown belong to all major groups. Most important ones are species of *Alternaria, Aspergillus, Chaetomium, Fusarium, Glomerella, Penicillium* and *Trichoderma*.

Pectic substances, the primary constituents of the middle lamella and primary cell wall are brokendown also by microbial pectinases. Pectinases belong to three major categories (i) hydrolytic that carry out hydrolytic cleavage of the pectin polymers (ii) transeliminases that bring about transeliminative cleavage of glycosidic links by the action of lyases, and (iii) pectinesterases which hydrolyse the methoxyl groups of the pectins, converting them to pectinic acids. Pectinases are produced by various species of *Rhizopus, Aspergillus, Penicillium, Monilia, Fusarium, Rhizoctonia, Geotrichum* etc.

Lignin, chiefly in woody parts is degraded by basidiomycetes, such as aphyllophorales and agaricales. Lignolytic enzymes of these microbes include phenol oxidases, laccases and peroxidases.

Besides fungi, bacteria are also involved in decomposition of organic matter. These include mostly aerobic bacteria, such as pseudomonads, bacilli and actinomycetes. Some anaerobic bacteria, including the characteristic methanogenic bacteria, growing in primitive conditions are also involved in carbon cycle of nature.

Algae being photoautotrophic are predominantly present in areas where sunlight can penetrate the very surface of the soil. They could be found at a depth of 1 m as some green algae and diatoms can grow hetero-as well as photoautotrophically. Generally, their populations are highest in the surface 10 cm of soil, where their populations range from 5,000 to 10,000/g of soil. A visible algae bloom developing in surface soil may contain millions of algal cells/g of soil. Algae are pioneer colonisers of barren volcanic areas, desert soils and rock surfaces. Their populations may exhibit seasonal variations. Four major groups of algae are found in surface soils. The green algae are most common in acidic soils. Members of chrysophycophyta (yellow-green algae) as *Navicula* and *Botrydiopsis* are also found. Red algae, as *Prophyridium* may also be present. Besides these species of *Nostoc* and *Anabaena,* more appropriately classified as bacteria are also common in soils. In tropical soils the cyanobacteria predominate among algal groups.

Protozoa, the unicellular eukaryotic microbes range up to 5.5 mm in length, though most are much small (Table 4.5). Most of them are heterotrophic on bacteria, yeast, fungi and algae. Due to their large size and predatory on smaller microbes protozoa are found mainly in the top 15 to 20 cm. of the soil. They are usually concentrated near root surfaces that have higher densities of bacteria and other microbes. Soil protozoa are flatter and more flexible than aquatic protozoa. Protozoa of each of three major groups—the flagellates, amoebae and the ciliates are found, but amoebae are usually the most abundant in soil environment. Numbers of protozoa reported range from 30,000/g of soil of nonagricultural land to 350,000/g of that from a maize field to 1.6×10^6/g soil from a subtropical area.

TABLE 4.5 Average dimensions of soil protozoa compared with bacteria (Ingham, 1998)

Group	*Length (μm)*	*Volume (μm^3)*	*Shape*
Bacteria	< 1–5	2.5	Spherical to rod-shaped
Flagellates	2–50	50	Spherical, pear-shaped, banana-shaped
Amoebae			
Naked	2–600	400	Protoplasmic streaming, pseudopodia
Testate	45–200	1000	Build oval tests or shells made of soil
Giant	6000	4×10^9	Enormous naked amoebae
Ciliates	50–1500	3000	Oval, kidney-shaped, elongated and flattened

Roles of Microorganisms in Surface Soils

Microorganisms in surface soil environment are involved in several vital processes, including the management of quality of deteriorating environment. Some of the important activities of these microbes are as follows:

(i) Soil formation. Besides physical and chemical processes, microbiological processes are also important in soil formation. Microbes participate in the formation and stabilisation of soil aggregates in soil structure. This occurs in several ways. Soil particles are bound together by filamentous microbes that grow over the surfaces of adjacent particles to create an extensive network of hyphae that connect the soil particles involved. Microbes also cause small clay particles to reorient along the microbial surface, that results into compaction of clay particles helping in aggregate formation. Aggregate formation is also facilitated by the production of exo-polysacharides by microbes as well as plant roots.

(ii) Nutrient cycling. In nature the balance between uptake of different nutrients and their return back to the environment is maintained through respective nutrient cycles involving different categories of the microorganisms. For instance since the life originated on earth, the delicate balance between carbon fixation and oxygen production, on one hand, and consumption of both fixed carbon and oxygen, on the other hand is maintained by the carbon cycle. Not only carbon, other elements-oxygen, nitrogen, sulphur, phosphorus and iron are also actively cycled (Chapter 9). It is interesting from an environmental microbiology viewpoint that these nutrient cycles not only depend largely on microbial activity but parts of these cycles are highly relevant to several economically important areas, including bioremediation, municipal waste disposal, sustainable agriculture and mining. These aspects are discussed in detail in Chapters 10, 11, 12, 13, and 15.

(iii) Bioremediation. In their structure many of the organic pollutants spilled or improperly disposed of in the environment are more or less similar to naturally occurring substances. However, the success of any bioremediation is site specific depending on the soil type, the indigenous microbes, environmental conditions and the type and quantity of the pollutant. Quite recently there has grown much interest in the use of microbes in bioremediation of contaminated sites. This topic is discussed in detail in Chapter 11.

(iv) Municipal waste disposal. The practice of disposal of biosolids produced at wastewater treatment plants on agricultural land is called **landfarming**. The sludge left after treatment of municipal waste is treated by indigenous microbes to breakdown more recalcitrant organics of the sludge. The sludge contains solid as well as liquid wastes. Role of microorganisms in municipal waste treatment and disposal is discussed in detail in Chapter 15.

The Vadose Zone

The vadose zone is defined as the unsaturated oligotrophic environment lying between the surface soil and the saturated zone. This zone contains mostly unweathered parent materials and has a very low organic carbon content (generally $< 0.1\%$). Thus when compared with the surface soils, availability of carbon and micronutrients is very limited in this zone. This zone is narrow in shallow or near the surface habitats or even absent as in wetlands, whereas many be up to hundreds of meters thick in several arid or semiarid areas of the world. This region is of microbiological interest because the pollutants originating from surface environments must pass through this zone before they can reach groundwater. These zones are receiving more attention of environmental microbiologists for their possible use in future to alter pollutants moving through them or to restrict their transport to groundwaters and potential drinking water supplies.

Microbial activities in vadose zone

Until recently it was thought the deep subsurface environments, both the unsaturated (vadose) as well as saturated ones, contain few if any microorganisms due to extreme oligotrophic conditions found there. Interest in this area began in the 1920s when increased oil demand led to increased oil production and exploration.

An examination of water extracted from deep within oil fields by geologists in U.S.A. revealed significant levels of hydrogen sulphides and bicarbonates, and it was confirmed that sulphate-reducing bacteria were responsible for their presence in such waters. Microbiologists could also detect the presence of such bacteria in cultures made from such water samples. During 1980s with the growing concern over groundwater quality, more information could be obtained about the presence and activity of microorganisms in samples from deep cores in both unsaturated and saturated zones of soil.

Several studies have now been made on microbial complex of deep cores in the unsaturated vadose zone of soils, including 70-m core from the semiarid high desert areas. Table 4.6 shows a comparison of bacterial counts in the surface and subsurface samples from such a site. It may be seen that direct as well as viable counts, when compared to surface samples declined in deep subsurface samples. The majority of the subsurface isolates are Gram-positive and strictly aerobic. In contrast, in surface soils Gram-negative bacteria were more in numbers. Thus, like ambient air, subsurface atmosphere was found aerobic. In general, microbial numbers and activities are higher in paleosols (buried sediments) that had exposure to earth's surface and plant production. As compared to surface soils, the present understanding of deep vadose zone is limited. It is very likely that there may exist areas in vadose zones differing in microbial numbers and their activities. But, when compared with surface soils, rates of metabolic activities are much lower in deep vadose zones.

TABLE 4.6 Bacterial counts in surface and 70-m unsaturated subsurface environments

Sample site	*Direct counts (counts/g)*	*Viable counts (CFU/g)*
Surface (10 cm)	2.6×10^6	3.5×10^5
Sub-surface basalt-sediment interface (70.1 m)	4.8×10^5	50
Sub-surface sediment layer (70.4 m)	1.4×10^5	21

Saturated Zone or Underground Aquifers

This zone occurs below the vadose zone and remains saturated with water. The characteristics of the environment and microbial communities of this zone have already been briefly considered in Chapter 3 in section on groundwater environments.

5

AEROMICROBIOLOGY

Chapter Outline

- *Allergic Disorders by Air Microflora*
- *Distribution of Microbes in Air*
- *Aeromicrobiological Pathway*
- *Outdoor Aeromicrobiology*
- *Indoor Aeromicrobiology*
- *Bioaerosol Control in Laboratory*

The term **aerobiology** was coined in the 1930s by F.C. Meier in relation with a project concerned with study of life in air. Aerobiology has been defined in various ways as "the study of aerosolisation, aerial transmission, and deposition of biological materials" or more specifically as "the study of diseases that may be transmitted via respiratory route". This relatively a new science has become increasingly important because of its concern with diverse scientific fields like public health, environmental science, industrial engineering, biological warfare and space exploration. **Aeromicrobiology** in its wider sense is the study of all forms of microbial life in air. To an environmental microbiologist, aeromicrobiology is "the study of various aspects of intramural (indoor) and extramural (outdoor) aerobiology in relation with the airborne transmission of environmentally relevant microorganisms including viruses, bacteria, fungi, yeasts, and protozoans". Besides these microbes, other biological materials like pollens, insect debris, animal danders (chief source of allergic disorders) are also found in air.

It is well known that airborne microbes are important pathogens causing diseases in plants, animals and humans. These pathogens spread by aeromicrobiological pathway (AMB pathway). Some of the important airborne human pathogens are as follows:

Diseases	**Pathogens**
(a) Viral	
Influenza	Influenza virus
Chicken pox	Varicella virus
Small pox	Variola virus
Measles	Rubeola virus
German measles	Rubella virus
Dengue fever	Flavivirus
(b) Bacterial	
Diphtheria	*Corynebacterium diphtheriae*
Pertussis (Whooping cough)	*Bordetella pertusis*
Meningitis (Meningococcal)	*Neisseria meningitidis*
Tuberculosis	*Mycobacterium tuberculosis*
Klebsiella pneumonia	*Klebsiella pneumoniae*
Serratia pneumonia	*Serratia marcescens*
Q fever	*Coxiella burnetii*

Pneumococcal pneumonia	*Streptococcus pneumoniae*
Strep throat, scarlet fever	*Streptococcus pyogenes*
(c) ***Fungal***	
Aspergillosis	*Aspergillus fumigatus*
Blastomycosis	*Blastomyces dermatidis*
Coccidioidomycosis	*Coccidiodes immitis*
Cryptococcosis	*Cryptococcus neoformans*
Histoplasmosis	*Histoplasma capsulatum*
(d) ***Protozoal***	
Pneumocystosis	*Pneumocystis carinii*

ALLERGIC DISORDERS BY AIR MICROFLORA

Microflora of air is responsible for several allergic disorders. A range of airborne particles, such as pollen, fungal spores, insect debris, animal danders, mites etc. are recognised allergens. They bring about such allergic disorders as bronchial asthma, allergic rhinitis and atopic dermatitis. In most countries pollen has been implicated and studied in relation to these allergic disorders. However, it has now been established that fungal spores are also equally important allergic agents. Fungal aerosols are present in much higher densities than pollen in some environments under some climatic conditions. A thorough screening of diurnal, seasonal and annual variations in both indoor as well as outdoor environments is necessary for diagnosis and therapeutic management of these disorders. In indoor environment, such as home and occupational surroundings, persons are exposed to fungal allergens. Fungal aeroallergens in the outdoor environments originate from such sources as cereal crops, decaying vegetable matter and organic debris. In indoor environment damp walls, dustbins, mattresses, window frames, humidifiers etc. are chief source of fungal allergens. Since first report of a fungus implicated in asthma in 1726 by J. Floyer, spores of a range of fungal taxa have been shown of clinical interest and being studied world over for their implications in respiratory allergic reactions. **Aeromycology** has now become an established field of research, involving scientists of basic sciences as well as professional areas.

Several molds (species of *Aspergillus, Penicillium, Mucor, Rhizopus*), yeasts, and rust and smut spores are shown to be implicated in occupational allergy. On the basis of prik and intradermal skin tests, several fungi have been identified as agents of Type I hypersensitivity disorders. These include species of *Alternaria, Aspergillus, Candida, Cladosporium, Curvularia, Drechslera, Epicoccum, Fusarium, Mucor, Nigrospora, Penicillium, Rhizopus* and *Trichoderma.*

Besides skin test, fungal allergy is diagnosed by the use of inhalation challenge, RAST, ELISA, immunodor and immunoblot techniques. These immunological techniques have proved to be useful in the diagnosis and treatment of pollen and fungal allergic disorders.

India is not only a sub-continent of rich biodiversity but also has various geo-climatic zones. In a given zone there are experienced marked seasonal fluctuations in temperature and relative humidity. Consequently the sources and nature of aero-allergens differ in different areas and in different seasons of a given area. For instance, pollen becomes important inhalant allergens chiefly in two seasons- (i) months of September- November, dominated by pollens of weeds and grasses, and (ii) months of March-May, dominated by pollens of trees.

Airborne microbes produce potent toxins. Very small doses of these toxins may become lethal to humans. The botulinal A toxin produced by *Clostridium botulinum* is a potential biological warfare agent. Though this neurotoxin is normally associated with consumption of contaminated food, a lethal dose is so small that aerosolisation can also be a means of its transmission. The lethal dose for this toxin by inhalation is 0.3 µg of toxin (death expected 12 hr. after exposure). Other examples of airborne toxins are lipopolysaccharides (LPS). These, also referred to as endotoxins of Gram-negative bacteria, cause respiratory distress syndromes.

Bioaerosols

Bioaerosols vary cosiderably in size (Fig. 5.1). Their composition depends on several factors including the type of microbe or toxin, type of particles they are associated with (as mist or dust), and the gases in which the bioaerosol is suspended. In general, bioaerosols range from 0.02 to 100 µm in diameter and are classified on the basis of their size into different modes as follows:

Mode type		Size range (µm in diameter)
Nuclei	Fine particles	< 0.1
Accumulation		0.1 to 2.0
Coarse		> 2.00

The composition of bioaerosols can be liquid or solid or a mixture of the two and should be thought of as microorganisms associated with airborne particles or as airborne particles containing microorganisms. It is rare to find microbes not associated with dust or mist or other airborne particles in the atmosphere.

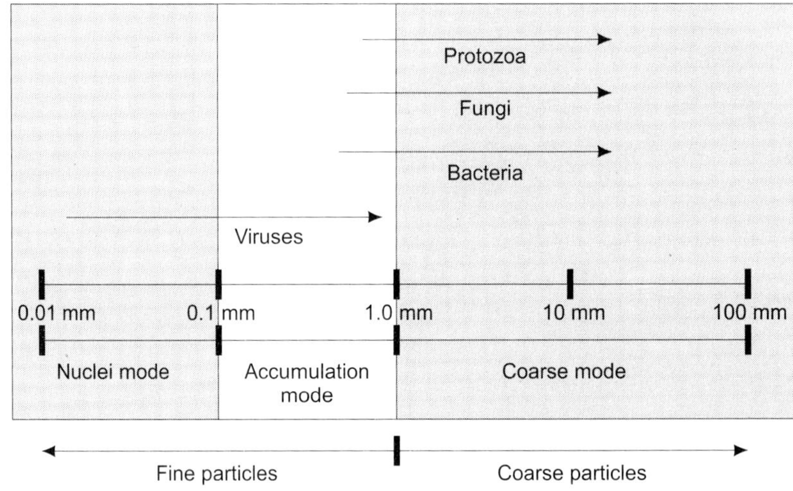

FIG. 5.1

Diagrammatic sketch showing the relative sizes of bioaerosols. The depictions of the various kinds of microbes are indicative of their potential sizes when associated with airborne particles (rafts). The terminologies used for the various sizes of the bioaerosols are also indicated

DISTRIBUTION OF MICROBES IN AIR

The **atmosphere** as a habitat is characterised by high light intensities, extreme temperature variations, low amount of organic matter and a scarcity of available water making it a non-hospitable environment for microorganisms and generally unsuitable habitat for their growth. Nevertheless, substantial number of microbes are found in the lower regions of the atmosphere (Table 5.1).

TABLE 5.1 Microbes found in the atmosphere

Types of microbes	*Percentage*
Bacteria	
Gram-positive pleomorphic rods, such as *Corynebacterium*	20
Gram-negative rods, such as *Achromobacter, Flavobacterium*	5
Endospore-forming genera, like *Bacillus*	35
Gram-positive cocci, like *Micrococcus*	40
Fungi	
Cladosporium	80
Alternaria	5
Penicillium	2
Others (*Aspergillus, Chaetomium, Fumago, Fusarium, Helminthosporium, Drechslera, Sclerotinia, Stachybotrys, Memnoniella, Trichoderma, Verticillium*)	13

The atmospheric layer of most interest in aeromicrobiology is the **boundary layer,** the earth's atmosphere extending to a height of about 0.1 km. from the earth's surface. Though, it is not uncommon to find microbes associated with the layers of troposphere above the turbulent boundary layer, it is the one that is largely responsible for the transport of particles over both short and long distances. The boundary layer consists of three parts: the laminar boundary layer, the turbulent boundary layer, and the local eddy layer. The **laminar boundary layer** is a layer of still air associated with the earth and projecting surfaces (solid or liquid), and is 1 µm to several meters thick. The **turbulent boundary layer** is always in motion and responsible for horizontal transport phenomena (wind dispersion). The **local eddy layer** is the actual zone of interaction between the still laminar boundary layer of surface projections and the turbulent boundary layer.

No microbes are indigenous to the atmosphere, rather they represent allochthonous populations transported from aquatic and terrestrial habitats into the atmosphere. Microbes of air within 300-1,000 or more feet of the earth's surface are the organisms of soil that have become attached to fragments of dried leaves, straw or dust particles, being blown away by the wind. Species vary greatly in their sensitivity to a given value of relative humidity, temperature and radiation exposures. More microbes are found in air over land masses than far at sea. Spores of fungi, especially *Alternaria, Cladosporium, Penicillium* and *Aspergillus* are more numerous than other forms over sea within about 400 miles of land in both polar and tropical air masses at all altitudes up to about 10,000 feet.

Microbes found in air over populated land areas, below altitude of 500 feet in clear weather include spores of *Bacillus* and *Clostridium,* ascospores of yeasts, fragments of myceilium and spores of, molds and streptomycetaceae, pollen, protozoan cysts, algae, *Micrococcus, Corynebacterium* etc. In the dust and air of schools and hospital wards or the rooms of persons suffering from infectious diseases, microbes such as tubercle bacilli, streptococci, pneumococci and staphylococci have been demonstrated.

These respiratory bacteria are dispersed in air in the droplets of saliva and mucus produced by coughing, sneezing, talking and laughing. Viruses of respiratory tract and some viruses of enteric tract are also transmitted by dust and air.

Pathogens in dust are primarily derived from the objects contaminated with infectious secretions that after drying become **infectious dust**. Droplets are usually formed by sneezing, coughing and talking. Each **droplet** consists of saliva and mucus and each may contain thousands of microbes. It has been estimated that the number of bacteria in a single sneeze may be between 10,000 and 100,000. Small droplets in a warm, dry atmosphere are dry before they reach the floor and thus quickly become **droplet nuclei** (Fig. 5.2).

Many plant pathogens are also transported from one field to another through air and the spread of many fungal diseases of plants can be predicted by measuring the concentration of airborne fungal spores.

AEROMICROBIOLOGICAL PATHWAY

Aeromicrobiological (AMB) pathway describes the launching of bioaerosols into air, the subsequent transport via diffusion and dispersion of these particles, and finally their deposition. An example of AMB pathway is the liquid aerosol containing influenza virus launched into the air through a cough, sneeze or even talking. The virus-containing aerosols are dispersed by a cough or sneeze, transported through the air, inhaled, and deposited in the lungs of a nearby person initiating new infection (Fig. 5.2).

FIG. 5.2

As shown in this figure, a cough or sneeze launches infectious microbes into the air. Anyone in the vicinity may inhale the microbes resulting in a potential infection

Launching is the process whereby particles become suspended within the earth's atmosphere. Though launching of bioaerosls occurs from both, terrestrial and aquatic sources, atmosphere becomes more loaded from the former one. Launching into the surface boundary layers occurs by various mechanisms, such as air turbulence created by the movement of humans, animals and machines; generation, storage, treatment and disposal of waste materials; natural mechanical process as action of water and wind on contaminated solid or liquid surfaces; and the release of fungal spores in natural life cycles. Airborne particles can be launched from either point, linear or area sources. Point source, as a pile of biosolid matter, is an isolated and well-defined site of launching, whereas linear and area sources involve larger, less well-defined areas.

Transport or dispersion is the process by which kinetic energy provided by the movement of air is transferred to airborne particles, with resultant movement from one point to another. Transport of bioaerosols can be defined in terms of time and distance. Submicroscale transport involves short periods of time, under 10 min. as well as well as relatively short distances, under 100 m. It is very common in buildings and other confined spaces. Microscale transport ranges from 10 min. to one hr. and from 100 m to one km and is the most common type. Mesoscale transport refers to transport in terms of days and distances up to 100 km. and in macroscale transport, both time and distance are still larger. As bioaerosols travel through time and space, different forces as diffusion, inactivation and ultimately deposition act upon them.

Deposition is the final step in the AMB pathway. The airborne bioaerosol eventually leaves the turbulence of the suspending gas and ultimately becomes deposited on a surface by one or combination of interrelated mechanisms. These mechanisms include gravitational settling, downward molecular diffusion, surface impaction or rain and electrostatic deposition.

It has now become possible to predict the airborne bioaerosol concentrations in the vicinity of a contaminated source through mathematical modelling. Appropriate equations have been developed to simulate the aerobiological pathways i.e. how far the bioaerosols (launched into air) will travel and in what concentration.

Atmosphere-an Inhospitable Niche for Microbes

Due to desiccation and other stress conditions, atmosphere is too harsh a habitat for majority of microorganisms. Some of them could evolve specific traits that help their survival in atmosphere. For instance thick-walled spores of bacteria and fungi and cysts of protozoa protect them from harsh gaseous environments. Many environmental factors have been shown to influence the survival of microbes in atmosphere.

Relative humidity has been of major importance in the survival of airborne microbes. In general, most Gram-negative bacteria associated with aerosols tend to survive for longer periods at low relative humidity. The opposite holds true for gram-positive bacteria which tend to remain viable longer in association with high relative humidities. Lipid bilayers of cell membrane help survival under low relative humidities, changing from crystalline structure to a gel phase. Viruses with envelops have longer survival in air at relative humidity below 50%. **Temperature** is a major factor in inactivation of microbes. In general, high temperatures promote inactivation and lower temperatures promote longer survival times. Effects of temperature are closely linked with relative humidity. **Radiation**, such as shorter UV wavelengths, and X rays cause damage to most bacteria, fungi, viruses and protozoa. UV and X-rays cause DNA damage. Carotenoids, high relative humidities and clouds tend to shield bioaerosols from radiation. Some bacteria make use of DNA repair mechanism for the same. Besides these, **oxygen**, **open air factors (OAFs)** and **ions** in combination also inactivate many species of airborne microorganisms. Oxygen converted to more reactive forms like superoxide radicals, H_2O_2 and OH radicals, becomes toxic to microbes causing their inactivation. **Open air factor** is a term coined to describe an environmental effect that cannot be replicated in laboratory conditions. It is closely linked to oxygen toxicity and is usually defined as a mixture of factors produced when ozone and hydrocarbons (generally related to ethylene) react. Their reaction causes inactivation of microbes. In addition to these, ions containing Cl, N or S, produced through some natural processes (as action of lightning, shearing of water and action of various forms of radiation) also cause inactivation of cell surface proteins and DNA damage of microbes.

OUTDOOR AEROMICROBIOLOGY

In outdoor or extramural environment, the expanse of space and the presence of air turbulence are the two controlling factors in the movement of bioaerosols. Other environmental factors, such as radiation, temperature and relative humidity are important in limiting the duration of viability of aerosolised microbes and hence modifying the effect of bioaerosols. This is beyond the scope of this text book to include all aspects of outdoor aeromicrobiology. Rather a brief account of areas of this new science that are relevant to public in general have been included.

Airborne Crop Pathogens

Bioaerosols are of direct relevance to agriculture. Airborne microbial pathogens are responsible for a range of important diseases of crop plants. Bioaerosols contaminate the crops and thus have significant economic impact worldwide. Major airborne pathogens of crops like wheat are responsible for causing the outbreak of wheat rusts in different areas of the world. There are similar examples of other such epidemics caused by bioaerosols in other crop plants including rice. Not only in crop, but in vegetable plants like potato also, airborne pathogens are responsible for outbreaks of late blight disease.

Besides plants bioaerosols are also important in animal husbandry. The occurrence of foot- and-mouth disease is an example of the role of bioaerosols in the spread of airborne disease. Besides respiratory pathogens bioaerosols are involved in transmission of gastrointestinal pathogens also. For instance the gastrointestinal pathogen of calves, *Salmonella typhimurium* is spread through bioaerosols. Similarily aerosolised *Solmonella enteridis* could infect laying hens.

Waste Disposal

A range of pathogenic microbes, viruses-bacteria, protozoa and helminths associated with waste effluents bring about health hazard during their treatment and disposal handlings. For instance, wastewater treatment plants utilising activated sludge and trickling filters create large amounts of aerosols during treatment processes. These aerosols are very rich in pathogenic microbes. Aerosols containing pathogenic microbes are also generated during other treatment processes, such as composting and land disposal etc.

Germ Warfare

Biological warfare has become the most dangerous hidden, inhuman weapon these days. However, such a strategy is wars in very old and as early as 1346 A.D., Tartars besieging the walled city of Kaffa used catapults to launch plague-infested bodies into the city. In 1950s in U.S.A. field-scale experiments were made using inert substances like fluorescent dyes to simulate biological warfare agents. These aerosols were released into air circulation system of a subway system and into the air off the coast of San Francisco. An accident at a biological warfare research institute in Russia caused the widespread exposure of nearby populations to a genetically modified strain of *Bacillus anthracis*. In 1998, in Japan, Tokyo police detected large quantities of *Clostridium botulinum* toxin during a raid on a terrorist-controlled facility. In Iraq there were investigations in 1990s for mass scale production facilities for biological warfare agents. Detection of biological warfare agents is an area that requires intensive training and sophisticated equipment to develop an advanced antibiological warfare defense.

INDOOR AEROMICROBIOLOGY

Indoor or intramural aeromicrobiology involves home and workplace environments in which airborne microbes create major public health concerns. When compared with outdoor ones, the indoor environments have much less radiation exposure and limited air circulation. Moreover, the controlled temperature and relative humidity are in the ranges that allow extended survival of microorganisms. All these conditions are thus suitable for the buildup and survival of microbes within many enclosed environments, including office buildings, hospitals, laboratories, and even spacecraft. Some such indoor environments have been briefly described here.

Private Homes and Office Buildings

Extent of bioaerosols development determines the health of any building. There are several factors that influence the formation of bioaerosols. These include the presence and/or efficiency of air filtering systems designed and fitted in the building, the health and hygiene of the occupants, the amount of clean outdoor air circulated through the building, the type of lightning, the ambient temperature in the building, and the relative humidity.

Inspite of all precautions, some pathogens could develop mechanisms for survival and transmission in indoor environments. One example is *Legionella pneumophilia*, the causative agent of both Legionnaires disease and Pontiac fever, usually in U.S.A.

Hospitals and Laboratories

These two indoor environments have much potential for the aerosolisation of pathogenic microbes. A high percentage of individuals, including patients and staff are active carriers of infectious, airborne pathogens in hospitals. Neonatal wards, surgical transplant wards and surgical theaters are of particular concern, Microbiological laboratories are also a breeding centre for pathogenic microbes. There are risks of aerosolisation even under all precautions taken during their culture and handling in laboratories.

Spaceflight

Microbes have been detected even from most harsh and extreme environments. They are associated with almost every aspect of life on earth, including the spacecraft. Microbes, though pathogenic, are also necessary and beneficial to us. They can be harnessed for food production, oxygen production, waste treatment, and air purification.

Air purification is an example of a beneficial use of microbes in association with the AMB pathway. **Biological air filtration (BAF)** is a method of air purification, currently being investigated for use during spaceflight.

This system uses microorganisms to remove organic contaminants from the airstream. The BAF is a closed two-phase system. The airborne organic contaminants are passed into the closed system, where they pass through a membrane filter into a liquid phase. The liquid phase is composed of microbes and nutrients that develop into an associated biofilm on the membrane surface. Different strains of bacteria or consortia of bacteria have been isolated that can completely mineralise most of the contaminants. This includes a greater than 99% reduction of toluene, chlorobenzene, and dichloromethane in the airstream. The BAF system has been found effective in removal of methane, acetone, toluene, and isopropanol.

Public Health

AMB pathway is used for immunisation against some diseases. For instance injectable influenza (flu) vaccines are being currently used. However, they are not widely used because their injection is very painful. Currently, a flu vaccine has been developed that is delivered by nasal spray. Aerosol delivery onto mucous membranes increases the levels and specificity of the immune response.

BIOAEROSOL CONTROL IN LABORATORY

Bioaerosols containing airborne microbes can be controlled at every point of spread-launching, transport and deposition by using different mechanisms which include ventilation, filtration, biocidal agents and isolation.

Ventilation is the most common method to check buildup of airborne particles, that involves creating a flow of air through contaminated areas. This can be achieved by open windows or use of air-conditioning and heating units that pump outside air into room. Though less effective, but is very important, provided that outdoor air is free from harmful microbes. The entry of extra air indoors at least would reduce the concentration of pathogenic microbes inside the room.

Unidirectional airflow **filtration** is also a simple and effective method for bioaerosol control. High-efficiency particulate air (HEPA) filters remove virtually all infectious particles. Since they are expensive, their use in building filtration systems could not become common. Baghouse filtration (a baghouse works on some principle as a vacuum cleaner bag) system has become common in buildings. The typical rating (dust-spot percentage) for the filters used in most buildings is 30 to 50%. A rating of 97% is required for removal of viruses.

Biocidal agents are used for superheating, superdehydration, ozonation and UV irradiation to eradicate the airborne microorganisms. The most commonly used method is ultraviolet germicidal radiation (UVGI) that controls majority of the pathogens.

Isolation is the enclosure of an environment through the use of positive or negative pressurised air gradients and airtight seals. Isolation chambers used in TB wards in hospitals provide protection to others present inside the isolation area (outside TB ward). Air from these rooms is exhausted into the atmosphere after passing through a HEPA filter and biocidal control chamber. This system works on negative-pressurised air. Positive-pressure isolation chambers, working on the opposite principle force air out of the room, and thus protect the occupants of the room from outside contamination. In this way, a TB ward is a negative-pressure isolation room while the rest of the hospital, or at least the nearby anterooms, are under positive-pressure isolation. Similar isolations are done in case of HIV-infected and chemotherapy patients.

Biosafety in the Laboratories

Microbiological laboratories are equipped for scientific experimentation or research on pathogenic microorganisms. Centrifuges and vortexes, so commonly used in these laboratories can promote aerosolisation of microbes. Specially-designed cabinets (biosafety cabinets) are used to check the spread of airborne microbes. **Biological safety cabinets** are the most effective and commonly used devices in the laboratories to check such microbes. Two basic types of these cabinets, class II and class III are used, each having specific designing characteristics. The levels of control provided are termed biosafety levels 1-4 in order of increasing efficiency of control.

Biosafety laboratories are carefully designed environments where infectious or potentially infectious agents are handled and/or contained for research or education. The purpose of such laboratories is to prevent the exposure of workers and the surrounding environment to biohazards. There are four levels of biohazard control, which are designated as biosafety levels 1 through 4. Different levels have been defined by concerned agency (as in USA by Centres for Disease Control, US Dept. of Health and Human Services). These are as follows:

Biosafety 1 – A teaching laboratory.

Biosafety 2 – An area where work is done using agents that are moderately hazardous to humans and the environment.

Biosafety 3 – Laboratories where such agents are handled that can cause fatal diseases as a result of AMB pathway.

Biosafety 4 – Laboratories with highest level of control of microbes causing life threatening diseases in association with aerosolisation.

Various methods have been developed for the collection of bioaerosols. One can choose any of these sampling methods depending upon the site of collection, environmental conditions of sampling, and mobility and sampling efficiency. Different kinds of samplers have been designed, from the very simple type, such as Durham's gravitational chamber to the most sophisticated one, the Andersen six-stage impaction sampler. Several types of samplers are commonly used, which are based on different methods of sampling-impingement, impaction, centrifugation, filtration, and deposition. **Impingement** is the trapping of airborne particles in a liquid matrix; **impaction** is the forced deposition of airborne particles on a solid surface; **centrifugation** is the mechanically forced deposition of airborne particles using internal forces of gravity; **filtration** is the trapping of airborne particles by size exclusion; and **deposition** is the collection of airborne particles using only naturally occurring deposition forces.

An account of these various devices for sampling and processing of **airborne microbes** by different types of samplers using different sampling methods has been given in Chapter 7 "Devices for Sampling and Processing of Microbes".

6

EXTREME ENVIRONMENTS

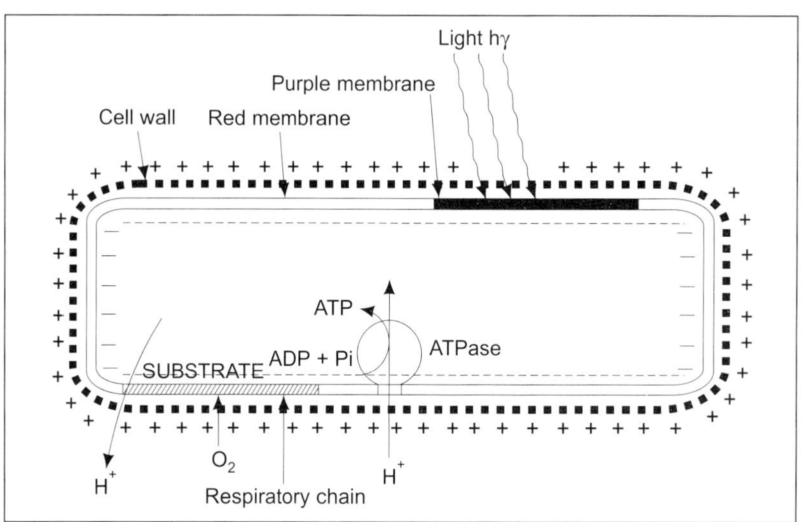

Chapter Outline

- *What is an Extreme Environment?*
- *Types of Extreme Environments*
- *Archaebacteria-the Heros of Extreme Environments*

In nature microorganisms are found in all sorts of habitats characterised by different environmental conditions. Features of three main types of environment —aquatic, terrestrial, and aerial, and the diversity of microbes occupying each of these environments have been presented in Chapters 3, 4, and 5. Conditions in any of these environments may turn to extremes (low or high). These environmental conditions can include pH, temperature, salinity, pressure, and nutrients. In a given habitat, normally any of the abovesaid conditions can reach to extreme levels. Interestingly, there also exist a few habitats where more than one of these conditions turn extreme at a given time, operating upon the microbial life simaltaneously. One such unique habitat is the air-water interface in water bodies. In this chapter we briefly describe some such types of environments.

WHAT IS AN EXTREME ENVIRONMENT?

An extreme environment can be defined in two ways: (i) an environment is said to be as extreme if the environmental conditions (pH, temperature, salinity, pressure, nutrients etc.) are at one of the two extremes (low or high), and (ii) an extreme environment is the one in which conditions select for extremely low microbial diversity. **Extremophile** is a term applied to any organism that has successfully adapted to environments where it is difficult or impossible for other organisms to survive. In general survival of an organism depends mainly on its growth and multiplication. In this way an extremophile must be able to grow and multiply in any such environment. Thus extremophiles have been selected over time for characteristics that allow them to grow and multiply in a variety of extreme environments. Terrestrial and submarine hot springs, salt lakes, antarctic desert soils, and some trenches are examples of extreme environments.

The conditions of extreme environments greatly restrict the range of microbial species that can grow in such habitats. The extremes of environmental conditions that the microbes must be able to tolerate include high temperatures (nearby to boiling water); low temperature (nearby to freezing levels), low acidic pH values, high alkaline pH values, high salt concentrations, low water availability, high irradiation levels, low concentrations of nutrients and high concentrations of toxic compounds. Many microbes that inhabit extreme environments, possess specialised adaptive physiological features that permit them to survive and function within the physiochemical constraints of these ecosystems. The membranes and enzymes of microbes of extreme environments often have also distinct modifications that allow them to function under conditions that would inhibit active transport and metabolic activities in organisms lacking these adaptive features.

Characteristic features of such extreme environments alongwith the structural and physiological adaptations used by extremophiles to compete or survive in their particular niches will be described briefly in the following section of this chapter.

TYPES OF EXTREME ENVIRONMENTS

Extreme environments exist in all habitats - air, soil and water. Some examples are given below.

Air-Water Interface

Air-water interface is a unique habitat that is often considered an extreme environment for many reasons, including high levels of solar radiation; accumulation of toxic substances (heavy metals, pesticides); large temperature, pH, and salinity fluctuations; and competition. This interface, also referred to as **neuston** actually harbours higher concentrations of organisms than other layers of water column. But, most reports suggest that inspite of high numbers the ratio of metabolic activity to total counts is lower in this habitat than in the planktonic zone. The neuston accumulates nutrients and especially attracts nonpolar organic and inorganic molecules which form a film at interface. Since this layer concentrates nutrients, bacteria attach to it. But neuston accumulates toxins too, such as pesticides (DDT), petroleum hydrocarbons, as well as metals (Cd, Cu, Mn, Hg, Pb, Se, Cr). Thus microbes in this habitat have developed special adaptive features to survive. These include the pathways that catabolise toxic compounds and provide resistance to metals. Some have also developed DNA repair mechanisms.

High Temperature

There are many examples of environments with extreme high temperatures (> 70°C) which include terrestrial and submarine hot springs, some of which can reach temperatures of 100°C, and hydrothermal vents, which can reach in excess of 300°C. Such high temperature are inhospitable for most forms of life, and only some eubacteria and archaebacteria are known to survive there. The growth of microbes of hot springs is also limited by low organic matter, oxygen and depending on a particular hot spring with either acid or alkaline pH values. The water outflowing the hot spring, flows down channels, establishing a temperature gradient, with a clear zonation of microorganisms, occupying habitats of differing maximal temperatures along this temperature gradient; bacteria being most tolerant of extreme temperatures. At temperatures above 75°C only a few bacterial species, as members of the genera *Thermus* appear to grow.

Bacillus stearothermophilus is dominant in hot springs in temperature zone of 55-70°C, but many other microbes as cyanobacteria and algae also occur there. Cyanobacteria occur as layers of growth within specific zones of thermal ponds. Cyanobacteria grow in higher temperatures zones than algae which are restricted to growth below 55°C.

Thermal vent communities are located at depths of 800-1000 m, where seawater percolates deeply into the crust to react with hot core materials. These regions receive no sunlight and there is minimal nutrient input from the water above. In these areas the community is supported energetically by the chemoautotrophic oxidation of reduced sulphur primarily by *Beggiatoa, Thiomicrospira,* and additional sulphide or sulphur-oxidisers of great morphological diversity.

Many of the organisms of hot springs and thermal vents are **obligate thermophiles** and are restricted to growth at high temperatures. They have several adaptive features to carry out metabolism at temperatures of over 60°C. Many these microbes produce enzymes that are not readily denatured at high temperatures. Sometimes unusual amino acid sequences occur in the their proteins, stabilising at high temperatures. Their membranes possess a major proportion of high molecular weight. and branched fatty acids that allow them to maintain semipermeability at high temperatures. Thermophiles have high proportions of guanine and cytosine in DNA that raise the melting point and add stability to the DNA molecule.

Most notable of these microbes are the genera, *Thermus, Methanobacterium, Methanobrevibacter, Methanococcus, Methanogenium, Methanosarcina, Sulfolobus, Pyrodictium,* and *Pyrococcus*. The last two genera can live at > 100°C. *Thermus aquaticus* is especially known for its thermotolerant DNA polymerase. This enzyme has been patented and used world over in PCR. Thermophiles have increased number of salt bridges (cations that bridge charges between amino acid residues), which help the protein to remain folded even at high temperatures. Thermophilic eubacteria have increased saturated fatty acids in cell membranes that allow the membranes to remain stable at high temperatures. Archaebacteria, the extreme thermophiles, have an entirely different cell-membrane. The details of the characteristics of their membrane and other physiological features will be described later in this chapter. Thermophiles contain special DNA binding proteins that arrange DNA into globular particles, more resistant to melting. Moreover DNA gyrase acts to induce positive supercoils in DNA.

There are numerous biotechnological applications for enzymes isolated from thermotolerant microbes, especially in commercial industry. Besides use of thermotolerant DNA polymerase in PCR, the proteases, lipases, amylases, and xylanases are used in agriculture, paper, pharmaceutical, water purification, bioremediation, mining and petroleum recovery industries.

High Salt Concentrations

Halotolerant or high-salt-tolerant organisms require salt concentrations for growth that are quite higher than those found in seawater. Dead sea, lying between Israel and Jordan and Utah's Great Salt Lake in U.S.A. are examples of such environments. The notable examples of halotolerant bacteria belong to the genera *Halobacterium* and *Haloanaerobium*. Some algae and fungi are also known to be halotolerant. Halotolerance is not a requirement for high salt concentration but is specific for Na^+. The main mechanism of salt tolerance in bacteria is internal sequestration of high concentrations of a balancing solute to equal the salt concentration found external to the cell. K^+ balancing mechanism occurs in halotolerant bacteria, whereas glycerol is balanced in eukaryotes. Acidic proteins with low proportions of nonpolar amino acids also help in salt tolerance. Due to these characteristics, halotolerant bacteria are usually unable to survive in the environments lacking high salt concentrations. Thus, many are considered obligate halophiles.

Low pH

Acidic environments, such as acid hot springs, the gastrointestinal tract, mining waste streams, acid mine waste water, and various mineral oxidising environments are inhabited by bacteria, such as *Thiobacillus*. Some members of the genus *Thiobacillus* are acidophilic, and grow only at pH values near 2.0. *Thiobacillus thiooxidans* has a minimum 1.0, optimum 2.0-2.8 and maximum 4.0-6.0 pH values. Acidophiles possess physiological adaptations that allow enzymatic and membrane transport activities at low pH. The cell membrane of such bacteria breaksdown and can not function at neutral pH value.

Some species of *Thiobacillus*, oxidise only sulphur compounds, whereas others, as *T. ferrooxidans*, also oxidise ferrous to ferric iron for ATP generation (chemolithotrophs). *Thiobacillus* spp. can be used in the recovery of minerals, including uranium, and their oxidation of reduced iron and sulphur compounds mobilises various metals so that they can be extracted from even low-grade ores. *T. thiooxidans* used in biological metal recovery is acidophilic with optimum growth in the pH range of 1.0-3.5. The metabolic activities of this bacterium, found often in association with waste coal heaps, produce mine drainage, a serious ecological problem associated with some coal mining operations.

These microbes are discussed extensively in relation to acid mine drainage in Chapter 10. Other examples of acidophiles are *Clostridium acetobutylicum* and *Sarcina ventriculi*, the obligate anaerobes that ferment sugars. Besides bacteria, some fungi, algae and protozoa are also known to tolerate acidic conditions. Modifications in architecture of cell membranes allow their survival under such conditions.

Thermoacidophiles are a heterogenous group of archaebacteria which are able to grow at high temperature and low pH. Several subgroups have been identified. A brief account of this group will be presented later in this chapter.

High Pressure

Microbes living in high pressure and cold temperatures of deep-sea environments are called **barophiles**. They are known to tolerate a pressure of more than 1000 bars found in deep-sea trenches. Thus besides being pressure tolerant, barophiles are also psychrophilic. They are adapted to darkness also. Various mechanisms allow them to survive there. These include high concentration of long-chain polyunsaturated fatty acids in the cell membranes that would maintain its fluid state under extremes of pressure and temperature.

No Nutrient

The ultrapure or nutrient-free water, used in semiconductor, medical and other industries is an example of extreme environment. Microbial contamination of such waters can be devastating. It can cause flaws in crystal design of computer chips. Very few microbes are known to survive in nutrient-free extreme environments. *Caulobacter* and *Pseudomonas fluorescens* are known to be present there. In distilled water, even the limited exchange of nutrients (CO_2) from the atmosphere provides enough nutrients to allow limited growth. *Pseudomonas aeruginosa*, an opportunistic pathogen has been detected in springwater bottles.

ARCHAEBACTERIA—THE HEROS OF EXTREME ENVIRONMENTS

A perusal of the above account of extreme environments and the kind of microorganisms inhabiting them makes it apparent that in most of such harsh, inhospitable conditions, archaebacteria are able to survive. What is unique with these microbes? How they are able to survive under such conditions? A brief account of the various structural and physiological characteristics of this unique group of microbes is being presented below.

Members of this group are shown to be phylogenetically related to each other on the basis of analysis of their 16s rRNA molecules, and to be distinct from eubacteria. They are unique in the sense that their cell walls lack murein and there are unusual ether linkages in their cell membrane lipids. Archaebacteria appear to be primitive bacteria, and as members of their own independent kingdom they are considered to be distantly related to other prokaryotes. They are important in evolutionary processes and other phylogenetic relationships.

Three major groups of archaebacteria are distinguished on the basis of metabolic or ecological features: the methanogens, the halophiles, and the thermoacidophiles. The methanogens are distinguished by their unique energy metabolism in which methane is a prominent end product. The halophiles and the thermoacidophiles are distinguished by their habitats: highly saline environment for the former and high temperature and low pH for the latter. However, two of the groups are internally heterogeneous.

Methanogens (Methane-Producing Bacteria)

Members represent a highly specialiscd physiological group. They belong to Methanobacteriaceae (*Methanobacterium, Methanobrevibacter*); Methanococcaceae (*Methanococcus*); Methanomicrobiaceae (*Methanomicrobium, Methanogenium, Methanospirillum*) and Methanosarcinaceae (*Methanosarcina*). They form methane by reducing CO_2 (chemolithotrophs). Methanogens are very strict obligate anaerobes. For methane production, they utilise electrons generated in the oxidation of hydrogen or simple organic compounds such as acetate and methanol. These microbes are unable to use carbohydrates, proteins or other complex organic substrates. The other microbes associated with methanogens maintain the low O_2 tensions and provide CO_2 and fatty acids to methanogens. Such associations are extremely important in the rumen of animals like cows- a major source of atmospheric methane. Some common methanogens are, *Methanospirillum hungatei, Methanobacterium thermoautotrophicum, M. soehngenii, Methanobrevibacter ruminatium, Methanococcus mazei,* and *Methanosarcina* sp. Methanogens comprise at least three major groups: Group I contains *Methanobacterium* and *Methanobreviabacter,* Group II contains *Methanococcus,* and Group III contains several genera, including *Methanospirillum* and *Methanosacina.*

At least three different types of **cell wall** are found among the methanogens. The most complex is that of Group I, which is rigid and composed chiefly of **pseudomurein**, a peptidoylycan similar to the murein of eubacteria. Pseudomurein contains N-acetyl- talosaminuronic acid instead of N-acetylmuramic acid, and lacks D-amino acids. In appearance the wall resembles those of Gram-positive eubacteria. In *Methanococcus,* of Group II, the wall is flexible and composed chiefly of proteins with traces of glucosamine. *Methanospirillum,* of Group III has the most complex cell wall. It is flexible, composed of at least two layers: an inner, electron-dense of unknown chemistry and the outer one appearing like a membrane in cross section but composed entirely of protein. This protein is resistant to hydrolysis by proteinases (as trypsin) and to solubilisation by detergents as sodium dodecyl sulphate, SDS.

Methanogens contain several cofactors not found in other bacteria. Three of them (methanopterin, methanofuran, and CoM) are carriers of the C_1 unit during its reduction from CO_2 to CH_4. $Factor_{420}$ functions as hydrogen carrier and F_{430} is the prosthetic group of methyl-CoM reductase, the last enzyme in reduction pathway. Methanogens are ubiquitous in highly reducing habitats (E*h* < – 0.33V). The impact of methanogens on their environments is substantial.

Halobacteriaceae (Halophiles)

The genera *Halobacterium* and *Halococcus* are included in this family and they have characteristics of archaebacteria. They possess unique metabolic features. Light energy is converted to ATP by bacteriorhodopsin, a red pigment located in the purple membrane portion of the red membrane of this bacterium. Thus in addition to respiration, there is also light-dependent ATP generation.

Besides the above unique metabolic features, Halobacteriaceae are obligate halophiles, growing only in media of at least 15% NaCl concentration. Members are found in ecosystems that have extremely high NaCl concentrations, such as salt lakes, the Dead Sea, and salt-preserved foods.

Salt lakes occur in arid regions where evaporation exceeds freshwater inflow or where a lake is fed by a salt spring. Since high concentrations of salt dehydrate cells and denature enzymes, relatively few organisms can grow in highly saline waters. Very often the biota of salt lakes is restricted to a few **halophiles** (salt-requiring) and salt-tolerant bacteria. Halophiles have high internal concentrations of KCI and their enzymes must have greater tolerance of salt than those unable to tolerate salt. In many cases high salt concentrations are required by halophiles to maintain their enzymatic activities. Many halophiles have unusual membranes, such as the purple bilayer membranes of *Halobacterium*. The cell wall of this bacterium lacks murein and appears to be stabilised by sodium ions. The principal component of the cell wall is a large (MW = 200,000) acidic glycoprotein, other component being nonglycosylated protein and glycolipid. The cell wall of *Halococcus* is different and resembles that of Gram-positive eubacteria. The ribosomes of *Halobacterium* require high concentrations of potassium for stability. These are the adaptive features that are used by halophiles to live in saturated brine environments of salt lakes.

Thermoacidophiles

They are a heterogeneous group defined by their ability to grow at high temperature and low pH. Several subgroups have been identified: *Sulfolobus, Thermoplasma* and *Thermoproteus* group.

Sulfolobus is a facultative chemoautotroph found extensively all over the world in hot acid springs and soils. The cells are irregularly lobate spherical. Temperature optima for growth vary among isolates from 63° to 80°C; pH optima normally around 2.0. The range of conditions over which it grows is fairly wide temperature of 55° to

85°C, pH of 1 to 5.9. Although *Sulfolobus* can grow on organic compounds (perhaps fermentatively), in its natural habitats it probably grows as a respiratory chemoautotroph. Geothermal steam or hot water leaches much amounts of iron and sulphide, which is rapidly oxidised to elemental sulphur by oxygen or ferric ion, either chemically or biologically. This bacterium rapidly oxidises H_2S. The principal substrate in hot spring and may be also in hot soils, is elemental sulphur and it grows on the surface of droplets or crystals of sulphur oxidising it to H_2SO_4, responsible for the acidic habitats.

Thermoplasma is a facultative anaerobe that uses small number of mono- and disaccharides as carbon and energy source. This microbe is found only in the refuse piles from coal mines which contain residual coal and substantial iron pyrite (FeS). Oxidation of FeS by chemoautotrophic bacteria acidifies and heats the piles, creating a favourable environment for *Thermoplasma* (pH 0.5 to 4.5; temp. 37° to 65°C). Small organic molecules in coal spoils (produced by pyrolysis of larger ones) are utilised by *Thermoplasma*. We still do not know the original habitat of this microbe, as attempts to isolate it from hot acidic springs or animal stomach have failed. Perhaps the true natural habitat is coal veins exposed to atmosphere by geological processes. *Thermoplasma* lacks a cell wall but cell membrane contains large amounts of lipopolysaccharide and glycoprotein, both of which contain mannose.

Thermoproteus group consists of thermoacidophiles whose metabolism is chiefly or exclusively respiratory with elemental sulphur serving as terminal electron acceptor. The product is H_2S, which is a required nutrient for organisms in this group. They all are strict anaerobes. The only habitats that could yield these microbes are geothermal areas of Iceland, where they are widespread. Two genera of this group are, *Thermoproteus* and *Desulfurococcus*. *Thermoproteus* are long thin rods, with strictly respiratory metabolism, growing over a temperature range of 78° to 95°C, and a pH range of 2.5 to 6. *Desulfurococcus* are spherical cells, with fermentative as well as respiratory metabolism, and grow over a temperature range of 75° to 95°C and a pH range of 5 to 7.

Structural and Physiological Features of Archaebacteria

Archaebacterial lipids

Two types of lipid structure are found: glycerol diethers and diglycerol tetraethers. Their hydrocarbon chains are normally the C_{20} phytane or the C_{40} biphytane, respectively. Membranes that contain diglycerol dibiphytanyl tetraether probably consist of a monolayer rather than a bilayer, with each lipid molecule spanning the entire membrane. In general, thermoacidophiles contain mainly dibiphytanyl tetraethers; the halophiles mainly diphytanyl diethers. Two different patterns of distribution occur in methanogens; coccoid cells contain only diphytanyl diethers, whereas the rest contain both diphytanyl diethers and dibiphytanyl tetraethers.

Cell walls

Cell walls of archaebacteria do not contain peptidoglycan (murein); rather their wall structures show great biochemical diversity. Some archaebacterial cell walls have pseudomurein which resembles peptido glycan of eubacteria but contain N-acetyl-talosaminuronic acid instead of N-acetyl-muramic acid and lack the D-amino acids found in eubacterial cell walls. Other archaebacteria have walls composed of protein subunits; still others have walls with different biochemical composition. Although, variable in chemical structure, the walls are able to protect the cell membrane even in hot, acidic and saline environments where many archaebacteria live.

Cell membrane

Archaebacteria have a cytoplasmic membrane with structures very different from true bacteria as well as eukaryotes and this feature distinguishes them from all other organisms. In extreme environment which they inhabit, unusual physiologically specialised membranes are needed for survival. In contrast to eubacterial and eukaryotic cell membranes that contain straight chain fatty acids linked to glycerol by ester linkages, the cell membranes of archaebacteria contain branched lipids. The lipids are diethers in which a glycerol unit is connected by an ether link to phytanols, the branched chains in which carbon atoms at regular intervals carry a methyl sroup. Moreover, glycerol has two optical isomers distinguished by the configuration of the molecule around the central carbon atom; the optical isomers rotate polarised light in opposite directions (Fig. 6.1). *Sulpholobus,* an archaebacterium of high temperatures in acidic environments - has a cell membrane that contains long chain branched hydrocarbons twice the length of the fatty acids found in the cell membranes of eubacteria. The lipid chains are long enough to extend from one side of the membrane to the other giving it the appearance of a monolayer while concealing its true bilipid structure (Fig. 6.2). Similar unusual membrane structures occur in other archaebacteria of other extreme environments, including *Thermoplasma* living in high temperatures and *Halobacterium* in habitats of high salt concentrations. The structure of these membranes makes them very resistant to conditions that would disrupt them and interrupt the function of a normal bilipid layer, enabling them to remain as semipermeable barriers in extreme environments.

Purple membrane of halobacteria

Halobacterium, an archaebacterium can generate ATP, both by using an organic substrate and by using light energy (Fig. 6.3). In presence of oxygen, these bacteria use oxidative phosphorylation for ATP synthesis (aerobic respiration). The electron

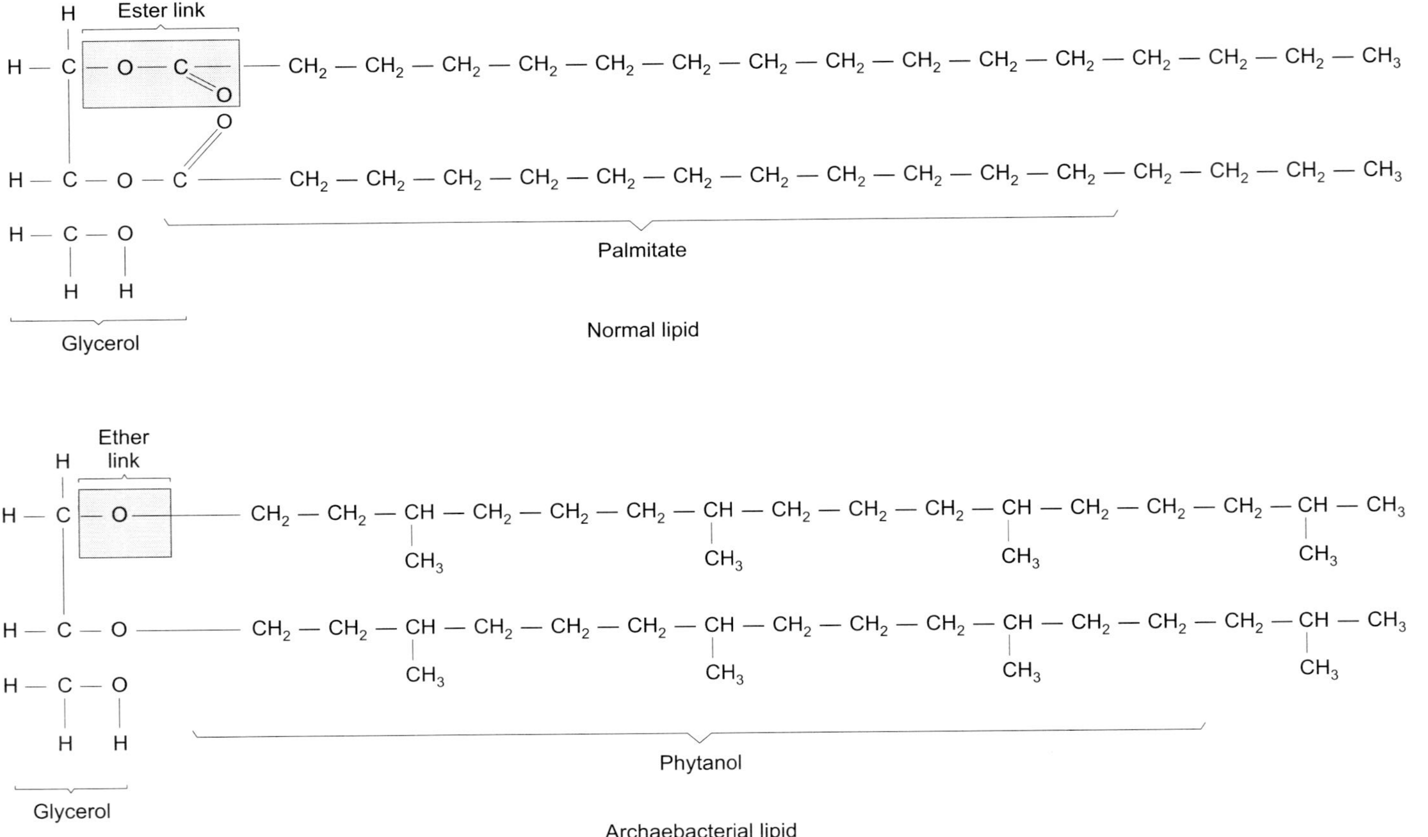

FIG. 6.1 Molecular structure of lipids of normal (eubacteria and eukaryotes) and archaebacterial cell membrances

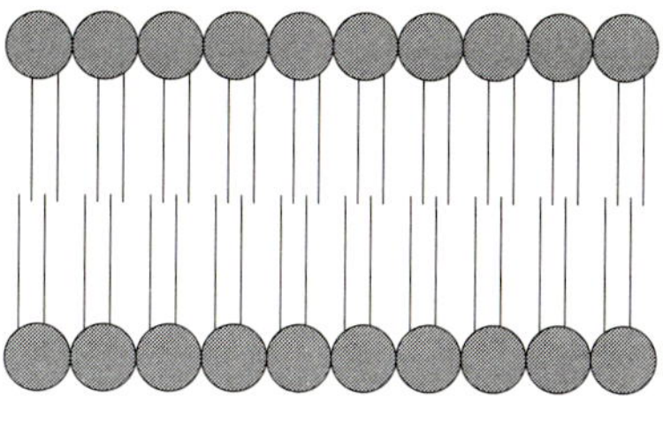

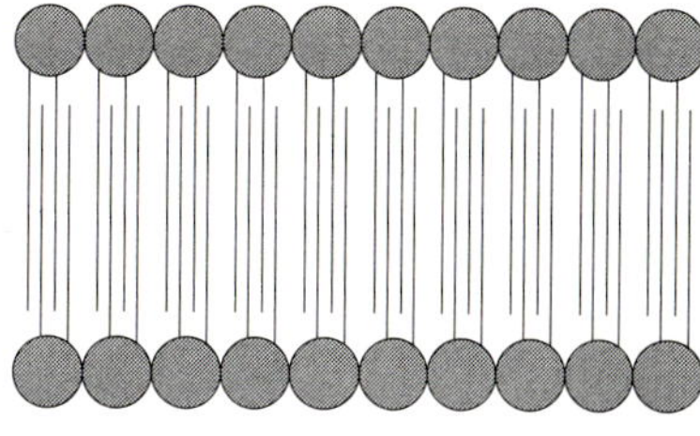

FIG. 6.2

Arrangement of bilipid layers of cell membranes of eubacteria and archaebacteria

transport chain for this pathway is located in a portion of cytoplasmic membrane that is red in colour and hence called the **red membrane.**

In the absence of oxygen, they turn to photophosphorylation for ATP synthesis. The mechanism here is different from other bacteria. It is based upon a **purple membrane** portion of the cytoplasmic membrane that contains **bacteriorhodopsin**, a protein of a chemical structure similar to that of rhodopsin pigment of human eye. When excited by light this pigment pumps protons to the outside of the membrane establishing a hydrogen ion gradient. The counterflow of hydrogen ions results in ATP

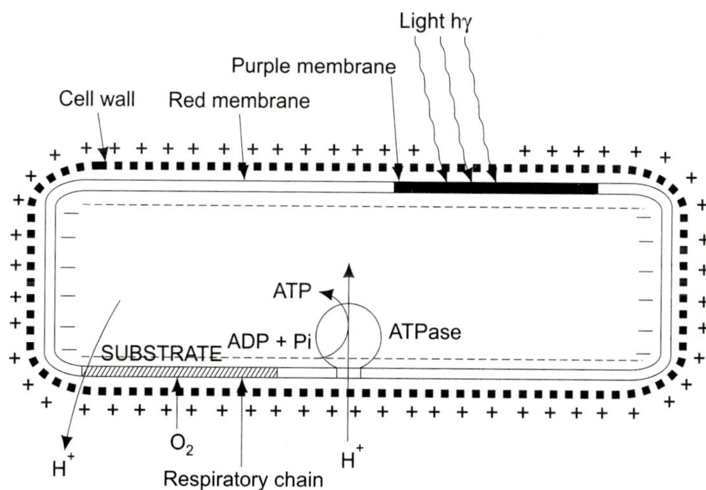

FIG. 6.3

The relationship of proton transport to ATP formation in *Halobacterium*. Regions of the cell membrane contain bacteriorhodopsin and have a purple colour. The purple membrane is involved with light-coupled ATP formation. A large portion of the cell membrane is red in colour and contains ATPase and mediators of the respiratory electron transport chain

synthesis. The enzyme required for phosphorylation is contained in a separate red membrane fraction so that the same, ATP-synthesising system used for photophosphorylation is used for oxidative phosphorylation (respiration). The halobacterial membrane system for using light energy to establish a hydrogen ion gradient that can drive ATP synthesis provides firm evidence for the essential role of chemiosmosis in the ATP synthesis.

The role of purple membrane in photophosphorylation was accomplished largely through the work of Walter Stoeckenius and co- workers. Through experimental work they could show that *Halobacterium* can carry out both, aerobic respiration and light-coupled photophosphorylation of ATP synthesis. These observations were consistent with the hypothesis that respiration and light generate an electrochemical proton gradient across the cell membrane and this gradient drives the ATP synthesis by membrane ATPase (Fig. 6.4). These workers made artificial vesicles from the purple membrane and designed experiments to establish role of bacteriorhodopsin in such process. This work was chiefly done by Stoeckenius and Efraim Racker.

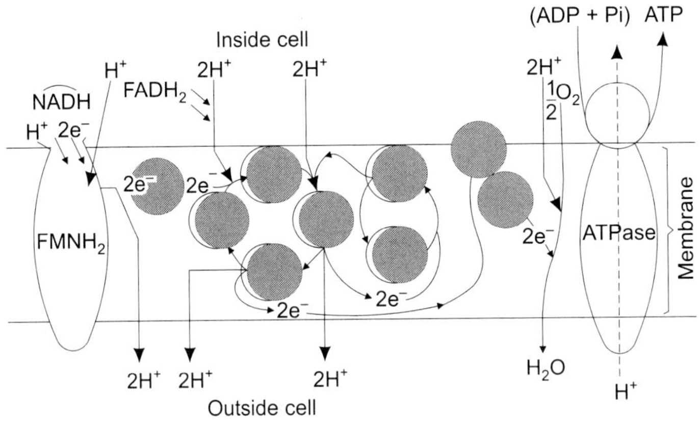

FIG. 6.4

The process of chemiosmosis, generation of hydrogen ion gradient across a membrane needed to generate ATP

7

DEVICES FOR SAMPLING AND PROCESSING OF MICROBES

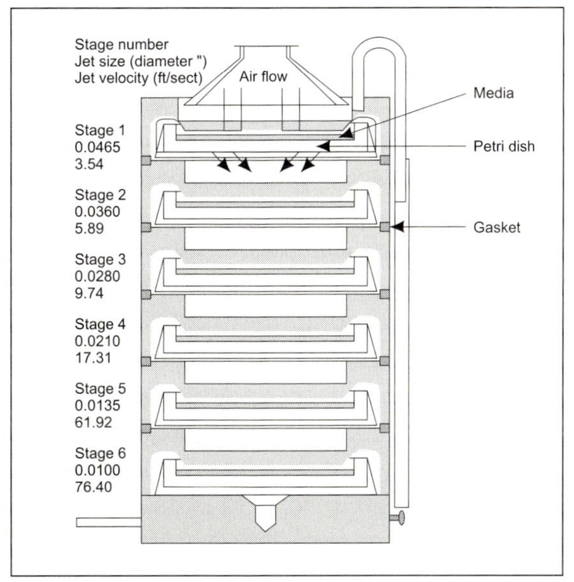

Chapter Outline

- *Soils and Sediments*
- *Water*
- *Air*

An account of the sampling devices for the collection and processing of microorganisms from **three** main types of environment: soils and sediments, water, and air has been presented in this chapter

SOILS AND SEDIMENTS

Sampling methods and their processing may differ for surface and subsurface soils and sediments.

Surface Soils

Bulk soil samples are obtained with a shovel or a soil auger. A simple hand auger is suitable for taking shallow soil samples from unsaturated areas. Samples are taken from a depth of 6 feet. Sterile spatula is used to scrape away the outer layer of the core (to avoid contamination) and inner part of the core is used for analysis. Auger must be washed with water and rinsed with 75% ethanol with final rinse in water at each sampling. Several samples are taken and pooled to make a composite sample (because soils may represent a heterogenous habitat). Composite samples are taken over a wide area and placing them in a bucket or plastic bag. The composite soil or its sample is then stored for processing.

In some instances, a series of experimental plots or fields need to be sampled to test the effect of a soil amendment, such as fertiliser, pesticide or sewage sludge on microbial populations. In this case a soil sample must be taken from each of several plots or field to compare with control untreated plots or fields. In such comparative studies, multiple samples or replicates are taken. In such cases two-dimensional sampling plans can be used to determine the number and location of samples taken. In two-dimensional sampling, each plot is assigned spatial coordinates and set sampling units are chosen according to an established plan. Some typical two-dimensional sampling patterns, including random, transect, two-stage, and grid-sampling are shown in Fig. 7.1. **Random sampling** involves choosing random points within the plot, which are then sampled to a defined depth. **Transect sampling** involves collection of samples in a single direction. In is useful in riparian areas. In **two-stage sampling** an area is broken into regular subunits (primary units). Within each unit, subsamples can be taken randomly or systematically. This method is useful in hillside slopes and a level plain. In **grid sampling**, samples are taken systematically at regular intervals at a fixed spacing.

For **rhizosphere** studies, normally roots are carefully excavated and shaken gently to remove bulk or nonrhizosphere soil. Soil adhering to the plant roots is then considered to be rhizosphere soil. The soil is then processed for microbial analysis. Soil suspension in water and therafter serial dilutions are made. Rhizosphere soil exists as

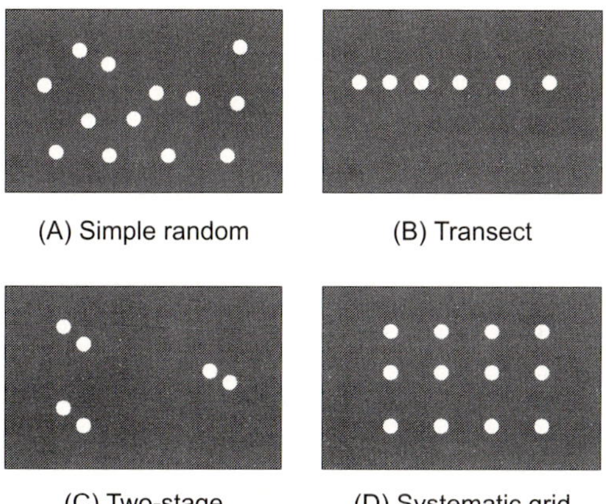

FIG. 7.1
Alternative spatial sampling patterns

a continuum from the root surface (rhizoplane) to a point where the root has no influence on microbial characteristics (normally 2-10 mm).

Subsurface Soils

Mechanical drill rigs are used for sampling the subsurface environments. For unsaturated systems, air rotary drilling can be used to collect samples from depth up to several hundred meters. In such drilling a large compressor is used to force air down a drill pipe, out the drill bit, and up outside the borehole. As the core barrel cuts downward, the air serves to blow the borehole cuttings out of the hole and also to cool the core barrel. In normal air drilling small amounts of water containing a surfactant are injected into the air stream to control dust and help cool the drill. The samples are immediately frozen and shipped to the laboratory.

Saturated subsurface environments are sampled in different way. For sampling of depths down to 100 feet, hallow-stem auger drilling with push-tube sampling is used. The auger consists of a hollow tube with a rotating bit at the tip that drills the hole. There is reverse threading outside of the hollow auger casing that pushes the cuttings upward and out of the hole. The casing acts as a sleeve into which a second tube, the core barrel, is inserted to collect the sample, when the desired depth is reached. For deeper cores (more than 100 feet), mud rotary coring is used. Mud rotary drilling has been used to obtain sediment samples to 3000 feet beneath the soil surface.

Sample Processing and Storage

The samples should be analysed for microbial composition as soon as possible after collection to minimise the effects of storage on microbial populations. The soil is

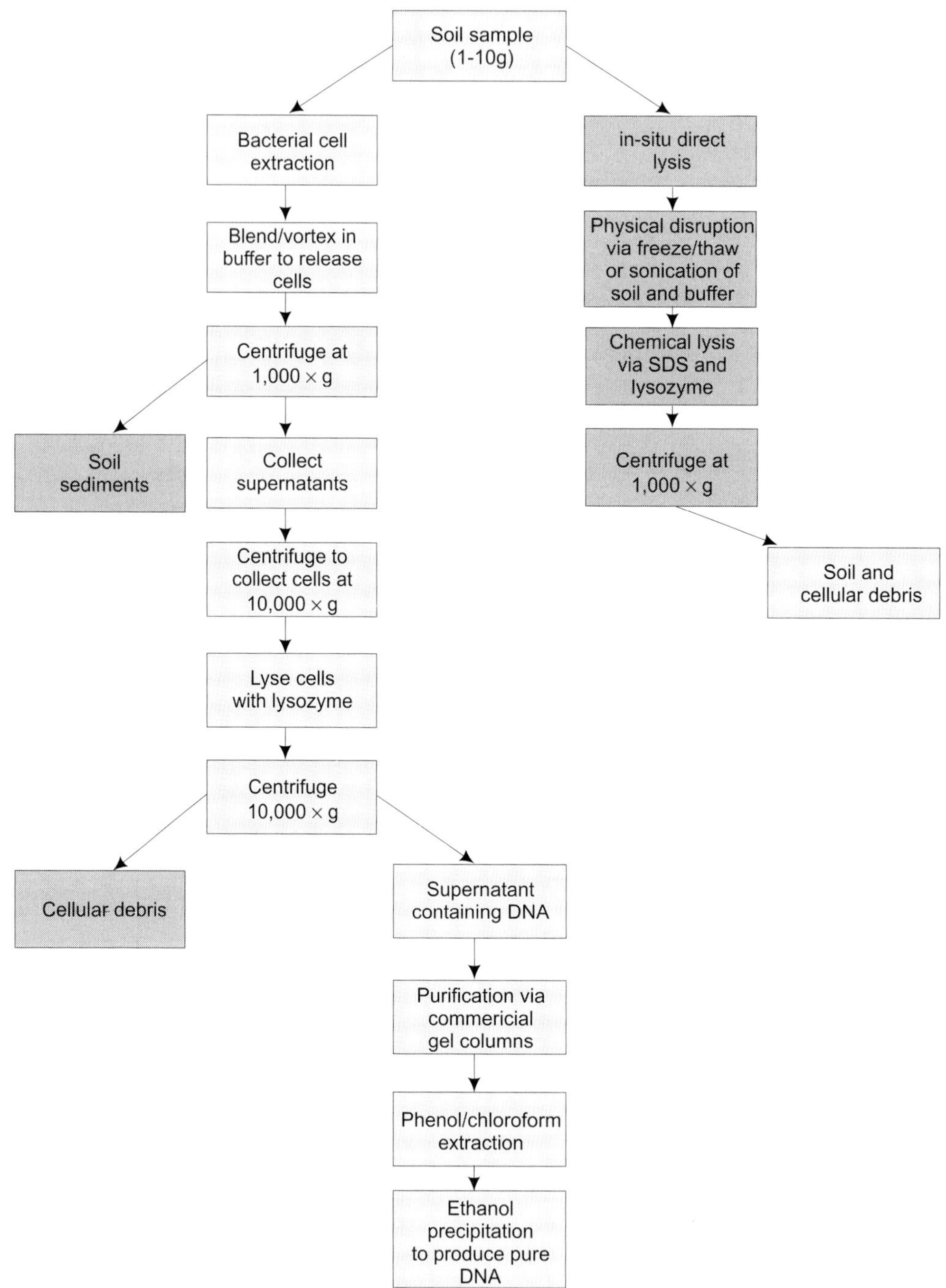

FIG. 7.2

Methods for obtaining community DNA from soil

passed through a 2-mm mesh to remove large stones and debris. Samples may be dried to facilitate sieving. A short-term storage should be done at 4°C before analysis. A 3-week storage normally does not alter microbial composition. Subsurface samples obtained by coring are either immediately frozen and brought to laboratory as intact core or processed at the coring site. Outside of the core is scrapped off with sterile spatula. The sample is placed in sterile plastic bag and analysed immediately or frozen for future analysis.

Bacterial populations of soil and sediment samples are analysed by traditional methods. Cultural assays utilising dilution and plating methodology on selective and differential media or direct count assays are done. These are described in Chapter 8. Plate counts provide more information on diversity of bacteria. Type of media is also important factor. Culture counts may vary with media. It has been found that most soil bacteria exist under nutrient-limited conditions.

A recently developed technique for study of bacterial population is the **estimation of total DNA** of bacteria within a soil sample. It is desirable to combine this technique with direct and cultural counts to have complete spectrum of bacterial populations in the soil sample. There are **two approaches** to the extraction of bacterial DNA (Fig. 7.2): (i) **extraction** of cells from soil followed by cell lysis and DNA extraction, and (ii) **in situ** lysis of bacterial cells within the soil matrix followed by DNA extraction released from the cells.

Fungi in soil samples are also detected by using direct as well as cultural methods. However, though possible, direct observational methods are labour intensive and rely on trial and error. For soil fungi, therefore, cultural methods are followed. Such methods which involve proper dilutions and plating are not much different from those used for bacteria. These have been described in Chapter 8.

Viruses in sludge, soil and sediment samples are detected by appropriate methods. Since viruses are important pathogens of the environment, their occurrence in sewage sludge (biosolids), soils to which wastewater or sludge is applied, or marine sediments affected by sewage outfalls or sludge disposal, is fully understood. The most common procedure for sludge involves adding $AlCl_3$ and HCl to 500-1000 ml of sludge to adjust the pH to 3.5. The viruses bind to the sludge solids, which are then removed by centrifugation and resuspended in a beef extract solution at a neutral pH to elute the virus. The eluate is then reconcentrated by flocculation of the proteins in the beef extract at pH 3.5, resuspended in 20-50 ml, and neutralised. The details of the procedure are shown in Fig. 7.3.

Procedure	Purpose
500 - 2000 ml sludge	
Adjust to pH 3.5 0.005M $AlCl_3$	To adsorb viruses to solids
Centrifuge to pellet solids	
Discard supernatant	
Resuspend pellet in 10% beef extract	To elute (desorb) viruses from solids
Centrifuge to pellet solids	
Discard pellet and filter through 0.22 μm filter	To remove bacteria, viruses are in supernatant
Assay using cell culture	

FIG. 7.3

Procedure for recovery and concentration of viruses from sludge

WATER

Sampling waters for microbial analysis is relatively easier than sampling soils. Known volumes of water can be collected from known depths easily. Normally 1 ml to 1000 litres of water are collected. As compared to soils, number of microbes in water tends to be lower. It is, therefore, desirable to achieve proper concentrations of microbes in water to ensure their detection in the samples. For larger microbes, including bacteria and protozoan parasites, samples are often filtered to trap and concentrate the organisms. For bacteria, the sample is filtered through 0.45 μm membrane filter, whereas for protozoa coarse woven fibrous filters are used. A different procedure is used for viruses as they are too small in size.

Virus Analysis

Since low numbers of viruses are encountered, their detection and analysis in water samples involves a longer procedure. There are **four basic steps** in virus analysis: sample collection, elution, reconcentration, and virus detection. The overall procedure for sampling and detecting viruses in water is shown in Fig. 7.4. Virus analysis is performed on a variety of waters: potable water, ground and fresh surface waters, marine waters and sewage. For **sample collection**, large volumes of water (100 to 1000 litres) are passed through a filter. The viruses are then **concentrated** by adsorption onto the filter. The class of filters most commonly used for virus collection

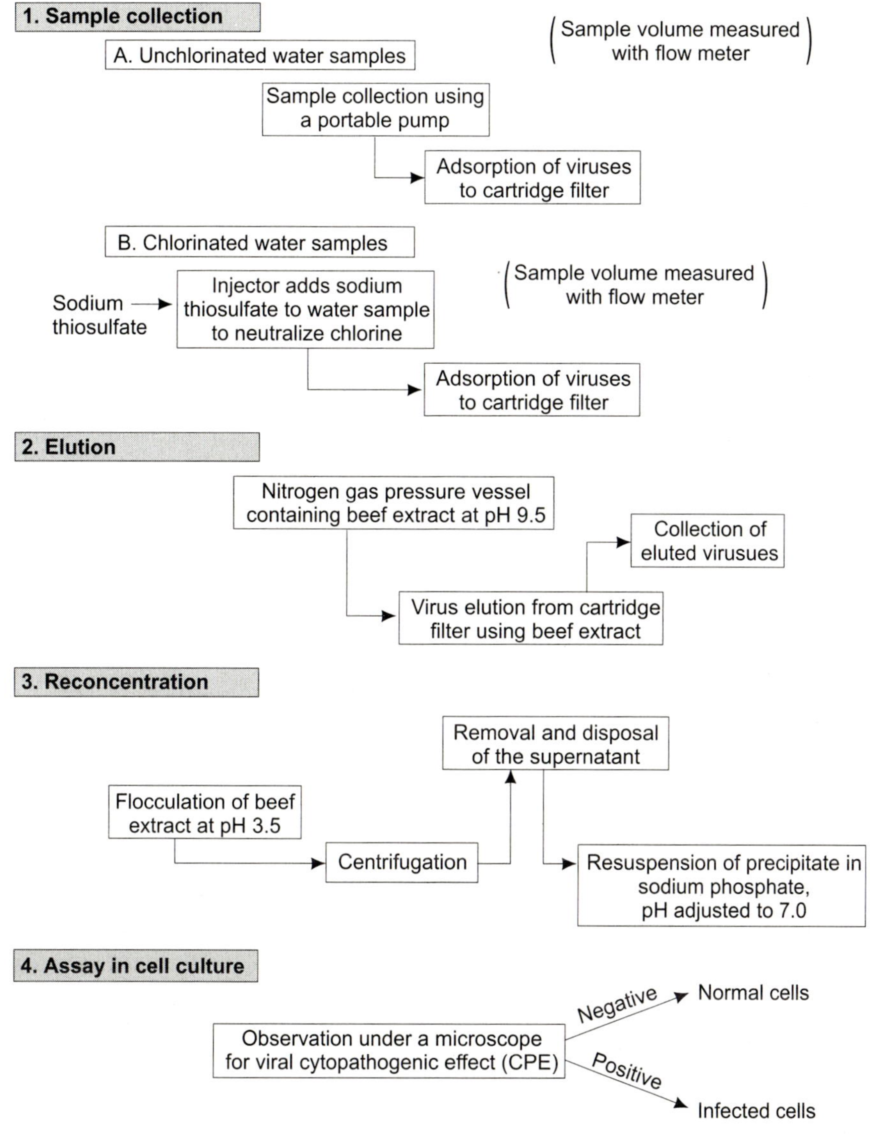

FIG. 7.4

Procedure for sampling and detection of viruses from water

and concentration is adsorption-elution microporous filters, more commonly known as **VIRADEL** (Virus adsorption-elution). Both electronegative and electropositive filters are available. Recovery of virus from the filter involves **elution** of the virus from the collection filter, and reconcentration to reduce the sample volume before assay. Adsorbed viruses are eluted from filter surfaces by pressure filtering 1-2 liters of an eluting solution through the filter. Elution fluid is 1.5% beef extract at pH 9.5. This volume is further reduced to 20-30 ml in reconcentration step. Viruses are then detected using cell culture or molecule methods, such as PCR. Virus can be detected by inoculation of a sample into an animal cell culture followed by observation of the cells for cytopathogenic effect or by enumeration of clear zones or plaque-forming units (PFU) in cell monolayers stained with vital dyes. PCR can also be used to detect viruses directly in either the sample concentrates or the animal cell culture.

Bacterial Analysis

Bacteria are collected and enumerated by one of the two procedures: (i) membrane filtration, and (ii) most probable number (MPN) methods. The former relies on collection and concentration of bacteria through filtration, whereas in the latter samples are generally not processed before the analysis. In both, bacteria are detected by cultural methods that are described in Chapter 8.

Protozoan Parasites

Appropriate methods have been developed for the detection of *Giardia* cysts and *Cryptosporidium* oocysts, as well as other protozoan parasites in water samples. Very often hundreds of liters of water are collected and filtered through 1 μm porosity spun fiber filter. The cysts and oocysts are extracted by cutting the filter in half, unwinding the yarn, and placing the fibers in an eluting solution of Tween 80, sodium dodecyl sulphate, phosphate-buffer. The 3-to 4-litre resulting volume is then centrifuged to concentrate the oocysts, which are then resuspended in 10% formalin or 2.5% buffered potassium dichromate (for *Cryptosporidium*). The pellet is further purified by density gradient centrifugation with Percoll-sucrose, sucrose or potassium citrate solutions. Aliquots of the concentrated samples are then filtered through membrane cellulose acetate filter, which collects the cysts and oocysts. These are stained with fluorescent monoclonal antibodies and the filter placed under an epifluorescence microscope. Fluorescent bodies of the correct size and shape are identified and examined by differential interference contrast microscopy for the presence of trophozites or sporozites. The overall procedure is shown in Fig. 7.5.

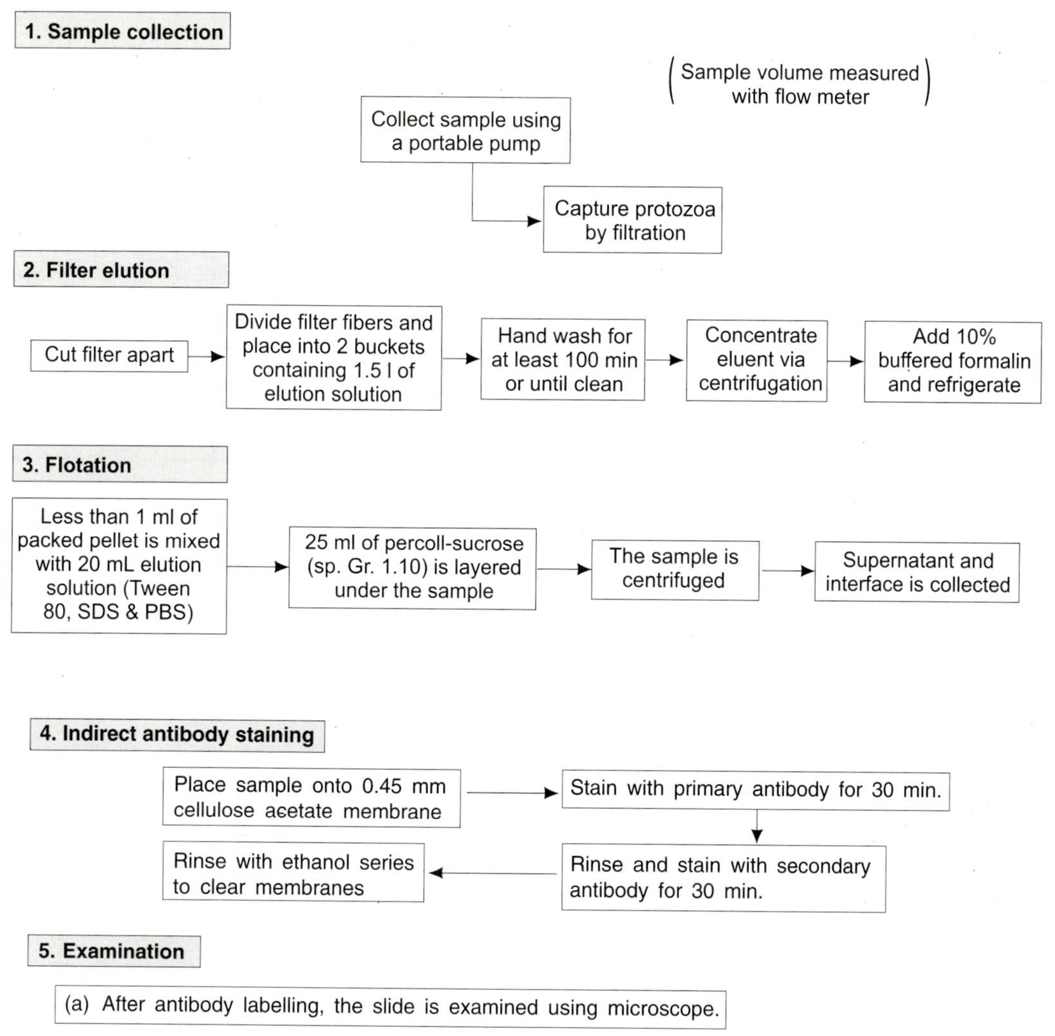

5. Examination

(a) After antibody labelling, the slide is examined using microscope.

(b) Examination is performed to determine characteristic size and shape of fluorescing microbes using UV epifluorescent microscopy.

Cryptosporidium sp	*Giardia* sp
4-6 μm	8-12 μm

(c) Examination is performed to determine characteristic internal structure using differential interference contrast (DIC) microscopy.

(i) up to 4 sporozites can be observed within an **oocyst** of *Cryptosporidium.*

(ii) Nuclei, axonemes, and median bodies are characteristically observed in **cysts** of *Giardia.*

FIG. 7.5

Procedure for collection and sampling protozoa from water, antibody staining, and examination of *Giardia* and *Cryptosporidium*

AIR

Many devices have been designed for the collection of bioaerosols. As pointed out in Chapter 5, several types of samplers are used to collect, quantify and detect the bioaerosols. These samplers work on the basis of different principles and sampling methods: impingement, impaction, centrifugation, filtration, and deposition. Most samplers designed for collection of samples are based on **impaction** and **impingement**.

Impaction

Impaction is the forced deposition of airborne particles on a solid surface. Several types of samplers have been developed that work on suction of air and subsequent impaction of the particles on a solid surface. **Durham's gravitational sampler**, developed in 1940s is the simplest and most commonly used sampler even today in some parts of the world including India. It is a simple device consisting of two metallic discs separated by three struts. The microscope slide coated with adhesive is mounted an inch above the centre of the lower disc, upper disc acting as rain shield. The slides are exposed for 24 hr. The sampler can be used for qualitative analysis of various kinds of spores, particularly larger ones, giving a rough idea of seasonal trend.

However, some **sophisticated samplers based on suction and impaction devices** (battery-as well as power-operated) have now been developed. These provide more accurate data on species-composition as well as density of aeroallergens. Based on suction device with definite amount of air sucked in and impacting the bioparticles on the adhesive material on the sampler is the most acceptable device world over. The Burkard volumetric traps and Rotorod aeroallergen models are such samplers. Burkard personal slide sampler (suction), battery-operated, suitable for indoor environment is 10 cm high having a rectangular orifice at the top (Fig. 7.6). The microslide is coated with glycerine jelly. The sampler sucks air at flow rate of 10 l/minute through the orifice. The particles get impacted on the slide forming a band. Slide is mounted and scanned. Burkard personal Petri dish sampler (Fig. 7.7) is similar to the slide sampler except that it has a stage to hold a Petri dish and a sieve to cover it. The orifice is circular. Petri dish containing suitable growth medium becomes exposed to a definite volume of air through the sieve on to the Petri dishes. These two samplers are suitable for indoor environment. Burkard continuous 7 day trap (continuous impaction) uses a glycerine jelly coated tape mounted on a rotating drum. The drum is connected to a timer and rotates at a constant speed. Tape exposed for a week is cut into seven strips, one for each day, mounted on a slide in glycerine jelly and scanned. Each strip can also be divided into 24 vertical bands, each band representing the particular hour of the day.

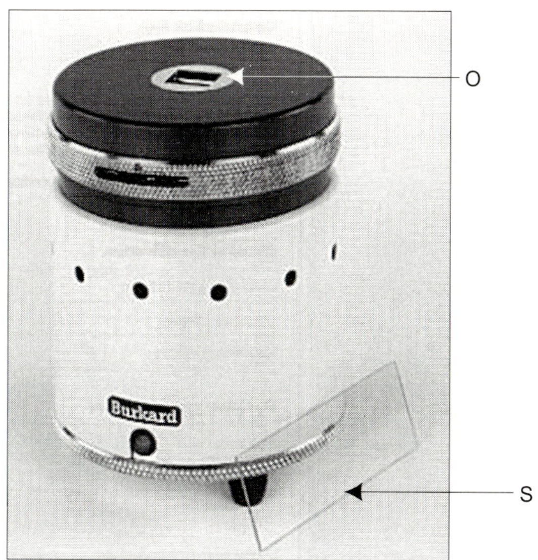

FIG. 7.6
Burkard personal slide sampler. O—orifice; S—slide

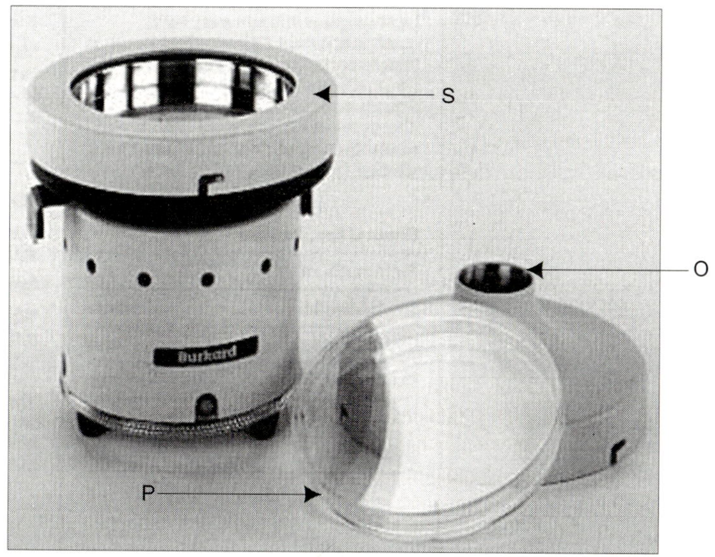

FIG. 7.7
Burkard personal Petri dish sampler. P-Petri dish; S-sieve; O-orifice

FIG. 7.8

Rotorod aeroallergen intermittent sampler. S-strip below cover (courtesy Dr. A.B. Singh)

The most accurate samplers to be used these days are the following two samplers.

Rotorod aeroallergen intermittent sampler

This sampler has an intermittent rotating impactor, is power-operated and ideal for both indoor and outdoor studies (Fig. 7.8). It is light weight, portable sampler. It has an arm that holds two cubical rods coated with silicon grease. Unexposed rods remain folded. Arm has a shield at top for protection from rains. Exposure time can be regulated. The rods are mounted on a grooved slide in a basic Fuschin stain and scanned.

Andersen six stage volumetric sampler

This sampler (Figs. 7.9, 7.10), developed by Anderson in 1958 is power-operated and ideal for indoor environment, though can also be used for outdoor studies. The general operating principle is that air is sucked through the sampling port and strikes

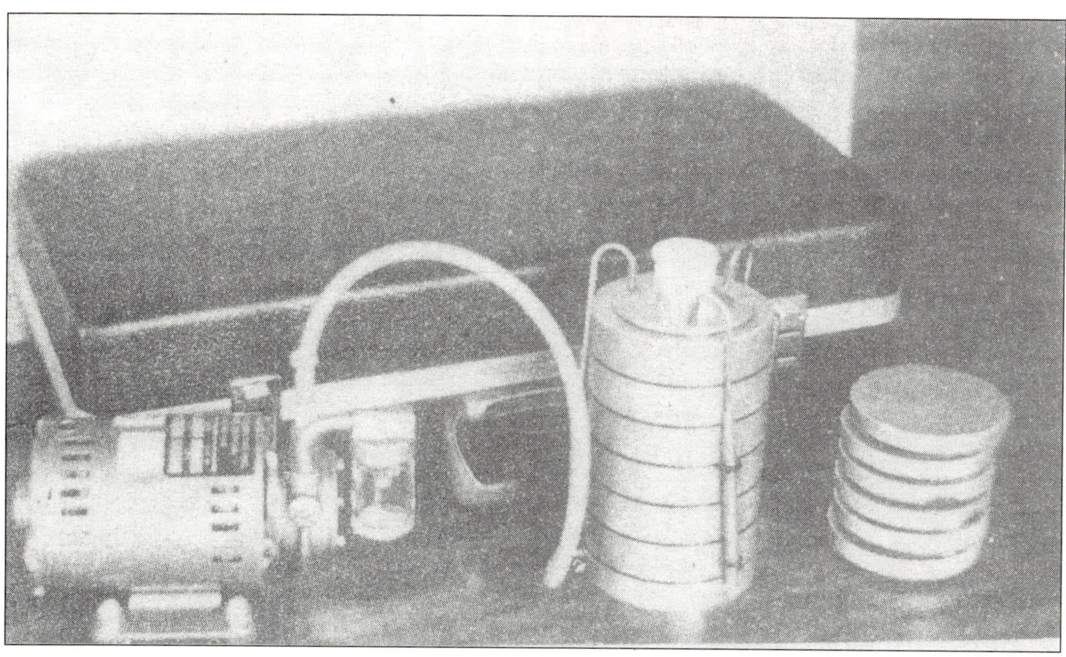

FIG. 7.9
Andersen six stage volumetric sampler shown attached with the motor. The six Petri dishes containing nutrient agar media are placed below each stage. The orifice through which the air is taken is covered with a plastic cap (courtesy Dr A.B. Singh)

agar plates. There are six Petri dishes containing suitable growth medium, kept under sieves of different pore size. Each sieve has 400 pores. The circular orifice at the top sucks air at flow rate of 28.3 l/min. passing it through the sieve arranged in gradually decreasing order of pore size. Air impacting on last Petri dish goes out. Larger particles are collected on the first layer, and each successive stage collects smaller and smaller particles by increasing the flow velocity and consequently the impaction potential. Particles of similar size impact on a given Petri dish only. In this way spores/pollens of six dimension orders are impacted on agar surface of individual Petri dish.

Impingement

Impingement is the trapping of airborne particles in a liquid matrix. Alongwith the Anderson six-stage impaction sampler, the all glass AGI-30 impinger based on impingement (Fig. 7.11) is the most commonly used device for microbial air samplings. This operates by drawing air through an inlet, similar in shape to human nasal passage. The air is transmitted through a liquid medium where the air particles become associated with the fluid and are trapped. The sampler is usually run at 12.5

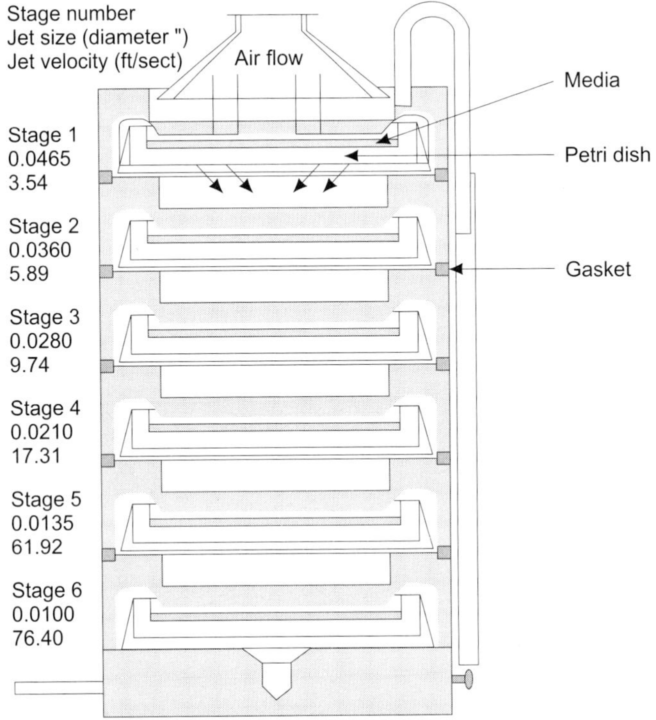

FIG. 7.10

A schematic representation of the Andersen six stage impaction air sampler. Air enters through the top of the sampler and larger particles are impacted upon the surface of the Petri dish on stage 1. Smaller particles, which lack sufficient impaction potential follow the air stream to the subsequent levels. As the air stream passes through each stage the air velocity increases thus increasing the impaction potential so that particles are trapped on each level based upon their size. Therefore larger particles are trapped efficiently on stage 1 and slightly smaller particles on stage 2 and so on until even very small particles are trapped on stage 6. The Andersen six stage thus separates particles based upon their size

l/min flow rate at a height of 1.5 m, the average human breathing height. It is easy to use, inexpensive, portable, easily sterilised and has good bioefficiency. It is very efficient for particles in the range of 0.8 to 15 μm. The usual volume of the collection medium is 20 ml and duration of sampling approximately 20 minutes. Sampling media may be a simple one as 0.85% NaCl, or more complex as 1% peptone. Enriched or defined growth media can also be used for specific microbes.

Centrifugation

Centrifugal samplers use circular flow patterns to increase the gravitational pull within the sampling device to deposit the particles. The cyclone, a tangential inlet and

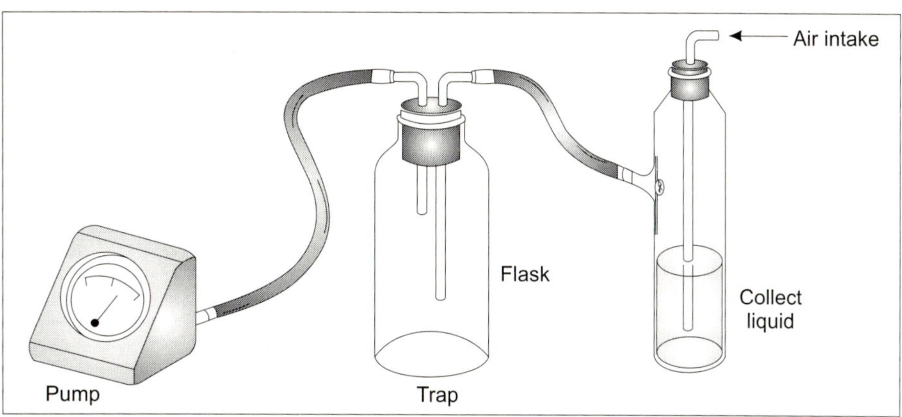

FIG. 7.11
Liquid impingement device for collecting bioaerosols

return flow sampling device, is the most common type. These samplers can sample a wide range of air volumes (1-400 l/min.), depending on the size of the unit. The unit operates by applying suction to the outlet tube, which causes air to enter the upper chamber at an angle. The flow of air falls in a tangential pattern, that effectively circulates air around and down along the inner surface of the glass housing. Due to increased centrifugal forces imposed on particles in the air stream, the paticles are sedimented out. The comical shape of the upper chamber opens into a larger bottom chamber, where most of the particles are deposited.

Filtration and Deposition

These two methods are widely used for microbial sampling for cost and portability reasons. For filtration device, there is a vacuum source involving passage of air through a filter, where the particles are trapped. Membrane filters of different pore sizes are used. After collection, the filter is washed to remove microbes for analysis. This device, however, has a low overall sampling efficiency. Deposition sampling is the easiest and most cost-effective sampling device. An agar plate is exposed to air, resulting into direct impaction, gravity settling, and other forces of deposition. But here also the efficiency of sampling is low.

8

METHODS OF ENUMERATION AND ACTIVITY OF MICROORGANISMS

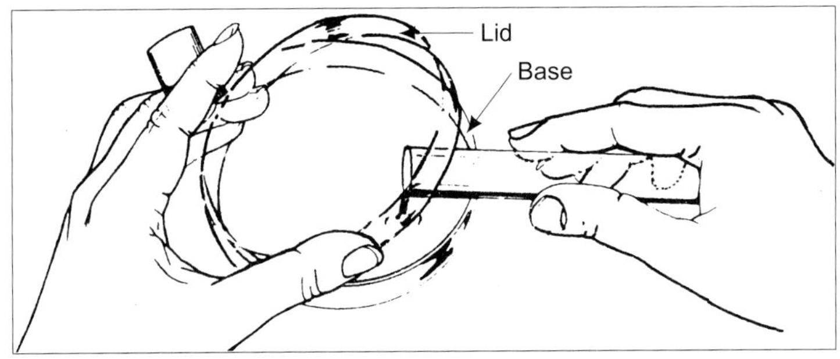

Chapter Outline

- *Cultural Methods*
- *Physiological Methods*
- *Immunological Methods*
- *Nucleic Acid-Based Methods*

I do not intend to describe the details of each of the various methods employed for enumeration of microorganisms and evaluation of their activities in different kinds of environments. Rather an attempt is made to consolidate and compile all such methodologies, presenting only broad outline of each method for the convenience of the reader. The detailed methodologies can be found in standard textbooks on general microbiology and other disciplines of biology, as well as in standard manuals of analysis of different components of environment—the soil, water and air. References to such literature will be cited in this chapter.

Methods employed in environmental microbiology are as diverse as the microbial world. Some methods can be employed for more than taxonomic groups of microbes as cultural methods for bacteria and fungi with necessary modifications. The expanded text of various cultural, physiological, immunological, and nucleic acid-based methods has been present in a condensed form in this chapter.

CULTURAL METHODS

To determine the appropriate culture method, it is necessary to define which microorganism or group of microorganisms is to be enumerated. The methods for enumerating bacteria, fungi, algae, protozoa, and viruses are very different. Some specific microbe may need a selective method also, as in case of indicator microbes like faecal coliforms or pathogens as *Salmonella* spp.

Cultural Methods for Enumeration of Bacteria

Two basic methods are used for viable counts of bacteria: (i) standard plate count, and (ii) most probable number (MPN) techniques.

Beterial cells from **soil** are extracted by appropriate methods described in Chapter 7. Methods often include hand or mechanical shaking with or without glass beads, mechanical blending, and sonication. Different extraction solutions may be used depending upon the pH and texture of the soil. A surfactant such as Tween 80 (Difco) may be used, often with a dispersing agent as sodium pyrophosphate.

Serial dilution of sample

After extraction, the sample is serially diluted to separate the microorganisms into individual reproductive units. Sterile water, saline, and buffered peptone or phosphate solution are often used for dilution.

The procedure is shown diagrammatically in Fig. 8.1. As shown in this figure, if we take one ml of the original sample (say a known weight of soil mixed in known volume of sterile water to have a soil suspension which contains a microbial mixture) and add it to 9 ml of steril water, it will give 1:10 or 10^{-1} dilution of the original

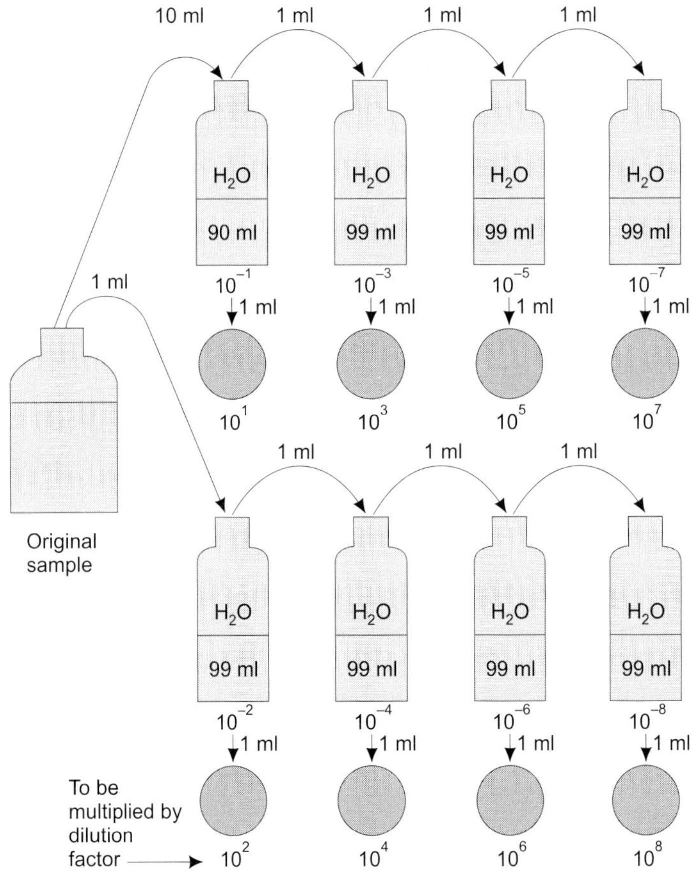

FIG. 8.1

Procedure of preparing different dilutions of an original sample of a microorganism and colony count

sample, i.e. the original sample has been diluted to 1/10th. Similarly we may prepare 1:100 (10^{-2}), 1:1,000 (10^{-3}), 1:10,000 (10^{-4}) and so on dilutions of the original sample as shown in Fig. 8.1. Finally one ml aliquot of any dilution is added to a sterile Petri dish to which are added 9 ml of sterile, cool, molten agar medium. The dishes are incubated at suitable temperature. Within few days colonies of each kind of microbes grow in the dish. The number of colonies of each kind is counted. This number is then multiplied by the **dilution factor** to find the total number of cells per ml of the original sample. This method is, therefore, also used for quantitative estimation of microbial cells in a known volume of original sample. Suppose, we use one ml of 10^{-3} dilution and 15 is the total number of cells per ml of this dilution. Then total number of cells per ml of the original sample will be 15 × 1,000 (dilution factor) i.e. 15,000 cells.

Water samples taken from marine or drinking sources require concentration rather than dilution because they contain low numbers of bacteria. A known volume of water is filtered through a membrane using a vacuum. The bacteria are trapped on the membrane, which is placed on the agar medium or a cellulose pad soaked in medium to allow growth of individual colonies. This is called **membrane filter technique**, referred in Chapter 7.

Plating methods

After dilution or concentration, the sample is added to Petri dishes containing a growth medium. Three different methods are used for application of the diluted sample to the growth medium: pour plate, streak plate and spread plate methods.

Pour plate method. As explained above, the sample is diluted successively with sterile water. The agar medium is maintained in molten state at 45°C. One ml of each dilution is added to each sterile Petri dish to which is then poured 9 ml of sterile, cool agar medium (Fig. 8.2). The contents are thoroughly mixed, and allowed to solidify. The dishes are incubated at suitable temperature. After few days, different kinds of microbes grow as separate colonies. Cells from individual colony may be picked up for a subculture.

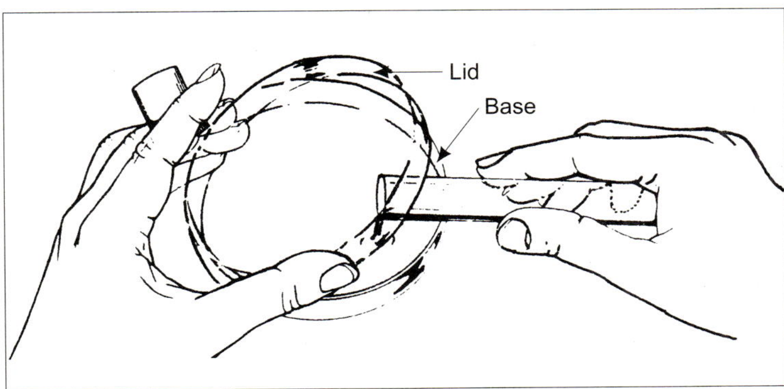

FIG. 8.2
Pour plate technique

Streak plate method. A small amount of sample is transferred onto the surface of a suitable, solid agar medium either by loop or transfer needle. This is then streaked in such a way as to provide successive dilutions arid ultimately to have well-

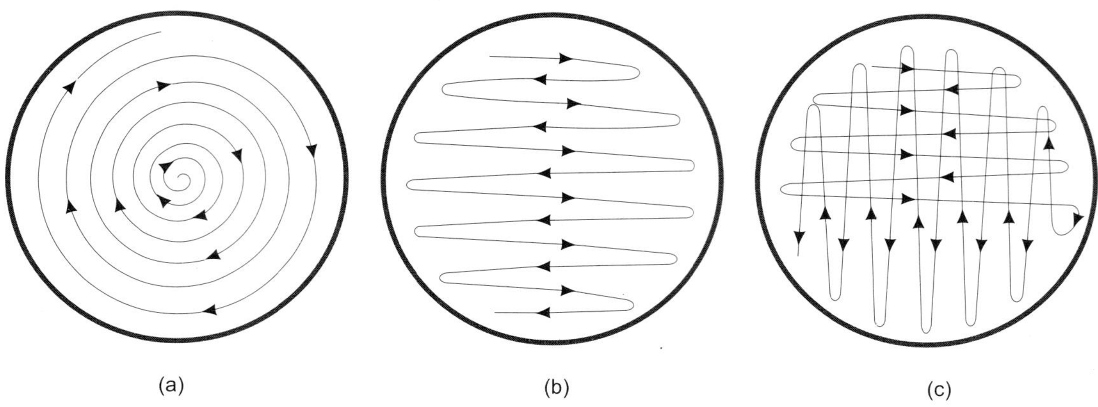

FIG. 8.3

Different ways of streaking on agar plates. The arrows indidicate the direction of streaking needle

isolated colonies. Streaking may be done in any of the ways shown in Fig. 8.3. In each case the sample becomes progressively diluted and at the end of streak one would expect properly isolated colonies.

Spread plate method. An aliquot of the diluted sample is placed onto the agar surface and is spread uniformly with a sterile, bent glass rod. Both streaking and spreading methods are useful particularly in separating aggregates of cells in the sample. There may be needed repeated streaking and spreading in case of pure culture of diplococci, sarcinae, streptococci and staphylococci.

Most probable number (MPN)

This technique is sometimes used in place of standard plate count method to estimate microbial counts. The sample is dispersed in an extracting solution and successively diluted as in plate count. This method relies on the dilution of the population to extinction followed by inoculation of 5-10 replicate tubes containing a specific liquid medium with each dilution. After incubation the tubes are scored as +/– for growth on the basis of such factors as turbidity, gas production or disappearance of a substrate (Fig. 8.4). Scoring a tube positive for growth means that at least one culturable microbe was present in the dilution used for inoculation. The number of positive and negative tubes at each dilution is used to calculate the number present in the original sample through the use of published statistical MPN tables (APHA, 1992).

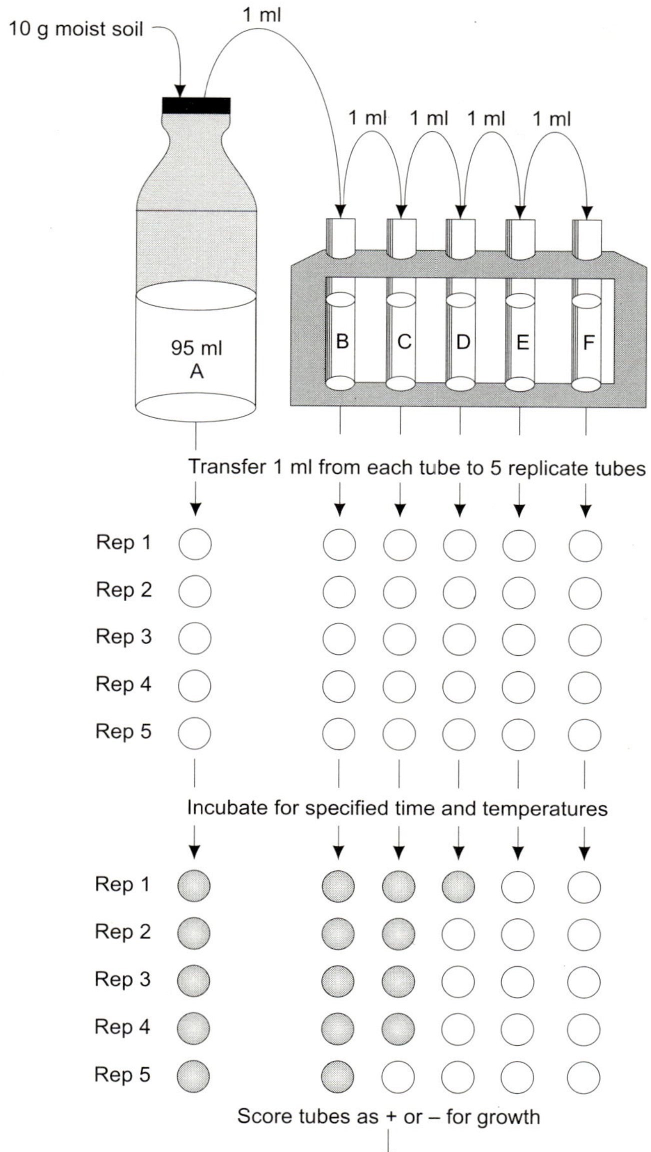

FIG. 8.4

Most probable number technique. An environmental sample is diluted to extinction. Diluted samples are used to inoculate replicate tubes at each dilution. The presence or absence of the microorganism of interest at a given dilution can be analysed statistically to estimate the original population in the environmental sample

Culture Media for Bacteria

Two major groups of culture media are used: **general media** for the common heterotrophic bacteria, and **enrichment media** for isolation of a specific particular type of microorganism from the mixture.

Heterotrophic plate counts

For common heterotrophic bacteria, two basic types of media can be used: (i) nutrient-rich, and (ii) nutrient-poor or minimal media. Examples of nutrient-rich media are, nutrient agar, peptone-yeast agar, and soil extract agar amended with glucose and peptone. Nutrient-poor media contain as much as 75% less of the ingredients of nutrient-rich media (peptone, yeast, and /or beef or soil extract). Examples of minimal media are R2A agar (BBL, Difco), m-HPC agar (Difco), and soil extract agar with no amendments.

Media for specific microbes

Enrichment medium is a liquid medium that promotes the growth of a particular physiological type of microbe present in a mixture of microbes while suppressing the growth of competitive flora. Such media may be elective or selective. **Elective enrichment medium** allows growth of a single or limited type of bacteria based on a unique combination of nutritional or physiological attributes. **Selective enrichment medium** involves the use of inhibitory substances or conditions to suppress or inhibit the growth of most organisms while allowing the growth of the desired one. Antibiotics are most often used for this medium. These include chloromphenicol, erythromycin, tetracycline, streptomycin, polymyxin, nalidixic acid, novobiocin, rifampicin, penicillin etc.

A **specialised isolation medium** contains formulations that meet the nutritional requirements of specific groups of microbes, such as *Staphylococcus* or *Corynebacterium*. A differential medium contains ingredients to allow distinction of different microbes growing in the same medium. It includes an indicator, usually for pH.

Fecal coliforms in water. The water is filtered using membrane filtration technique. The membrane is then inoculated on m-FC agar (Difco) and incubated at 45°C for 24 hours. There develop blue colonies of coliforms, whereas non-coliforms appear as gray to cream coloured. Fecal coliforms can be isolated on an enriched lactose medium with 1% rosalic acid to inhibit the growth of non-coliforms, and an elevated incubation temperature is critical in the selction for fecal coliforms. This is a standard method published in "Standard Methods for Examination of Water and Wastewater" (Method 9222 D). (APHA, 1995)

Salmonella from sewage. Culture methods for *Salmonella* are varied, complex and time consuming requiring final confirmatory tests (Fig. 8.5). The traditional method for detecting *Salmonella* relies on an enrichment and plating technique with subsequent estimation using MPN tables. Selected volumes of sewage, 10 to 0.01 ml are added to a pre-enrichment broth and incubated for 18-24 hours at 37°C. This is followed by selective enrichment and other steps as shown in Fig. 8.5.

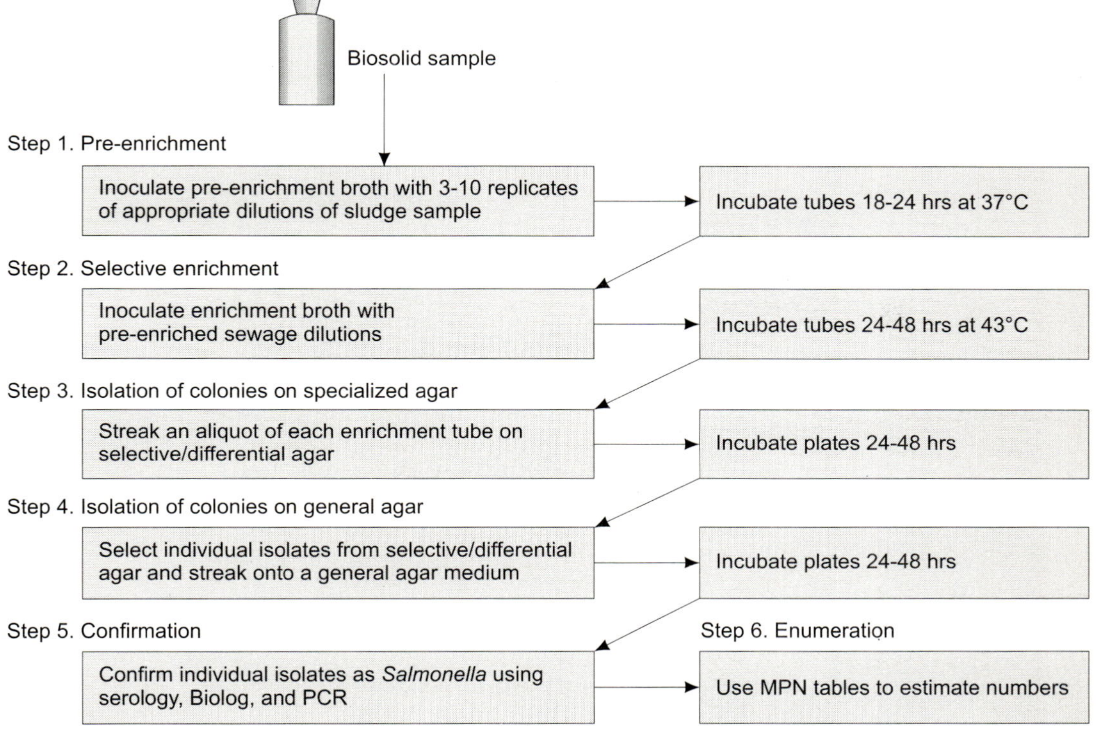

FIG. 8.5

Protocol for the detection of *Salmonella* from sewage sludge

Fluorescent Pseudomonas from soil. *Pseudomonas* are Gram-negative aerobic chemoheterotrophs. The soil sample is added to an extracting solution and a dilution series is followed. Spread plate technique is used to isolate individual colonies on selective media. One such medium is S1 medium, which contains sucrose, glucose, casamino acids, sodium lauryol sarcosine, trimethoprim, various salts and agar. Fluorescent colonies are identified on the medium with UV-light.

Nitrifying organisms from the soil. *Nitrosomonas* and *Nitrobacter* are isolated using a somewhat lengthy procedure. It includes soil enrichment, an initial culture enrichment and rigorous purity checks (Fig. 8.6). *Nitrobacter* spp can also be estimated using MPN technique. The broth in the tubes contains potassium buffers, Mg salts, trace elements, Fe, $NaNO_2$, and Bromthymol Blue as indicator. The colour changes from blue-green to yellow after oxidation of NH_4^+ to NO_2^-

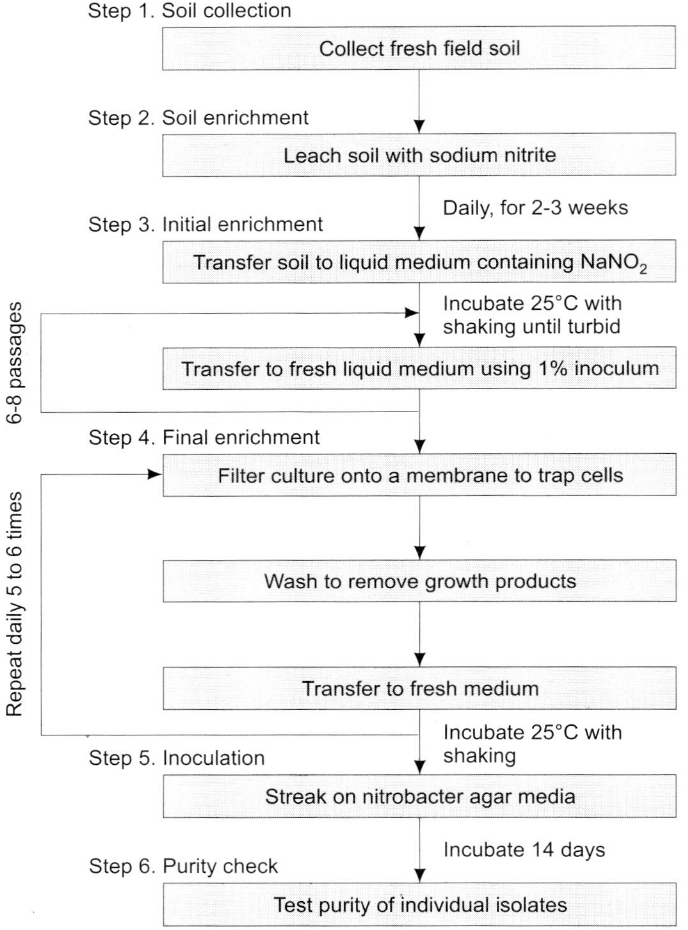

FIG. 8.6
Protocol for the isolation of nitrifying organisms

2,4-D degrading bacteria from soil. A selective, differential enrichment medium, called eosin-methylene-blue (EMB)-2,4-D agar is used for such bacteria. This contains minimal salts and 2,4-D as the carbon source. Indicators eosin B and methylene blue allow selection for gram-negative bacteria and differentiation of 2,4-D-degrading colonies which turn black.

Cultural Methods for Fungi

Cultural methods for fungi are similar to those for bacteria, but are to be amended with chemicals to restrict bacterial growth. Common chemicals used for the purpose are antibiotics or dyes as Rose Bengal. Plating analyses and quantitative dilution, however, do not provide accurate results. Unlike bacteria, where each colonyforming unit originates from a single cell, in fungi colonies can arise from a single spore, an aggregate of spores, or a mycelial fragment containing more than one viable cell. Inspite such limitations, plating techniques are still a valuable tool for the assessment of fungi in the environment. Fungi are cultured by pour plate, spread plate and the membrane filtration techniques. These protocols can be found in "Standard Methods for the Examination of Water and Wastewater" (Methods 9610 B, C, D). Fungal hyphae picked up from samples can be plated on nutrient agar. Plant roots are washed and placed on agar medium.

Cultural Methods for Algae and Cyanobacteria

Algae and cyanobacteria can be enumerated by dilution and plating techniques as well as MPN assay with some modifications. Filamentous algae require grinding the soil samples with a mortar and pestle prior to extraction as well as more vigorous homogenisation with blender or glass beads. Antibiotics like penicillin and streptomycin are added to check bacterial and fungal growth on solid agar media. Incubation for 1-3 weeks at 20-25°C with a photoperiod of either 12 hr. light-12 hr. dark or 16 hr. light-8 hr. dark are required. For cyanobacteria, a typical medium, BG-11 is used. This includes $NaNO_3$, 1.5 g; K_2HPO_4, 40.00 mg; $MgSO_4.7H_2O$, 75.0 mg; $CaCl_2.H_2O$, 36 mg; citric acid, 6.0 mg; ferric amm. citrate, 6.0 mg; EDTA (disodium salt), 1.0 mg; Na_2CO_3, 20.0 mg; trace metals, 1.0 mg; agar, 10.0 g; dist. water, 1.0 litre; cycloheximide, 20.0 mg.

Detection of Viruses by Cell cultures

Two main types of cell culture are used in virology: **primary cell culture**, where cells are removed directly from an animal and can be used to subculture viral cells for only a limited number of times (20-50 passage), and **continuous cell culture**, in which cell lines derived from animals or humans, are subcultured indefinitely. Such cell lines may be derived from normal or cancerous tissue. Cell cultures are initiated by dissociating small pieces of tissue into single cells by treatment with a proteolytic enzyme and a chelating agent, generally trypsin and EDTA. The dispersed cells are then suspended in cell culture medium composed of balanced salt solution containing glucose, vitamins, amino acids, a buffer, a pH indicator, serum (usually fetal calf), and antibiotics.

The two most common methods of detecting and quantifying viruses in cell culture are: the **cytopathogenic effect** (CPE) method, and the **plaque-forming unit** (PFU) method. the cytopathogenic effects are observable changes in host cells due to virus replication. These include rounding or formation of giant cells or hole formation in monolayer due to local lysis of infected cells. Different viruses may produce very individual and distinctive CPE. For instance adenovirus causes rounding of cells. Two procedures can be used to quantify viruses utilising CPE: (i) serial dilution endpoint or TCID50 method which involves adding serial dilutions of virus suspension to host cells and subsequently observing the production of CPE over time. The titer or endpoint is the highest viral dilution capable of producing a CPE in 50% of the tissue culture vessels or wells and is referred to as medium tissue culture infective dose or TCID50. (ii) the MPN method, which is similar to the one used for bacteria.

Viruses like enteroviruses produce plaques or zones of lysis in cell monolayers overlaid with solid nutrient medium. These plaques originate from a single infectious virus particle; thus the titer can be quantified by counting the plaques or plaque-forming units (PFUs). This can be thought of analogous to enumeration of bacteria using CFU.

PHYSIOLOGICAL METHODS

Various methods of measurements of the activity of microorganisms, both in pure culture and environmental samples, have been briefly described in this section.

Microbial Activity Measurements in Pure Culture

In a pure culture, activity of microorganisms can be measured in terms of appropriate parameters. These includ substrate disappearance, terminal electron acceptor (TEA) utilisation, cell mass increase, and carbon dioxide evolution.

Substrate disappearance

For heterotrophic activity substrate is carbon-based, and such substrate can be measured in different ways depending on the chemistry of substrate molecule. Accordingly, these **heterotrophic substrates** are measured by using UV spectrophotometer, fluorimeter, high-performance liquid chromatograph (HPLC), gas chromatograph (GC), and mass spectrophotometer. For **chemoautotrophic substrates**, as oxidation of ammonia (nitrification), and sulphur (sulphur oxidation), different methodologies are employed. For nitrification ^{15}N-labelled products are measured by mass spectrophotometer, whereas sulphates formed during sulphur oxidation are measured by atomic absorption spectrophotometer.

Terminal electron acceptors

A variety of TEAs are available for measuring microbial activity, which is reflected in their disappearance and the appearance of the reduced TEA. Majority of the environmental activities are measured under aerobic conditions, oxygen being the TEA. Measurement of oxygen in a pure culture is relatively rapid, routine and cheap to perform. This is most extensively used parameter in waste water treatment to determine the BOD.

Oxygen levels in an aqueous sample can be measured using an oxygen probe, by colorimetric assays or by manometry. An oxygen probe is basically an electrode covered by a gas-permeable membrane combined with a meter that converts electric signal into an analytical measurement. In colorimetry the sample is mixed with a series of reagents to produce an oxygen-dependent color. A titration method, Winkler method is also used for dissolved oxygen, which is precipitated using manganous sulphate and KOH-KI mixture. Manometry quantifies microbial respiration by measurement of O_2 depletion from the atmosphere within a sealed flask. The method is based on ideal gas law, $PV = nRT$. Warburg constant volume respirometer and Gilson manometer are used for this purpose.

Cell mass

Increase in cell mass in a pure culture is very often quantified by culturable plate counts or direct microscopic counts. Another common method to estimate cell mass is to measure the turbidity or the protein content in the culture. **Turbidity** of a bacterial solution can be measured by a colorimeter or spectrophotometer, using a standard curve and wavelength between 490 and 550 nm. A variety of methods are in use for measurement of **protein** content. The most commonly used methods are those of Lowry *et al* (1951), Bradford (1976) and Smith *et al* (1985).

Carbon dioxide evolution

An alkaline trap, usually composed of NaOH or any other strong basic solution can be used to trap CO_2 produced during mineralisation. The trapped biocarbonate can be detected by using titrimetric, gravimetric, or conductimetric measurements. The most common method is to quantify trapped CO_2 by titration with standardised acid solution.

The use of ^{14}C-radiolabelled substrates allows very sensitive and specific measurements of $^{14}CO_2$ evolution. This technique is often used to evaluate the rates of biodegradation of organic contaminants. A biometer flask can be used, and the alkali within the sidearm trap is assayed for radioactivity using liquid scintillation counting.

Microbial Activity Measurements in Environmental Samples

Microbial activity in a pure culture, though can be easily determined by measurements of cell number (culturable plate or direct counts) or biomass (turbidity or protein content), these values are not realistic for environmental samples. In environmental samples, any such value is obtained for all forms of organisms present in the sample rather than an individual as in case of pure culture. Realistic information on microbial activity is in fact provided by environmental samples. In samples, activity is quantified in terms of respiration or synthesis of cellular macromolecules, which provide more direct reflection of levels of microbial activity within a sample. Different tests to quantify different metabolic activities of microbes are listed in Table 8.1.

TABLE 8.1 Popular tests for the measurement of microbial activity in environmental samples

Test	*Basis of test*
1. Measurement of respiration gases	Measurement of O_2 utilised or CO_2 produced in sample. Respiration flux indicates overall metabolic activity (level of microbial activity)
2. Respiration of radio-labelled substrates	Metabolism of labelled substrate monitored measuring evolution of labelled CO_2
3. Microelectrodes	Probes (tips $< 20\ \mu m$ diam.) inserted into the sample provide continuous monitoring
4. Incorporation of radio-labelled thymidine into cellular DNA	Microbes will scavenge DNA precursors, as thymidine, from their environment. Its rate of incorporation into DNA is measured
5. Adenylate energy charge (AEC)	AEC values reflect a continuum between active microbial community (AEC > 0.8) and a community with a high proportion of dead cells (AEC < 0.4)
6. Dehydrogenase assay	Measurement of rate of oxidation-reduction reactions by monitoring the reduction of tetrazolium salt by respirating microbes
7. Hydrolysis of fluorescein diacetate	This is performed by several enzymes including esterases, proteases and lipases

Measurement of respiration gases (CO_2 and O_2)

As described earlier for pure cultures CO_2 and TEA (terminal electron acceptor) can be measured in environmental samples, both in laboratory as well as field conditions (i.e *in situ*). In controlled conditions of a laboratory, a sample of porous medium is typically incubated in a sealed, airtight enclosure, usually referred to as a **microcosm**. Field studies can be performed by placing a field chamber over a plot of surface soil and using this setup to make *in situ* experiments. In microcosm, flux of CO_2 and/or

O_2 within the headspace atmosphere is determined. Gas samples from headspace can be withdrawn using a gastight syringe and concentration of these gases measured by gas chromatography. Alternatively, CO_2 can be trapped in a basic solution using trap (biometer flask used for pure cultures). Gas chromatography is suitable method for field studies also.

Respiration measurements have several applications in environmental microbiology. These include the assessment of soil "health" or condition as reflected by basal rate determinations of microbial activity in the samples; indicators of rates of biodegradation of pollutants, as petroleum hydrocarbons in contaminated sites; microbial biomass determinations; biological oxygen demand (BOD); and better understanding of the working of mixed microbial mats communities (biofilms).

Incorporation of radio-labelled tracers into cellular macromolecules

Protein or nucleic acid estimations can be used to monitor the increase in biomass of a bacterial population. In environmental samples biomass, specifically of bacteria is measured by incorporating radiolabelled tracer molecules into cellular macromolecules. Such tracers include nucleoside thymidine labelled with tritium (^3H) which is incorporated into DNA, and the amino acid leucine, labelled with either ^3H or carbon-14 (^{14}C) which is incorporated into protein.

Adenylate energy charge (AEC)

In terms of determining microbial activity, the relative abundance of ATP compared with its precursors, ADP and AMP indicates a rapid rate of ATP formation. The ATP/ADP/AMP ratio provides a biochemical basis for assessment of physiological and nutritional status of organisms. A measure of AEC ratio is the weighted ratio of cellular adenylates:

$$\text{AEC} = \frac{\text{ATP} + 1/2\ \text{ADP}}{\text{ATP} + \text{ADP} + \text{AMP}}$$

High AEC values (> 0.8) reflect an active community, intermediate values (0.4 to 0.8) reflect cells in resting state, and low values (< 0.4) reflect a high proportion of dead or moribund cells. Estimation of AEC values from environmental samples can be used as an indicator of bacterial biomass. Quantification of ATP from samples needs an extraction procedure, followed by concentration of cellular components into a buffer. The ATP is then determined directly using the luciferin-luciferase assay. ADP and AMP in the sample are then converted to ATP by enzymatic reactions and quantified.

Enzyme assays

Many enzyme assays have been developed to detect either specific or general microbial activity in environmental samples. Some of these are as follows:

Enzyme	**Substrate**
Dehydrogenase	Triphenyltetrazolium
Phosphatase	p-Nitrophenol phosphate
Protease	Gelatin
Amylase	Starch
Chitinase	Chitin
Cellulase	Cellulose, Carboxymethyl cellulose
Nitrogenase	Acetyline
Nitrate reductase	Nitrate

The two commonly used assays for general microbial activity are, the dehydrogenase assay, and the esterase assay. **Dehydrogenase** reactions can be detected using a colourless, water soluble tetrazolium salt. When reduced this salt forms a reddish coloured formazan product that can be detected in several ways. These salts compete with other electron acceptors. Thus measurement of tetrazolium reduction will reflect the electron transport chain activity. A commonly used tetrazolium salt is 2-(p-iodophenyl)-3-(p-nitrophenyl)-5-phenyltetrazolium chloride (INT), which is transformed into an intensely coloured, water-insoluble formazan (INT-formazan). The samples are suspended in a solution containing INT and incubated for 1-12 hrs. INT-formazan can be detected by microscopic examination of the red deposits formed within the cell or by quantifying total INT-formazan production.

Esterase assays are based on hydrolysis of fluorescein diacetate (FDA), which is also susceptible to proteases and lipases. The product of FDA hydrolysis is the highly fluorescent molecule, fluorescein, which is water soluble and can be detected spectrophotometrically or fluorometrically. Like dehydrogenase assay FDA hydrolysis has been used to examine microbial activity in soil, subsurface soils, and activated sewage sludge. The sample is incubated with FDA in a buffer solution. The supernatant is then separated from the cell or environmental debris by centrifugation and the fluorescein quantified. Fluorescein can also be extracted from sample using acetone.

IMMUNOLOGICAL METHODS

Several of the basic immunological methods (immunoassays) are being widely used in environmental microbiology. Immunoassays are analytical methods used for the detection and/or quantification of the antigen-antibody interaction. For almost all types of immunoassays, attachment of a signal molecule to the antibody and/or antigen is very important. Many types of signal molecules are used in immunoassays, including iodine, enzymes, fluorochromes, and radioisotopes. **Antiglobulins**, the antibodies that are specific for another individual antibody are attached to a signal molecule and used as secondary or indirect detection molecules.

What are Antibodies?

Antibodies are protein complexes produced by the immune system of higher life forms that help depend the host against pathogen invasion. A great diversity of antibodies are produced by B cells. Antibodies are **specific** in recognising the particular antigen. They have thus great **affinity** with an antigen.

Antibody molecules represent about 17 percent of the total protein in the blood serum. Enormous variety of antibodies should be evident from only single fact that a single bacterium may have a thousand different antigens, and that each antigen might elicit its own highly specific antibody. The term **immunoglobulin (Ig)** is used synonymously with antibody because antibody molecules exhibit the characteristics of the globulin group of proteins.

It was only during late 1950s and early 1960s that details about the nature of antibodies began to emerge. Research focused on the Bence Jones proteins, identified a century earlier by Henry Bence Jones. Enough information and several theories were available about the origin and functions of these proteins, but little was known about their detailed structure. It was presumed that an understanding of the structure of these proteins (now known to be antibodies) would provide clue to their unique specificity for antigens. The answers were provided by Gerald M. Edelman and Rodney M. Porter who described the chemical composition and structure of proteins and received the 1972 Nobel Prize in Physiology or Medicine.

The basic antibody molecule consists of four polypeptide chains: two identical "heavy" (H) chains and two identical "light" (L) chains. These chains are joined together by sulphur to sulphur (disulphide) linkages to form a Y-shaped structure as shown in Fig. 8.7a. Each heavy chain consists of about 400 amino acids, while each light chain has about 200 amino acids.

Within each polypeptide chain there exist constant and variable regions. The amino acids in the constant region of both light and heavy chains are virtually identical among antibodies. However, the amino acids of the variable region vary among the hundreds of thousands of different antibodies. Thus the variable regions of a light and heavy chain combine to form a highly specific, three-dimensional structure somewhat analogous to the active site of an enzyme. This portion of the antibody molecule combines with the antigenic determinant (Fig. 8.7c). Moreover, the "arms" of the antibody are identical so that a single antibody molecule may combine with two antigen molecules. This combination may lead to a complex antibody and antigen molecules.

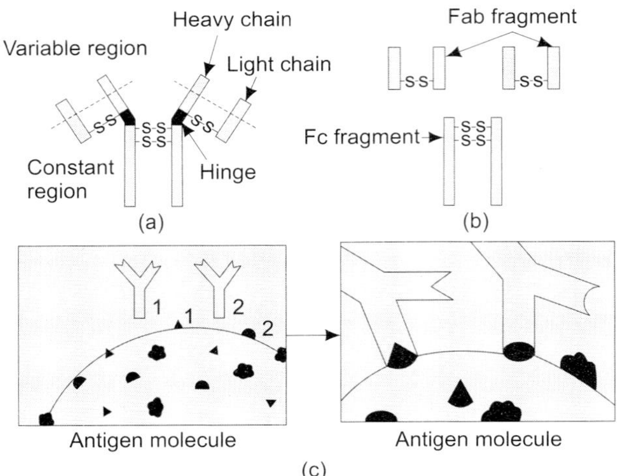

FIG. 8.7

Details of the antibody molecule. (a) the antibody molecule consists of four chains of protein: two light chains and two heavy chains connected by disulfide linkages. The heavy chains bend at a hinge point. The variable region is that portion of the molecule where the amino acids vary among antibodies. (b) on treatment with papain enzyme, cleavage occurs at the hinge point and three fragments result. Two are Fab fragments. and the third is the Fc fragment. (c) the reaction between antibody and antigenic determinants in an antigen molecule. Each antibody molecule has two reactive sites complementary to an antigenic determinant. Antibodies 1 and 2 each react with different antigenic determinants at the variable end

When an antibody molecule is sectioned with papain, an enzyme from papaya fruit, the molecule separates at the hinge point, and two functionally different segments are isolated (Fig. 8.7b): (i) the **Fab fragment,** for "fragment-antigen-binding", is the portion that will combine with the antigenic determinant, and (ii) the **Fc fragment,** for "fragment able to be crystallised", has various functions; it is the part of the antibody molecule that combines with phagocytes in opsonization; it appears to neutralise viral receptor sites; it attaches to certain cells in allergic reactions; and it activates the complement system in resistance mechanisms.

Till present time, five types of antibodies have been identified, based on differences in the heavy (H) chains. The five classes of antibodies (immunoglobulins, abbreviated as Ig), are designated as IgM, IgG, IgA, IgE and IgD. The structures of these antibodies are shown in Fig. 8.8. IgM consists of a pentameter (five subunits), whose segments are connected by a glycoprotein called a joining or J chain, IgA is a dimer (two subunits), with the segments also connected by a J chain. In addition,

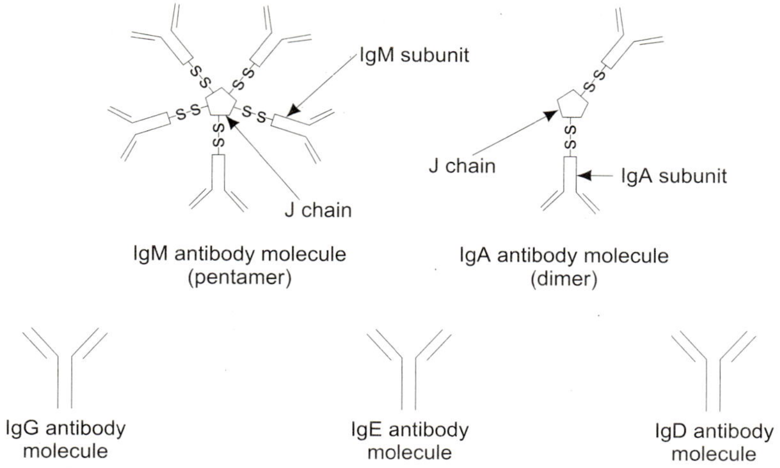

FIG. 8.8

The structures of five types of antibodies. Note the complex structures of IgM (a pentamer) and IgA (a dimer). IgG, IgE, and IgD consist of monomers, each composed of two heavy chains and two light chains of amino acids

there is a secretory component that enables the molecule to leave secretory epithelial cells. The remaining three antibodies are monomers (one subunit).

IgM is the first antibody to appear in the circulation after stimulation of B-lymphocytes. It is the principal component of the **primary antibody response** (Fig. 8.9), and the largest antibody molecule. Owing to its size, most IgM remains in the circulation. IgM is formed during fetal cases of rubella or toxoplasmosis, indicating that a certain immunological competence exists in the fetus. About 5 to 10 percent of the antibody component of normal serum consists of this antibody.

IgG is the classical gamma globulin and is the major circulating antiboody, comprising about 80 per cent of the total antibody component in normal serum. IgG appears about 24 to 48 hours after antigenic stimulation and continues the antigen-antibody interaction begun by IgM. It is thus the antibody of the **secondary antibody response** (Fig. 8.9). In addition, it provides long-term resistance to disease as a product of the memory B-lymphocytes. Booster injections of a vaccine considerably raise the level of this antibody in the serum. IgG is also the **maternal antibody** that crosses the placenta and renders immunity to fetus until child is able to make antibodies at about six months of age.

IgA is approximately 10 per cent of the total antibody component of normal serum. One form of this antibody, called **serum IgG**, exists in the serum and is similar to IgA. A second form accumulates in body secretions and is referred to as

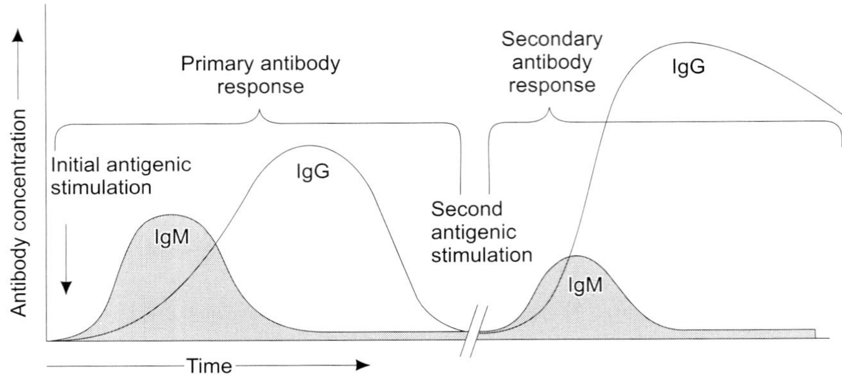

FIG. 8.9

The primary and secondary antibody responses. On initial antigenic stimulation, IgM is the first antibody to appear in the circulation. It is the principal component of the primary antibody response and is later supplemented by IgG. On second exposure to the same antigen, the production of IgG is more rapid and the concentration in the serum reaches a higher level than previously. The IgG antibody is more concentrated in the secondary antibody response

secretory IgA. This antibody provides resistance in the respiratory and gastrointestinal tracts. possibly by inhibiting attachment of the parasites to tissues. It is also located in tears and saliva, and in the colostrum, the first milk secreted by nursing mother. When consumed by child, the antibodies may lend resistance to gastrointestinal disorders.

The characteristics of the five types of antibodies are summarised in Table 8.2.

TABLE 8.2 Characteristics of five types of antibodies

Antibody type	*% of antibody in serum*	*Location in body*	*Mol. wt. (daltons)*	*No. of fourchain units*	*Crosses placenta*	*Characteristics*
IgM	5-10	Blood, Lymph	900,000	5	No	Principal component of primary response
IgG	80	Blood, Lymph	150,000	1	Yes	Principal component of secondary response
IgA	10	Secretions, Body cavities	400,000	2	No	Protection in body cavities
IgE	< 1	Blood, Lymph	200,000	1	No	Role in allergic reactions
IgD	0.05	Blood, Lymph	180,000	1	No	Unknown

IgE plays a major role in allergic reactions by sensitizing cells to certain antigens. The functions and significance of **IgD** are not known at present. Perhaps this antibody may stimulate B-lymphocytes to secrete other antibodies.

The commonly used immunoassays are as follows:

Fluorescent immunolabelling (Immunofluorescence)

This is the use of fluorescent signal molecules conjugated to antibodies to interact with and subsequently indicate the presence of a particular antigen by the production of fluorescent light. The sample antigen is attached to microscope slide and a fluorescent chemical/antibody conjugated specific to the antigen is added. This is then viewed under microscope equipped with a fluorescent light source. The labelled antigen appears bright green against dark background. This assay is commonly used in detection of protozoan parasites, *Giardia* and *Cryptosporidium* in waters.

Enzyme-linked immunosorbent assay (ELISA)

It is very sensitive method used to detect the presence of pathogens. Antigens or antibodies are atached to a solid surface and the coated surfaces are combined (immunosorbed) with the test material. An enzyme system is then linked to the complex.

The antigens, if necessary concentrated, and solibilised in buffer. Depending on the type of antigen and the antibodies, one can now perform direct or indirect assay (Fig. 8.10). For a direct ELISA, a primary antibody is attached to microtiter plate, microcentrifuge tube or other solid support. The antigen is then added and allowed to incubate for binding with the antibody. A second antibody (signal antibody) is now added. These secondary antibodies are conjugated to a signal molecule (enzyme, alkaline phosphatase). The secondary antibody then binds to the antigen, which is already bound to the primary antibody. Substrate is then added that causes a colour change due to the presence of the signal molecule. This color change is usually in proportion to the amount of the antigen present, thus quantifying the antigen present in a given sample.

This method is very useful in studies on degradation of pollutants in water treatment plants and bioreacotors, where the concentration of bacteria within biofilms is very important. This test has been found very useful for estimating specific biomass within biofilms. ELISA has also been used to quantify populations of *Dehalospirillum multivorans* (asociated with biodegradation of organic pollutants) in mixture culture biofilms.

Competitive ELISA

In this test, both a labelled control antigen and a sample, containing unknown quantity of unlabelled antigen are added to a sample well coated with antibody.

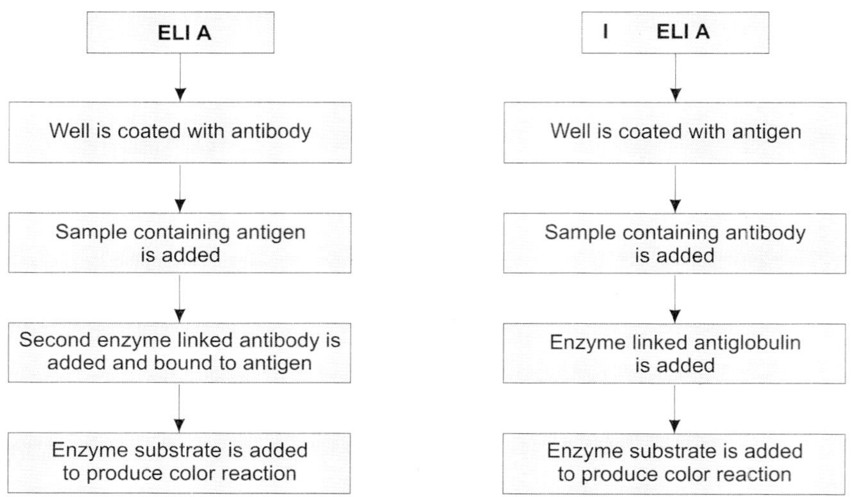

FIG. 8.10

Schematic representations of direct and indirect sandwich ELISAs. These reactions are usually carried out in a microtiter plate and the color change shown can be detected and quantified using a plate reader

Sample is added first and the bound antibody captures the unlabelled antigen. The labelled antigen is then added and captured by any remaining antibody binding sites. If all sites are taken, no labelled antigen will bind and no signal emitted. If there is no antigen in the sample, only labelled antigen will bind and the signal will be maximised. This test is very sensitive even to lowest concentrations of the antigen. This test has been found very useful in detection of even minute quantities of mycotoxins in cereals.

Immunomagnetic separation assays

This is an antigen capture using antibodies conjugated to paramagnetic beads to attach to, concentrate, and purifiy antigens. Antibodies coated on magnetic beads (100 nm to 20 μm) bind the antigens in solution and are then separated from the solution using a magnet. This assay has been used to recover thermophilic sulphate-reducing bacteria as *Thermodesulphobacterium mobile* from oil field waters below oil production platforms.

Western immunoblotting assay

This assay identifies the presence of target antigens in a complex mixture of many other nontarget antigens present in samples. The assay is performed with simple dot blot hybridisation with a labelled antibody or an electrophoretic separation followed by hybridisation with the labelled antibody. In dot blot hybridisation, samples are added directly to an immobilising nitrocellulose membrane, followed by immunolabelling and signal detection. In second method, a sample of antigen is added to a gel and separated by size using electrophoresis. After electrophoretic separation, the sample is transferred to an immobilising nitrocellulose membrane. The membrane is then incubated with enzyme-labelled or radio-labelled antibodies, that specifically bind to the antigen. After incubation, a substrate for enzyme-labelled or photographic film for radiolabelled substrate is used to detect the presence of target antigen. An outline of the assay is shown in Fig. 8.10. This assay has been found very useful for determining the presence and amounts of environmentally relevant proteins such as methane monooxygenase (MMO) produced by *Methylosinus trichosporium* OB3b, studied for bioremediation of trichloroethylene (TCE), a common environmental pollutant.

Immunoprecipitation assay

This assay uses the antigen-antibody reaction in a solution to determine the amount of antigen or antibody in a sample by determining the amount of precipitation or clumping of the antigen-antibody complex. A series of reaction tubes are set up, each containing a constant titer of antibodies. Antigen is then added in increasing concentration to consecutive tubes. There is no precipitation in initial tubes (as lowest concentration of antigen is added to them). With increase in antigen concentration formation of antigen-antibody complexes increases until a visible precipitate is formed. With further increase in antigen concentration (exceeding that of antibody) the amount of precipitate will decrease again. This assay has been used to determine the mechanism for inhibition of important fungal plant pathogens by another fungi. For instance to investigate the inhibitory methanism of *Talaromyces flavus* (that produces H_2O_2) against *Rhizoctonia solani, Verticillium dahliae* and *Sclerotinia sclerotiorum.*

NUCLEIC ACID-BASED METHODS

Since traditional cultural methods tend to underestimate the true soil microflora, several methods have been developed to secure more representative soil microbial populations. These are based on the total nucleic acid content of the community or **community DNA**. In **soil environment** community DNA is obtained by separating all cells from the soil and then lysing the cell and extracting the nucleic acids. Cell lysis is done by heat, chemical or enzymatic treatment. Another way of DNA extraction from soil is to lyse the bacterial cells *in situ* (within soil matrix) and then extract and purify the DNA. Though several methods of extraction of community DNA are available (More *et al*, 1994), the most reliable one is the use of lysozyme or sodium dodecyl sulphate as the lysing agent. After extraction the DNA is purified by cesium chloride density centrifugation or the the use of commercial purification kits. Purified DNA is processed further for molecular analysis.

Cells from **aquatic environments** are concentrated by filtration through membrane 0.2 μm filters. Viruses are concentrated by electronegative or electropositive filters. Larger parasites as *Giardia* are trapped in a simple fiber filter cartridge. Biomass of these aquatic microbes is lysed, purified and treated for molecular analysis.

Various nucleic acid-based methods commonly used in environmental microbiology are as follows:

Gene probing

Gene probes can be used to detect a target organism within a sample of mixture of organisms, to confirm the identity of a pure culture isolate, as well as to detect a specific gene sequence within bacterial colonies containing a mixed population of bacteria. Gene sequence is detected by **colony hybridisation** or **lifts**. Gene probe technology is also applied for identification of a target sequence of interest by Southern or Northern hybridisation, also known as blots. **Southern blotting** is a technique used to identity DNA sequence, whereas **Northern blotting** is the analogous process used to analyse RNA. **Dot blotting** or **dot hybridisation** is a technique used to evaluate the presence of a specific nucleic acid sequence without running a gel.

Polymerase chain reaction (PCR)

Discovered in 1985, PCR has become key protocol in most biological disciplines including environmental microbiology. The amount of target DNA is amplified upto 106-fold or more that makes the detection very sensitive and easy. Several new modified versions of PCR have been developed during last 10 years or so, that proved useful in detection of microbes in the environment, including *Rhizobium* spp. and *Salmonella*.

Recombinant DNA techniques

DNA cloning has been widely used in environmental microbiology to examine the genetic make up of bacteria.

Restriction fragment length polymorphism analysis

RFLP analysis, used for bacterial isolates, consists of designing smaller fragments of extracted DNA by action of restriction enzymes (endonucleases). These enzymes recognise specific DNA sequences, usually 4-6 bp long. These fragments are processed further for analysis.

Denaturing/Temperature gradient gel electrophoresis

This is a new technique of depth insight into the diversity of microbial communities. PCR-amplified DNA fragments are analysed by this method. Originally developed in medical research, the technique was introduced in microbial ecology by Muyzer *et al* (1993).

Plasmid analysis

Specific bacterial isolates are identified by detecting the plasmids. These have been particularly used for identifying *Salmonella* and *Rhizobium*.

Reporter genes

A reporter gene is a gene (or set of genes) that is inserted into a target gene of interest and subsequently signals or reports the activity of that gene. Thus nucleic acid sequence are not only used to detect specific organisms, but also to demonstrate a specific activity. The **lacZ** gene (coding for β-galactosidase), the **gusA** gene from *E. coli* (coding for β-glucuronidase), and **xylE** gene from TOL plasmid of *Pseudomonas putida* (coding for catechol 2,3-dioxygenase) are examples of reporter genes.

9

BIOGEOCHEMICAL CYCLES

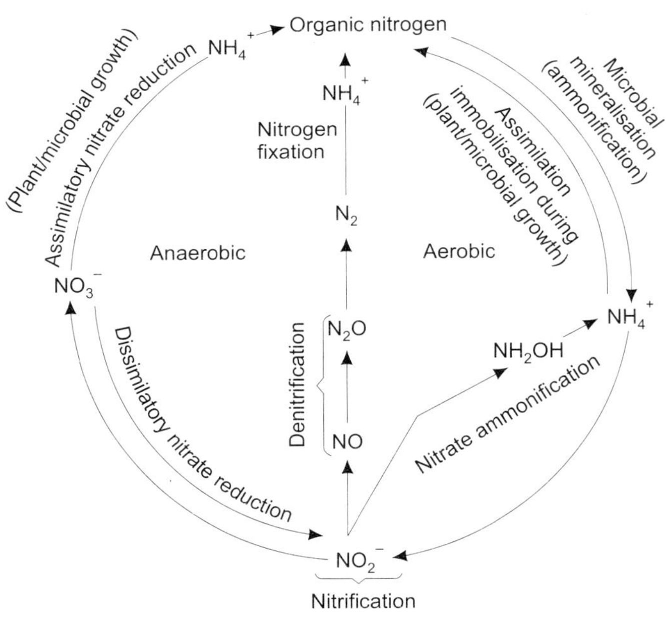

Chapter Outline

- *Carbon Cycle*
- *Methane-the Central Molecule of Carbon Cycle*
- *Nitrogen Cycle*
- *Sulphur Clycle*

The main building elements of organisms are carbon, hydrogen, oxygen, nitrogen, sulphur and phosphorus. Growth of an organism involves the conversion of these elements present in an inorganic form to the organic compound that make up the living matter. The energy for this elemental conversion is ultimately derived from solar sources of photosynthesis.

Inorganic forms of elements —Living organisms / Solar energy→ Organic forms of elements

If this were the only process, life would soon cease as the inorganic forms of elements, particularly carbon and nitrogen would be locked into organic matter. In fact, the reverse process - **mineralisation** must also occur and is brought about by the activity of living organisms so that cycles of elements and matter occur.

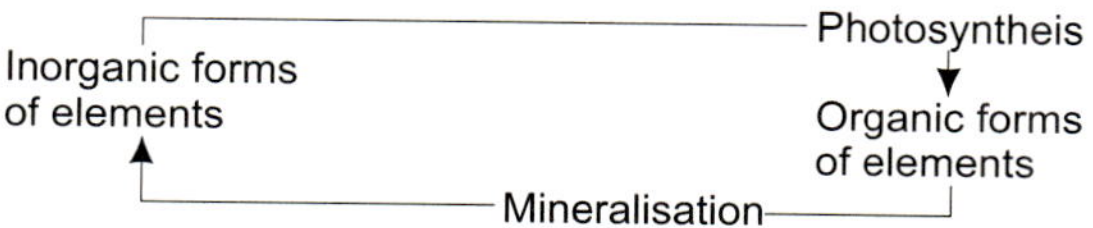

The situation is complicated in nature by the existence of oxidised and reduced states of most of the essential elements; organisms may only be able to use one or other form and further cycles therefore exist between them.

Oxidised form	CO_2	O_2	H_2O	NO_3^- N_2	SO_4^{2-}
Reduced form	CH_4	H_2O	H_2	$NH_4^+ \rightleftharpoons NH_3$	H_2S

The microbes interconvert these compounds for their own growth and each process is energy consuming. While the different elements are cycled, energy flows through the ecosystem and ultimately lost as heat. The organic matter produced by plants and other autotrophs does not keep accumulating rather it is consumed and degraded, and a delicate global balance of all elements (C, N, S, P, etc.) is maintained. This balance is a result of the biologically driven, characteristic cycling of elements between biotic forms and abiotic components. To stress the biological, geological and chemical nature of the processes, the cycles are termed as **biogeochemical cycles**. **Gaia hypothesis**, put in the early 1970s by James Lovelock that Earth behaves like a superorganism is of direct relevance to biogeochemical cycle. This concept developed into what is now known as the Gaia hypothesis. To quote Lovelock (1995) "Living organisms and their material environment are tightly coupled. The coupled system is a superorganism, and as it evolves there emerges a new property, the ability to self-regulate climate and chemistry."

A glance to a simplified carbon cycle (Fig. 9.1) shows that the two fundamental processes driving the cycle are: carbon fixation, and carbon respiration. The first half includes the **autotrophs**, both photo-as well as chemoautotrophs, also referred to as primary producers. These include plants as well as microorganisms such as algae, cyanobacteria, bacteria, and some protozoa. They fix carbon dioxide into organic matter. The carbon dioxide thus fixed into organic carbon is available for consumption or respiration, and this second half of the cycle includes **heterotrophs**, the animals and microorganisms.

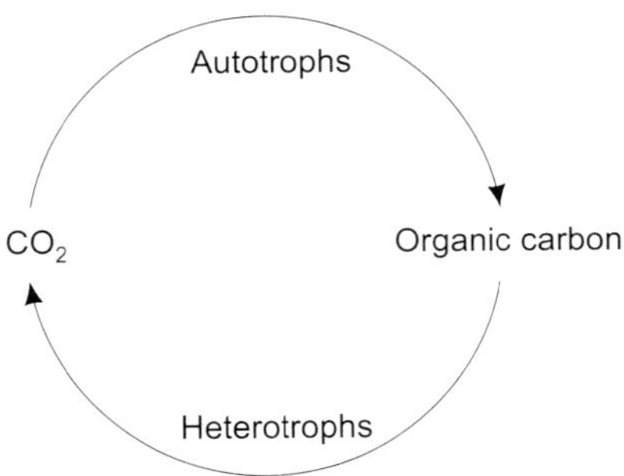

FIG. 9.1
A simplified carbon cycle

Microbial activities that drive the biogeochemical cycles are highly relevant to any biologist. However, these become relatively more relevant to environmental microbiologists due to the simple reason that there are many microbial activities that can be harnessed in a beneficial way (bioremediation of metal and organic pollutants, recovery of precious metals as copper or uranium), or may be detrimental to environmental quality (acid rains, acid mine drainage, metal corrosion, global warming, ozone depletion etc.). Thus the knowledge of these cycles is very much relevant and critical as the demographic pressure of increasing human population continues to grow with direct impact of human activity on earth's environment. We will cosider briefly the cycles of three major elements, carbon, nitrogen and sulphur in this chapter, though other elements, phosphorus, iron, calcium etc. also undergo similar cycling in nature.

CARBON CYCLE

A little extended view of the carbon cycle (Fig. 9.2) shows that carbon is actively cycled between inorganic CO_2 and the variety of organic compounds that compose living organisms and their dead organic matter. This cycle primarily involves the transfer of CO_2 and organic carbon between the atmosphere, where carbon occurs principally as inorganic CO_2, and the hydrosphere and lithosphere (Fig. 9.2). Autotrophs (photo-as well as chemolithotrophs) are responsible for primary production i.e. conversion of CO_2 to organic carbon. Once carbon is fixed (reduced) into organic compounds, it can be transferred from one organism to another, supporting a variety of heterotrophs. The respiration of live organisms and the degradation of organic matter, mostly in soil bring CO_2 back to the atmosphere.

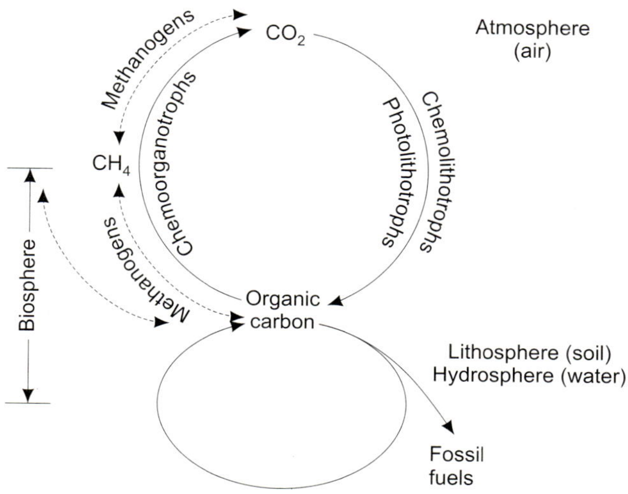

FIG. 9.2
A little extended carbon cycle

Organic matter in soil is the chief source of nutrients to plants. Moreover, these are the microbes which transform this organic carbon to various nutrients. Decomposition of organic matter by microbes is a central step in carbon cycle.

In environment, carbon dioxide is the major source of carbon for living organisms. Besides CO_2, carbon in nature is also present in the reduced form of methane (CH_4) in some specific habitat conditions. Interplay of methane in carbon cycle has generated much interest among environmental microbiologists. As shown in Fig. 9.2 this gas may undergo interconversions between inorganic (CO_2) and organic carbon compounds. Methane can be produced (methanogenesis) by special group of anaerobic archaebacteria, the methanogens, found in special habitat conditions, as well as can also be utilised as sole carbon and energy source by another special group of aerobic bacteria, methane-utilising or methane-oxidising bacteria (methanotrophs).

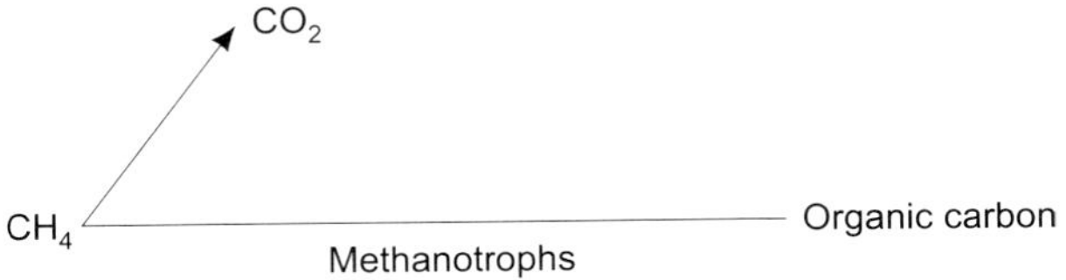

The total CO_2 in the atmosphere would be completely exhausted with this rate of conversion of inorganic to organic carbon. However, the reverse process of mineralisation of organic to inorganic carbon by the activity of heterotrophs prevents this exhaustion. The major end product of mineralisation is CO_2, though some methanogenic bacteria also produce CH_4 by anaerobic respiration and fermentation.

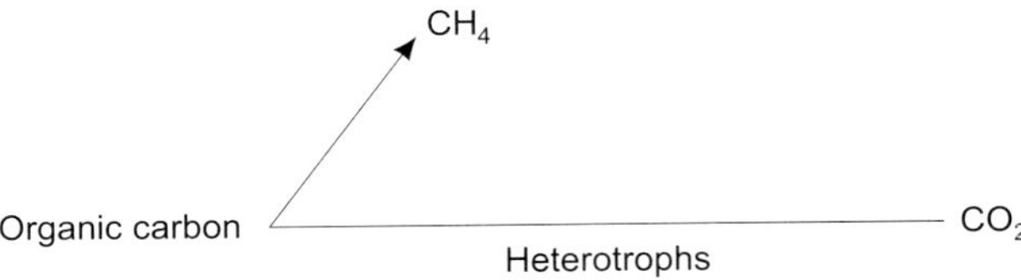

Thus, a carbon cycle is constructed to give an equilibrium with equal rates in both directions.

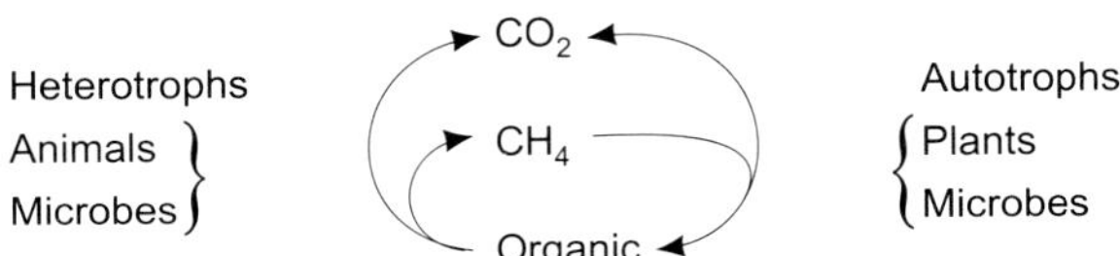

The conversion of inorganic to organic carbon by plants and by autotrophic microbes is relatively straightforward. However, the reverse process by the animals and heterotrophic microbes is more complex and will be considered in some details.

Heterotrophs-animals or microbes, obtain their carbon and energy by the metabolism of an organic source provided by another form of life.

$$\text{Organic carbon} \xrightarrow[CO_2\ (CH_4)]{\text{Heterotrophs}} \text{Organic carbon} \quad \begin{pmatrix}\text{Cells and} \\ \text{organic end} \\ \text{products}\end{pmatrix}$$

As a result, some of the organic carbon is mineralised whereas the rest is converted to further organic form in the form of new growth of the heterotroph or as end-products of metabolism. The efficiency of this conversion will vary according to the organism and to the aerobic or anaerobic conditions. The particular heterotroph and any organic carbon produced by it then serve as food source for other heterotroph and so on along a food chain. At each trophic stage of the chain, a percentage of the original organic carbon is mineralised until the process is complete.

$$\text{Autotroph} \xrightarrow[\text{Heterotroph}]{CO_2\ (CH_4)} \begin{matrix}\text{Organic} \\ \text{carbon}\end{matrix} \xrightarrow[\text{Heterotroph}]{CO_2\ (CH_4)} \begin{pmatrix} CO_2\ (CH_4) \\ \text{Organic carbon} \\ \text{Heterotroph}\end{pmatrix}$$

Microbes play an exceedingly important part in the process. Any compound that is a component of a living organism must be susceptible to mineralisation so that it does not accumulate, and all the carbon remains unavailable in that form. Microbes have a very wide ability to breakdown organic compounds. Dead plant and animal matter on and in soil and in water is rapidly degraded. A sequence of microbes do this, each kind of microbe utilising one or more component.

Microorganisms occupy the central position in the process of mineralisation of the organic carbon. The most common organic compounds found in the environment and acted upon by microbes include organic polymers (the most common being plant cellulose, hemicellulose, and lignin), humus, and C_1 compounds such as methane (CH_4). A brief account of the decomposition of these different components of organic matter, including humus by microorganisms has already been given earlier in Chapter 4.

METHANE—THE CENTRAL MOLECULE OF CARBON CYCLE

Interconnected microbial activities result into both, the formation (methanogenesis) as well as utilisation of methane in the environment.

Methanogenesis

Methane formation, termed as **methanogenesis** takes place by a group of obligately anaerobic archaebacteria, called the methanogens. Methane is released to atmosphere from abiogenic as well as biogenic sources (Table 9.1). A substantial amount of methane is released to the atmosphere as a result of energy harvesting and utilisation. Methane is also released through landfill gas emissions. In U.S.A. landfills alone generate more than 10×10^6 metric tons of methane per year.

Methane is an explosive gas at concentrations as small as 5%. Thus, to avoid accidents, the methane generated in landfills must be managed in appropriate way. If present in concentrations higher than 35% it can be collected and used for energy. Alternatively the methane can be burnt off at concentrations of 15% or higher. Most commonly, it is simple vented to the atmosphere to check its build up in higher concentration to ignite. Concentration of methane is increasing in the atmosphere at the rate of about 1% every year. Wetlands are regarded as significant biogenic source accounting for about 20% of the annual emmissions.

Although methane makes a relatively minor carbon contribution to the global carbon cycle, methane emission is of much concern to environment. Like carbon dioxide methane is a greenhouse gas contributing to global warming. In fact, it is the second most common greenhouse gas emitted to the atmosphere. Moreover, it is 22

TABLE 9.1 Estimates of methane released into the atmosphere

Source	*Methane emission (10^6 metric tons/year)*
Abiogenic	
Coal mining	10-35
Natural gas flaring and venting	10-35
Industrial and pipeline losses	15-45
Biomass burning	10-40
Methane hydrates	2-4
Valcanoes	0.5
Automobiles	0.5
Biogenic	
Ruminants	80-100
Termites	25-150
Paddy fields	70-120
Natural wetlands	120-200
Landfills	5-70
Oceans and lakes	1-20
Tundra	1-5
Total	350-820
Total abiogenic	48-155
Total biogenic	302-665

times more effective than CO_2 at trapping heat. Not only these, localised production of methane in landfills can create safety and health concerns.

The microorganisms responsible for methanogenesis are a group of obligately anaerobic archaebacteria called the **methanogens** which occur in special habitats such as saturated soils of natural wetlands, paddy fields, marshes, algal mats and landfills, and even inside the gut of termites and ruminants, and heartwood of trees. In saturated soils as paddy field, the methanogens produce methane autotrophically through CO_2 fixation, as well as heterotrophically growing on some C_1 and C_2 substrates including acetate, methanol and formate produced during decomposition (fermentation) of dead organic matter of aquatic plants and other debris (Fig. 9.3).

The basic metabolic pathway used by the methanogens is:

$$4H_2 + CO_2 \rightarrow CH_4 + 2H_2O$$

$$\Delta G^\circ = -130.7 \text{ kj}$$

This is an exothermic reaction where CO_2 acts as the terminal electron acceptor and H_2 as electron donor providing energy for the fixation of carbon dioxide. Such methanogens are capable of utilising CO_2 and are, therefore, autotrophic. Besides

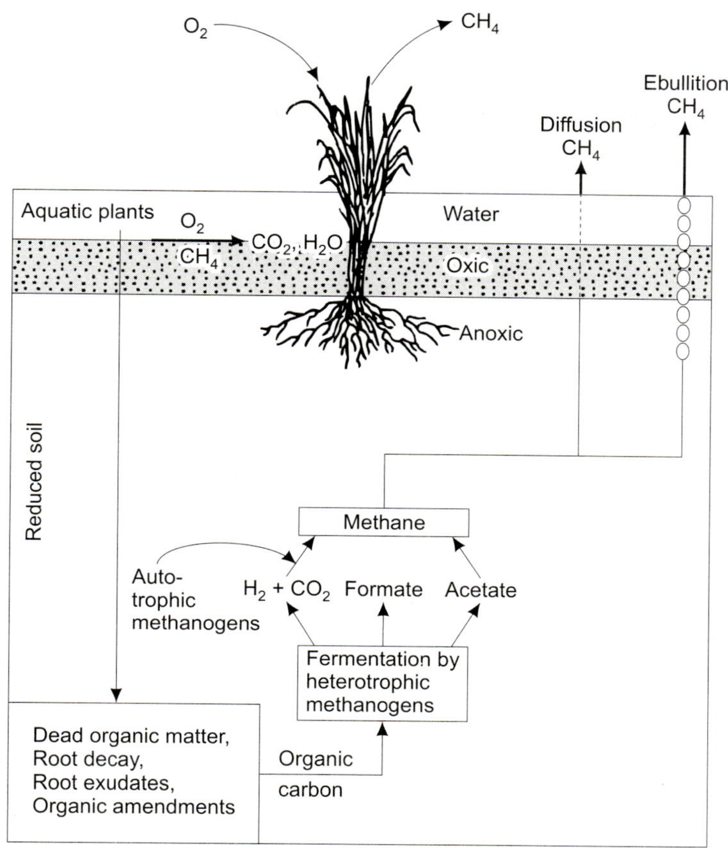

FIG. 9.3

Diagrammatic representation of production, reoxidation and transport of methane in a paddy field

autotrophic, there are heterotrophic methanogens also involved in methanogenesis. They can produce methane during heterotrophic growth on a limited number of other C_1 and some C_2 substrates including acetate, methanol and formate. Since there are very few carbon compounds that can be used by methanogens, these microbes depend on the production of such compounds by other microorganisms in the surrounding community. As such an interdependent community of microbes typically develops in anaerobic environments. In this community, the more complex organic molecules (dead organic matter accumulated, organic amendments etc.) are catabolised by heterotrophic populations that ferment or respire anaerobically, generating C_1 and C_2 carbon substrates as well as CO_2 and H_2 that are then used by autotrophic methanogens.

Methane Oxidation

Methane is thus found extensively in nature as the end product of CO_2 fixation by methanogens under anaerobic conditions. This is available as such to a group of bacteria called the **methanotrophs** which have developed to utilise methane as a carbon and energy source (food) for their growth. These methanotrophs are thus chemoheterotrophic and obligately aerobes. They metabolise methane as follows:

$$CH_4 + O_2 \xrightarrow[\text{monooxygenose}]{\text{Methane}} \underset{\text{methnol}}{CH_3OH} \rightarrow HCHO \rightarrow HCOOH \rightarrow CO_2 + H_2O$$

Oxygenases incorporate oxygen into the substrate and are important enzymes in the initial steps of degradation of hydrocarbons. This enzyme has gained importance as it is being exploited to biodegrade chlorinated solvents like trichloroethylene (TCE). Methanogens can thus be used for bioremediation of ground waters contaminated with TCE.

Carbon Monoxide and Other C1 Compounds

Bacteria that can utilise C1 carbon compounds other than methane are called **methylotrophs**. There are several C1 compounds produced from both natural and anthropogenic activities (Table 9.2). One of these is carbon monoxide (CO). The annual global production of CO is $3\text{-}4 \times 10^9$ metric tons. The two major CO inputs are abiotic: (i) atmospheric photochemical oxidation of carbon compounds, such as CH_4, accounting for roughly 1.5×10^9 metric tons/year, and (ii) burning of wood, forests and fossil fuels, accounting for 1.6×10^9 metric tons/year. A small proportion of CO, 0.2×10^9 metric tons/year results from biological activity in ocean and soil environments. This gas is highly toxic. Microbial activities have been found to contribute significantly to CO menace. They cause efficient destruction of this gas. Key microbes detected in terrestrial environments can metabolise CO aerobically as well as anaerobically. Under aerobic conditions *Pseudomonas carboxydoflava* oxidises CO to CO_2.

$$CO + H_2O \rightarrow CO_2 + H_2$$

The organism is a chemoautotroph and fixes CO_2, thus generated into organic carbon. The oxidation of the hydrogen produced provides energy for CO_2 fixation. Under anaerobic conditions, methanogenic bacteria can reduce CO to methane.

$$CO + 3H_2 \rightarrow CH_4 + H_2O$$

A number of other C1 compounds support the growth of methylotrophic bacteria (Table 9.2). Many, but not all, methylotrophs are methanotrophic also. Both types of these bacteria are widespread in environment because these C1 compounds are

TABLE 9.2 C1 compounds of major environmental importance

Compound	*Source*
Carbon dioxide (CO_2)	Combustion, respiration, and fermentation end product, a major reservoir of carbon on earth
Carbon monoxide (CO)	Combustion, common pollutant, plant, animal and microbial respiration end product
Methane (CH_4)	Anaerobic fermentation or respiration end product
Methanol (CH_3OH)	Fermentation end product, hemicellulose degradation
Formaldehyde (HCHO)	Combustion product, intermediate metabolite
Formate (HCOOH)	In plant and animal tissues, fermentation product
Formamide ($HCONH_2$)	From plant cyanides
Dimethyl ether (CH_3OCH_3)	From methane by methanotrophs, industrial pollutant
Dimethyl sulphide [$(CH_3)_2S$]	Most common in environment, generated by algae.
Dimethyl sulphoxide [$(CH_3)_2SO$]	Anaerobically from dimethyl sulphide

ubiquitous metabolites, and in response to their presence, microbes have evolved capability to metabolise them aerobically or anaerobically.

NITROGEN CYCLE

In contrast to carbon, both elements, nitrogen and sulphur are taken up in the form of mineral salts and these cycle in nature oxido-reductively. For instance, nitrogen can exist in numerous oxidation states, from-3 in (NH_4^+) to + 5 in (NO_3^-). These element cycles are referred to as mineral cycles. The best studied of these is the nitrogen cycle (Fig. 9.4).

Nitrogen is the fourth most common element found in cells, making up approximately 12% of cell dry weight and includes microbially catalysed processes of nitrogen fixation, ammonium oxidation, assimilatory and dissimilatory nitrate reduction, ammonification and ammonium assimilation.

As an inert gas, ntrogen has accumulated in the earth's atmosphere since the planet was formed. It is being continuously released into the atmosphere from volcanic and hydrothermal eruptions, one of the major global reservoirs of nitrogen.

Another major reservoir is found bound, nonexchangeable ammonium in earth's crust. Neither of these reservoirs is actively cycled. Other reservoirs, the organic nitrogen in biomass and deed organic matter tend to be actively cycled.

Nitrogen Fixation

Ultimately, all fixed forms of nitrogen, NH_4^+, NO_3^-, and organic nitrogen, come from

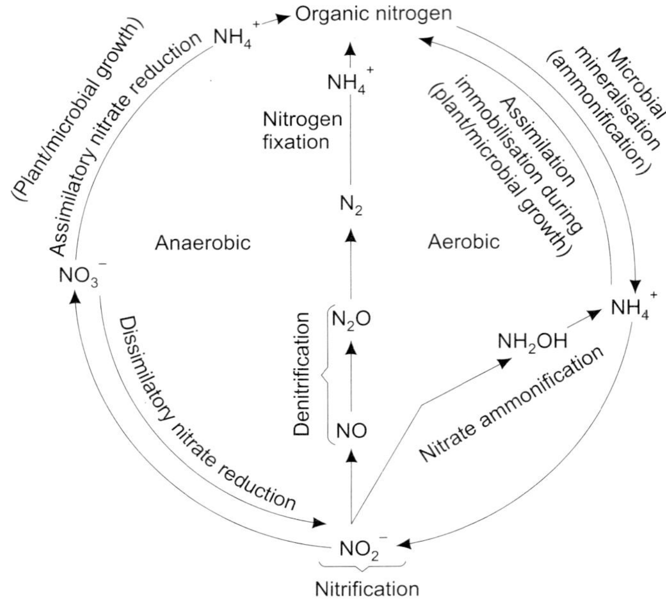

FIG. 9.4
Nitrogen cycle

atmospheric nitrogen. Approximately 65% of the N_2 fixed annually is from terrestrial environments (natural as well as managed agriculture system).

Whereas carbon, hydrogen and oxygen are actively cycled by microbes, plants and animals, the biogeochemical cycling of nitrogen is largely dependent on microbes (Fig. 9.4) Nitrogen fixation, the conversion of N_2 to ammonia or organic nitrogen, is restricted almost exclusively to microbes—bacteria, cyanobacteria and some actinomycetes. These occur in free-living state as well as in symbiotic associations with plants, both legumes and non-legumes. Few microbes and no plants or animals are able to use atmospheric nitrogen directly: plants, animals and most microbes depend on the availability of fixed forms of nitrogen for incorporation into their cellular biomass. It is estimated that microbes world over convert about 200 million metric tons of nitrogen to fixed forms of nitrogen every year compared to about 30 million metric tons produced by industrial plants as nitrogen fertilisers. The amount of N_2 fixed in each system depends on the environment (Table 9.3).

Ammonia is the first detectable product of nitrogen fixation. It is assimilated into amino acids, and subsequently synthesised into proteins and nucleic acids. Proteins, amino acids, and inorganic ammonium ions are used as a nitrogen source by many organisms that are unable to assimilate atmospheric nitrogen directly.

TABLE 9.3 Rates of nitrogen fixation

Nitrogen-fixing system	*Nitrogen fixation (Kg N/ha/year)*
Rhizobium-legume	200-300
Anabaena-Azolla	100-120
Cyanobacteria-moss	30-40
Rhizoshpere associations	2-25
Free-living	1-2

The fixation of atmospheric nitrogen depends on the nitrogenase enzyme system, composed of nitrogenase and nitrogenase reductase. The active site of nitrogenase, where reduction of nitrogen actually occurs, is associated with molybdenum- and iron- containing cofactors. Nitrogenase is very sensitive to O_2 and is irreversibly inactivated upon exposure to even low concentrations. In root nodules, the enzyme is protected by red pigment leghemoglobin. The fixation of atmospheric nitrogen requires a high energy input (approx. 30 ATP/N_2 fixed) and in terrestrial ecosystems largely dependent on the availability of relatively high levels of organic matter for use in respiratory generation of ATP.

In soil, microbial fixation of atmospheric nitrogen is carried out by free-living bacteria (asymbiotic) and those in symbiotic association with plants (symbiotic). In agricultural soils symbiotic nitrogen fixation by *Rhizobium* or *Bradyrhizobium* is most important, where these bacteria live in association with legume crops only. In forests, other symbiotic N-fixing bacteria and actinomycetes live in association with various trees. Rate of N-fixation by *Rhizobium* and *Bradyrhizobium* are 2-3 times more than *Azotobacter,* a free-living N-fixing bacterium. Free-living N-fixing members of the genera *Azotobacter, Azomonas* and *Derxia* are common in temperate regions in neutral or alkaline soils. They are sensitive to low pH. In tropics, *Beijerinckia* spp, more acid tolerant are prevalent in soil. The ability of microbes to fix nitrogen is readily detected by the acetylene reduction assay.

In aquatic habitats, cyanobacteria, such as *Nostoc* and *Anabaena* are important N-fixers. In these microbes nitrogenase is protected by thick-walled heterocysts where oxygenic photosynthesis does not occur. Rates of nitrogen fixation by cyanobacteria are 10 times higher than those of other free-living bacteria.

Symbiotic Nitrogen Fixation

Symbiotic relationship between *Rhizobium* and *Bradyrhizobium* and leguminous plants is very important in maintenance of soil fertility. These bacteria involve the plant roots and form nodules, within which atmospheric nitrogen fixation takes place.

Bradyrhizobium species nodulate soybeans, lupines, cowpeas and a variety of tropical legumes. *Rhizobium* spp nodulate alfalfa, peas, clover and a variety of temperate plants. The establishment of association between them and plants is very specific, and there is a mutual recognition between the bacterium and binding sites on the plant root surface. The interaction between these microbes and plant involves (i) attraction of bacteria by amino acids secreted by root, (ii) binding of. the bacteria to receptors (lectins) on the root, (iii) the activity of plant growth substances, leading to curling and branching of rootlets, (iv) entry of bacteria into root hairs, (v) development of an infection thread, (vi) transformation of plant cells to form tumorous growth, (vii) multiplication of bacteria within the nodule and, (viii) transformation of the invading bacteria into distorted, pleomorphic forms. Within the nodule, metabolites are transferred between bacteria and plant. Within the infected plant tissue the bacteria multiply forming unusually shaped pleomorphic cells- **bacteroids,** which are no longer capable of independent reproduction. The bacteroid cells contain active nitrogenase, not generally found in free-living *Rhizobium* cells. Bacteria provide fixed nitrogen to plant and plant in return organic compounds, for required ATP generation, to the bacteria. Leghemoglobin in nodule supplies O_2 to the bacteroids for respiration, maintaining it at a low concentration so as not to inactivate nitrogenase.

In addition to symbiotic association between these two bacteria and legumes, various other bacteria including cyanobacteria and actinomycetes also enter into similar mutualistic relationship with a restricted number of other types of plants. They also form nodules and fix atmospheric nitrogen. Leghemoglobin is not the specific O_2–carrier in these associations. Actinomycete, *Frankia alni* infects roots of a non-legume, alder tree to form nodules.

Ammonification (Mineralisation)

Microbes also carry out important transformations of organic and inorganic fixed forms of nitrogen. Nitrogen in organic matter is found predominantly in the reduced amino form, such as occurs in amino acids. Many microbes as well as plants and animals are able to convert organic amino nitrogen to ammonia- i.e. ammonification. Deaminases play an important role in this process. Microbial decomposition of urea results in the release of ammonia which is returned to atmosphere or may go to neutral aqueous environments as ammonium ions.

Nitrification

Relatively few organisms are capable of nitrification, the process in which ammonium ions (oxid. level = −3) are initially oxidised to nitrite ions (oxid. level = +3) and subsequently to nitrate ions (oxid. level = +5). Nitrification is an example of aerobic respiration. Both these steps are energy-yielding from which chemolithotrophs derive energy. The two steps of nitrification are carried out by different microbes (nitrifying bacteria) as shown below.

Genus	*Habitat*	*Converts*
Nitrosomonas	Soil, freshwater, marine	Ammonia to nitrite
Nitrosospira	Soil	Ammonia to nitrite
Nitrosococcus	Soil, freshwater, marine	Ammonia to nitrite
Nitrosolobus	Soil	Ammonia to nitrite
Nitrobacter	Soil, freshwater, marine	Nitrite to nitrate
Nitrospira	Marine	Nitrite to nitrate
Nitrococcus	Marine	Nitrite to nitrate

For the most part nitrification is carried out by autotrophic bacteria. Some heterotrophic bacteria and fungi are also able to oxidise inorganic N-compounds. In soil, *Nitrosomonas* is the dominant genus. Other autotrophic nitrifying bacteria are ammonia-oxidising, as *Nitrosospira, Nitrosococcus* and *Nitrosolobus,* and nitrite-oxidisers, such as *Nitrospira* and *Nitrococcus.*

Nitrate Reduction

Nitrate in environment can be leached to groundwater and surface waters. Besides this, nitrate can be taken up and incroporated into biomass by plants and microbes. It is first reduced to NH_4^+ and then incorporated into biomass. This process is called animilatory nitrate reduction or nitrate immobilisation. Finally, microbes can utilise nitrate as a terminal acceptor in anaerobic respiration. There are two pathways for this dissimilatory process: (i) dissimilatory nitrate reduction to ammonium (end product), and (ii) denitrification, where a mixture of gaseous products including N_2 and N_2O is formed.

Denitrification

The conversion of fixed forms of nitrogen to molecular nitrogen i.e. **denitrification** is another important but undesirable process in cycling that is mediated by microbes. Denitrifaction refers to microbial reduction of nitrate through various gaseous inorgamic forms, to N_2. Some aerobic bacteria can use nitrate in place of O_2 as a final electron acceptor, reducing nitrate as a result of anaerobic respiration. Some, as *E. coli* are only able to reduce NO_3 to NO_2, but a variety of other bacteria are able to carry out subsequent anaerobic respirations, where NO_2 ion is reduced to N_2O gas and subsequently to molecular N_2. Some species of *Pseudomonas, Moraxella, Spirillum, Thiobacillus* and *Bacillus* are capable of denitrification. The return of nitrogen to atmosphere by this process completes the nitrogen cycle.

Nitrite Ammonification

Some bacteria, particularly *Clostridium* spp. reduce nitrite to ammonium ions in a process called **nitrite ammonification.** This is competitive with denitrification. This does not remove nitrogen from the soil. Infact much of the nitrate added to soils is reduced to NH_3 by plants and fermentative bacteria rather than to N_2 by denitrifiers.

Genetic Engineering to Create Nitrogen-Fixers

Is it possible to introduce the ability of nitrogen fixation in plants, such as wheat, corn etc.? Yes, what genetic engineers claim in recent years. By use of recombinant DNA technology, microbiologists are involved in the feasibility of introducing nitrogen-fixing bacterial genes into plant cells. Genes for N-fixation are first inserted into the genome of yeast; plasmids from *E. coli* and from a yeast cell are cleaved and then fused to form a single hybrid plasmid, which can be recognised by the yeast cell and integrated into its chromosomal DNA (Fig. 9.5). In the next step, the genes that are to be introduced into the yeast are isolated from the chromosome of *Klebsiella pneumoniae,* a nitrogen-fixer. The genes, collectively designated **nif.** code for some 17 proteins. Another *E. coli* plasmid is cleaved, and the isolated **nif** genes are introduced to form a second hybrid plasmid. Beacause of the bacterial DNA already inserted into one of the yeast chromosomes, the yeast cell recognises the hybrid *E. coli* plasmid. The plasmid is then integrated into the yeast chromsome. Though it could be demonstrated that transfer of genes between prokaryotes and eukaryote (here yeast) could be possible, the N_2-fixing proteins are not expressed in the yeast. More studies are being done on expression of **nif** genes into eukaryotic cells.

SULPHUR CLYCLE

Sulphur is the tenth most abundant element in earth's crust. An essential element for organisms, it makes up approximately 1% of dry weight of a bacterial cell. Sulphate is the major form assimilated by microbes and plants. Sulphur is cycled between oxidation states of +6 for sulphate (SO_4^{2-}) and – 2 for sulphide (S^{2-}). From the earth's core sulphur is outgassed through volcanic activity. Released primarily as SO_2 and H_2S, it becomes dissolved in the oceans and aquifers. In these reservoirs H_2S forms metal sulphids and SO_2 and the metal sulphates with Ca, Ba and Sr. Thus outgassed sulphur becomes converted to rock. Thus there is very little sulphur in

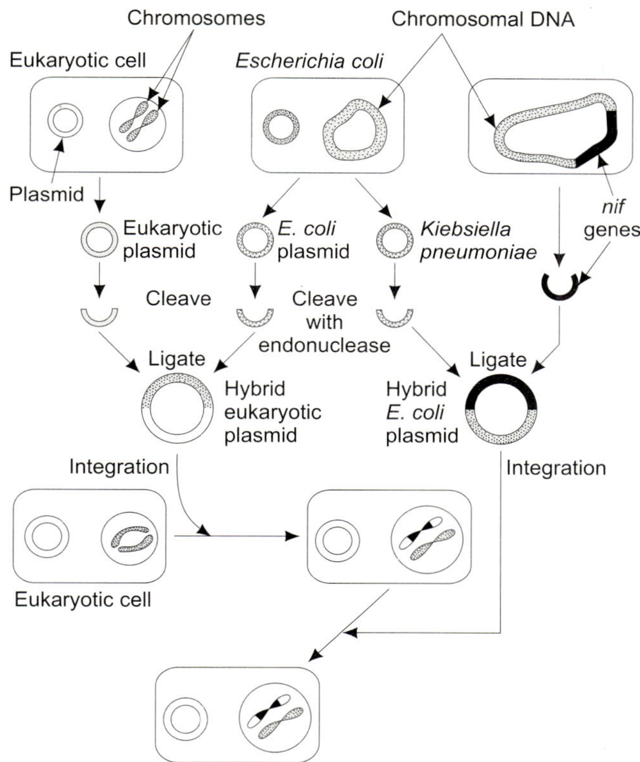

FIG. 9.5

Genetic engineering technique to create plants that contain the bacterial *nif* gene for nitrogen fixation

atmosphere. The largest reservoir of sulphur is found in earth's crust in the form of inert elemental sulphur deposits and sulphur-metal precipitates as pyrite (FeS_2) and gypsum ($CaSO_4$). The second large reservoir is the slowly cycled sulphate found in ocean. Smaller and more actively cycled reservoirs of sulphur include sulphur found in biomass and organic matter in terrestrial and ocean environments.

Sulphur thus can exist in a variety of oxidation states within organic and inorganic compounds. Each oxidation-reduction reaction, mediated by microbes changes the oxidation states of sulphur within various compounds, establishing the sulphur cycle (Fig. 9.6). Microbes are able to remove sulphur from organic compounds. Under aerobic conditions, the removal of sulphur results in the formation of sulphate, whereas under anaerobic conditions H_2S is produced from mineralisation of organic sulphur compounds. H_2S may also be formed by sulphate-reducing

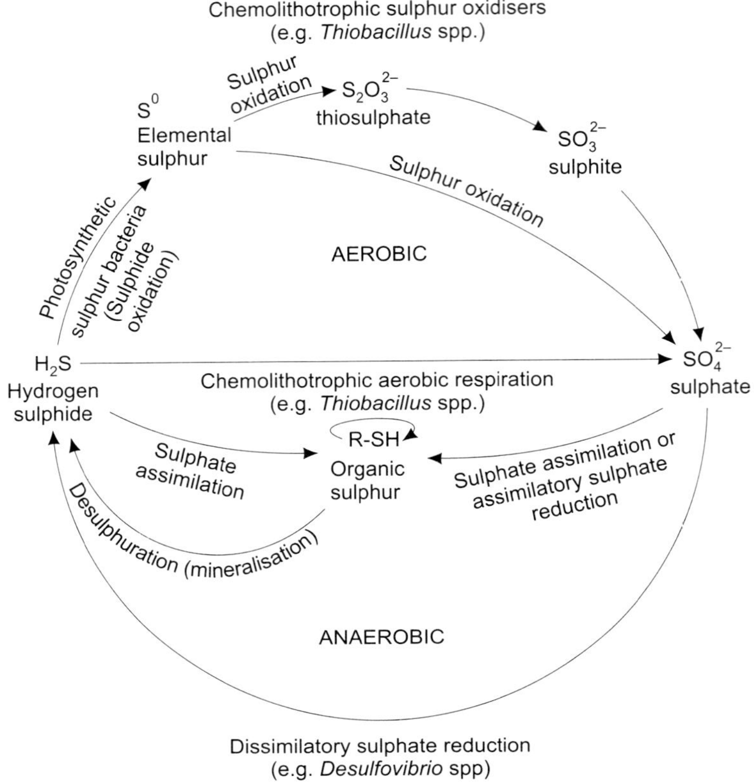

FIG. 9.6
Sulphur cycle

bacteria. H_2S can react with metals to form insoluble sulphides. In organic rich soils, most of the H_2S is generated from decomposition of organic sulphur compounds. In anaerobic sulphate-rich marine sediments H_2S is generated by sulphate-reducers like *Desulfovibrio*.

Although sulphur cycle is not as complex as the nitrogen cycle, the global impacts of this cycle are extremely important, including the formation of acid rain, acid mine drainage and corrosion of concrete and metal. These aspects will be considered in next chapter.

The main steps of sulphur cycle are follows: The primary soluble form of inorganic sulphur found in soil is sulphate. Whereas plants and most microorganisms incorporate reduced sulphur (sulphide) into amino acid or other molecules, they take up sulphur in the oxidised sulphate form and then reduce it internally. This is called **assimilatory sulphate reduction**. The release of sulphur from organic form is called **sulphur mineralisation**. It occurs under both aerobic and anaerobic conditions. Appropriate enzymes are involved. In marine environments one of the major products of algal metabolism is dimethylsulphoniopropionate (DMSP) used in osmoregulation of the cell. It degrades into dimethylsulphate (DMS). Both DMS and H_2S are volatile, released to atmosphere.

Sulphur oxidation is an important process in sulphur cycle. In presence of oxygen, reduced sulphur compounds can support the growth of a group of chemoautotrophic bacteria under strictly aerobic conditions (*Thiobacillus, Thiomicrospira, Achromatium, Beggiatoa, Thermothrix,* common in mud, hot springs, mining surfaces, soil) and a group of photoautotrophic bacteria under strictly anaerobic conditions (*Chlorobium, Chromium, Sctothiorhodospira, Thiopedia, Rhodopseudomonas,* common in shallow water, anaerobic sediments metor hypolimmion, anaerobic water). Chemoautotrophs are the predominant sulphur oxidisers in most environments. Besides them, several aerobic heterotrophic microbes, including bacteria and fungi oxidise sulphur to thiosulphate or to sulphate. Among chemoautotrophs, *Thiobacillus thiooxidans,* oxidising elemental sulphur, is extremely acid tolerant. This bacterium, in conjuction with iron-oxidising, acid-tolerant, chemoautotroph, *T. ferrooxidans* is responsible for acid mine drainage, an undesirable consequence of sulphur cycle activity. The same organisms can be harnessed for the acid leaching and recovery of precious metals from low-grade ore, also known as **biometallurgy**.

Photoautotrophic oxidation of sulphur is brought about by green and purple sulphur bacteria. This group of bacteria evolved on early Earth when atmosphere was devoid of oxygen. These microbes fix CO_2 using light energy, but instead of oxidising water to oxygen, they use oxidation of sulphide to sulphur.

Sulphur reduction occurs in three ways. One is assimilation of sulfate into cell components under aerobic or anaerobic conditions. In contrast, there are two dissimilatory pathways, both of which use an inorganic sulphur as a terminal electron acceptor (TEA). In both these reduction of sulphur occurs under anaerobic conditions. The two types of sulphur, used as TEA are elemental sulphur and sulphate. These two types of metabolism are differentiated as **sulphur respiration** and **dissimilatory sulphate reduction**. *Desulphuromonas acetooxidans, Desulphobacter, Desulphobulbus, Desulphococcus, Desulphosarcina* and *Desulphovibrio,* all collectively are known as **sulphate-reducing bacteria (SRB)**.

10

DETRIMENTAL IMPACTS OF DIVERTED BIOGEOCHEMICAL CYCLES

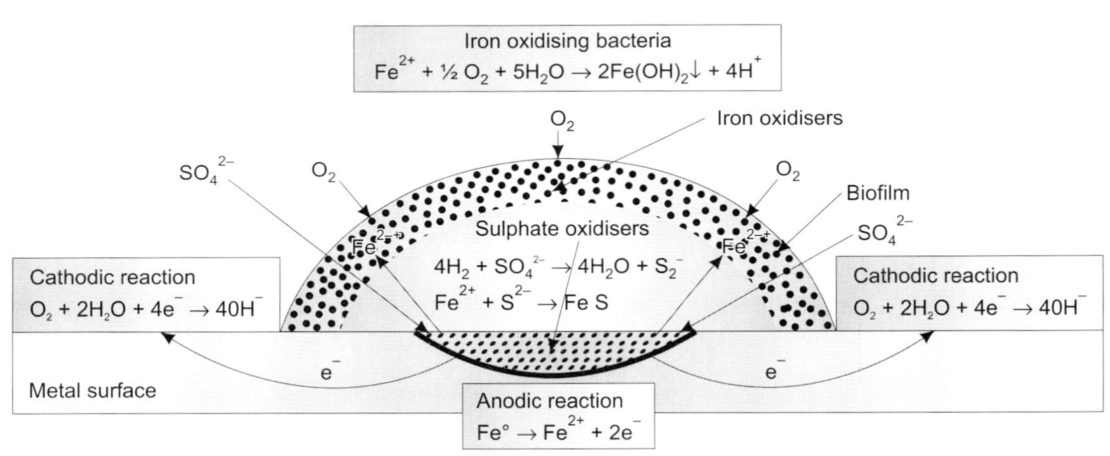

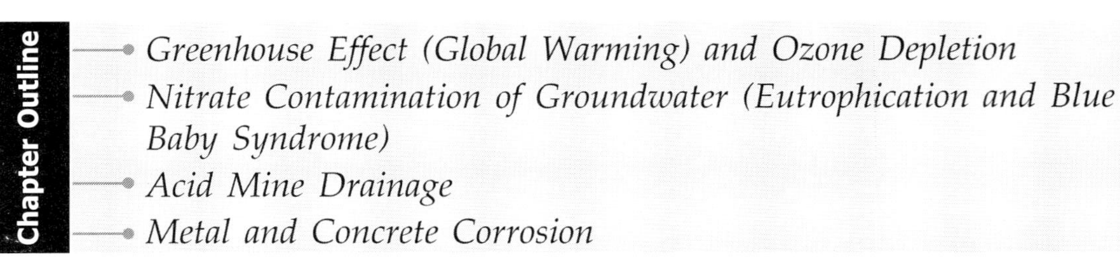

A brief account of the basic biogeochemical cycles for carbon, nitrogen and sulphur was given in Chapter 9. Since there was only little or no human interference, these cycles remained more or less stable for thousands to millions of years. But the stability of these cycles has been drastically threatened due to interference by the growing human population in need for food and energy. As a result, there have been detrimental impacts of such interference on the environment world over (Table 10.1). We would examine such detrimental impacts in some detail in this chapter.

TABLE 10.1 Detrimental impacts of increasing need for food and energy by growing human population on three basic biogeochemical cycles

Diverted cycles	*Chief activities causing diversions*	*Components of the cycle affected*	*Impact on global environment*
Carbon	Deforestation; Fossil fuels; Energy harvesting and utilisation; Waste disposal in landfills (gas emissions)	Atmospheric CO_2 and CH_4	Greenhouse effect-Global warming
Nitrogen	Chiefly agricultural practices of applying excess of chemical fertilisers to soil; Biomass fuels; Fossil fuels	Atmospheric nitrous oxide (N_2O)	Greenhouse effect-Global warming; Ozone depletion in stratosphere
	Excess of ammonia fertilisers; Human and animal waste; Septing tank	Nitrates in soil, transported with water flow contaminating groundwater supplies and oceans	Eutrophication leading to algal blooms causing depleted oxygen levels in waterbodies; Methaemoglobinaemia (Blue baby syndrome)
Sulphur	Metal mining	Sulphur- and iron-oxidising bacteria (e.g. *Thiobacillus ferrooxidans*) acting an pyrite	Acid mine drainage affecting water quality of rivers and streams
	Biofilm formation in waterbodies developing anaerobic microsites	Anaerobic sulphate-reducing bacteria (SRB), induced to become active	Corrosion of pipe line metals resulting into pipe line failures in water treatment, chemical, gas and oil industries
	Sewage load	Sulphur-oxidising bacteria	Corrosion of concrete sanitary sewer pipes and stone objects

GREENHOUSE EFFECT (GLOBAL WARMING) AND OZONE DEPLETION

The major gas in the atmosphere responsible for greenhouse effect is **carbon dioxide**. The other two greenhouse gases are methane and nitrous oxide resulting in concern about global warming. This has occurred because of direct human action, namely deforestation (trees, the natural CO_2-fixers disappearing), utilisation of fossil fuels and waste disposal in landfills.

Though, some of the excess carbon dioxide released into the atmosphere has been absorbed by the oceans (oceans acting as a buffer between the atmospheric and sediment carbon reservoirs), there has still been a net efflux of CO_2 into the atmosphere of approximately 7×10^9 metric tons/year. The problem with this imbalance is that because atmospheric CO_2 is a small carbon reservoir, the result of a continued net efflux over the past 100 years or so has been a 28% increase in atmospheric CO_2 from 0.026% to 0.033%. This increase consequently contributed to global warming through the greenhouse effect.

Major amount of carbon dioxide is released in the atmosphere from burning of fossil fuel (coal, oil etc) for domestic cooking, heating etc. and the fuel consumed in furnaces of power plants, industries, hot-mix plants etc. From fossil fuels alone more than 18×10^{12} tonnes of CO_2 is being released into atmosphere each year. In our country, on an average, thermal power plants are likely to release around 50 million tonnes of CO_2 each year in the atmosphere. Indian coals are notorious for their high ash content (20-30% and 45% in some cases) and for very bad ash qualities. The projected annual coal consumption for the four NTPC super-thermal power plants is eight million tonnes at Singrauli (low grade), five million tons at Korba (high grade), 8.7 million tonnes at Ramagundam and nearly five million tonnes at Farakka (high grade). The coal we burn was produced 250 million years ago, over a period of millions of years. If eight million tonnes of coal to be burnt at Singrauli, is mined over an area of 10 sq. km. then the deposit-formation period will he roughly 500 years and if mined over an area of 1 sq. km. it would be 5000 years. Can one afford this foolish act with nature? CO_2 is also emitted during volcanic eruptions. On a global time scale, the known amounts of CO_2 in lime stone and fossil sediments suggest that normal persistence period of CO_2 in the atmosphere is around 100,000 years.

To some extent an increase in CO_2 level in atmosphere increases the photosynthesis rate and consequently plant growth, acting as fertiliser especially in hot tropical climates. This potential of fertiliser effect may be exploited by using modified crop varieties and agricultural practices. However, an increase in CO_2 concentration in atmosphere may result into diastrous effects also. This is described below-the greenhouse effect.

Greenhouse Effect

Since CO_2 is confined exclusively to the troposphere, its higher concentration may act as a serious pollutant. Under normal conditions (with normal CO_2 concentration) the temperature at the surface of the earth is maintained by the energy balance of the sun rays that strike the planet and heat that is radiated back into space. However, when there is an increase in CO_2 concentration, the thick layer of this gas prevents the heat from being re-radiated out. This thick CO_2 layer thus functions like the glass panels of a greenhouse (or the glass windows of a motor car), allowing the sunlight to filter through but preventing the heat from being re-radiated in outer space. This is the so-called **greenhouse effect**. Thus most heat is absorbed by CO_2 layer and water vapours in the atmosphere, which adds to the heat that is already present. The net result is the heating up of the earth's atmosphere. Thus increasing CO_2 levels tend to warm the air in the lower layers of atmosphere on a global scale. Nearly 100 years ago the CO_2 level was 275 ppm. Today it is 350 ppm. and by the year 2035 and 2040 it is expected to reach 450 ppm. Imagine the earth's temperature. CO_2 increases the earth temperature by 50% while CFCs are responsible for another 20% increase. There are enough CFCs up there to last 120 years. What will be if we do not stop CFC release?

The heat trap provided by atmospheric CO_2 probably helped create the conditions necessary for the evolution of life and the greening of earth. Compared to moderately warm planet, Mars, with too little CO_2 in its atmosphere is frozen cold and Venus with too much is a dry furnace. The excess CO_2 to some extent is absorbed by the oceans. But with the industrailization of West and increased consumption of energy, CO_2 was released into atmosphere at a faster rate than the capacity of oceans to absorb it. Thus its concentration increased. According to some estimates CO_2 in air may have risen by 25% since the middle of 19th century. It may even be doubled by 2030 A.D.

There are some differences of opinions, however, about the extent of rise in earth's temperature due to increasing CO_2 levels. According to some, computerised models, doubling the CO_2 level will increase the global mean temperature (15°C) by 2 degrees C. But, some others say that this will be less than one quarter of a degree. There are other gases also which contribute to greenhouse effect. These are SO_2, NOx, CFCs discharged by industry and agriculture. Even a change of two degrees may disrupt the earth's heat budget, causing catstrophic consequences.

Some analysts believe that changes in the earth's mean temperature will be apparent by 2050, when the temperature would increase by 1.5 to 4.5°C. According to one projection, changes will be the least in the tropics and the most at the poles. So, Greenland, Iceland, Norway, Sweden, Finland, Siberia and Alaska will be among the most affected. The polar icecaps would melt. The floating Western Antarctica ice sheet could begin to melt. A rise of five degrees would raise the sea level by five meters within a few decades, threatening all the densely populated coastal cities from Shanghai to San Francisco. It is suggested that North America would be warmer and drier. The U.S. would produce less grains. On the other hand, North and East Africa, the Middle East, India, West Australia and Mexico would be warmer and wetter, enabling them to produce more grain. Rice-growing season as well as area under rice cultivation could increase. However, this may not happen as higher surface temperature will increase the evaporation of water, thus reducing grain yield. According to U.S. Scientist, George Woodwell, India's annual monsoon rains may even cease altogether.

According to an estimate, if all the ice on the earth should melt 200 feet of water would be added to surface of all oceans, and low-lying coastal cities as Bangkok and Venice would be inundated. A rise in sea level of 50-100 cm caused by ocean warming would flood low-lying lands in Bangladesh and West Bengal. Due to greenhouse effect, there may occur more hurricanes and cyclones and early snow melts in mountains causing more floods during monsoon. According to some, within next 25 years or so, there will be a rise in sea level by 1.5 to 3.5 meters and in Bangladesh alone 15 million people will have to move or drown. Low-lying cities of Dhaka and Calcutta may be inundated.

Besides, the five emerging environmental issues (new technologies, red tides, diesel pollution, acid fog and threats to Antarctica), that the UNEP has been able to identify, the one that has proved the most vexatious, and disquieting is the greenhouse effect of global warming. It is caused by the build-up of CO_2 and other toxic gases discharged by industry and agriculture in the atmosphere. If unchecked, it could alter temperatures, rainfall and sea levels of the earth. The UNEP has appropriately chosen the slogan **"Global Warming: Global Warning"** to alert the people on World Environment Day, June 5, 1989. The cost of defense (reduction of gas emissions and research to identify the hardest hit regions and plan of coastal defense) would be enormous: in the region of $ 100 billion or more for a one meter rise in sea level. The problem is that most vulnerable areas in developing world do not have economic resources. The hardest hit may be developing world, which discharge 2/5ths of the global carbon emissions each year which itself is increasing by over 100 million tonnes a year.

Methane, the second most common greenhouse gas directly affects the tropospheric chemistry and controls numerous chemical processes and constituents in the troposphere and stratosphere. The infrared spectrum of CH_4 makes it a strong greenhouse gas. Atmospheric concentration of CH_4 has more than doubled during the past 200 years, rising over the past 15 years by an average of 1%/year. Wetlands are primary natural sources of atmospheric methane accounting for about 20% of the current global annual emission of ca 450-500 Tg. It is 22 times more effective than CO_2 at trapping heat. A substantial amount of methane is released to the atmosphere as a result of energy harvesting and utilisation. Methane is also released through landfill gas emissions.

Nitrous oxide (N_2O) gas is released to the atmosphere from both industrial and biogenic sources. It can contribute to both global warming and the destruction of the protective ozone layer in the stratosphere. N_2O has a long residence time (> 100 years) in the atmosphere, and is highly efficient in absorbing long-wave radiations. One molecule of N_2O is equivalent in heat-trapping ability to about 200 molecules of CO_2.

A major source of N_2O released to atmosphere is the soil environment. N_2O is produced during denitrification. The agricultural practice of added nitrogen sources, either as chemical fertilisers or as animal manure, to soil is a major contributor to the gradual increase in N_2O emissions to the atmosphere. In fertilisers, the nitrogen source is mainly ammonia or ammonia-producing compounds such as urea. However, only about half of the total nitrogen applied to a field is assimilated by the crop. The remainder is lost through leaching, erosion, and gaseous emissions. Thus agricultural practices account for a large percentage of the N_2O released by human activities, although other sources of N_2O in the atmosphere are burning of biomass and combustion of fossil fuels.

NITRATE CONTAMINATION OF GROUNDWATER (EUTROPHICATION AND BLUE BABY SYNDROME)

Nitrates normally does not accumulate in the environment. However, in agricultural practices the use of ammonia fertilisers and production of large animal waste result in addition of excess nitrogen to soil and groundwater environment. Septic tanks are another source of elevated nitrogen. Moreover, conversion of ammonia to nitrate by aerobic, chemoautotrophic nitrifying bacteria also results in accumulation of nitrates in many agricultural systems.

Instead of returning the human waste nitrogen to soil, sewage systems pass it into rivers and so on to oceans. At this point, the nitrogen cycle has gone out of joint. Flush- and-forget sewage disposal systems improved sanitation, but have aggravated the damage to nitrogen cycle. Sewage systems take them to treatment plants, thus diverting the natural way of human waste to the soil. This now reaches to surface water bodies and oceans. About 32% of the nitrogen fertilisers applied in USA each year (about 3.6 million tonnes) reaches surface water and groundwater. According to an estimate at global level the amount of nitrogen added to environment between 1860 and 2000 has increased from 15 Tg (1 Tg = 1 million tonnes) of nitrogen per year to over 165 Tg of nitrogen per year. Global production of nitrogen will reach between 250 million tonnes to 900 million tonnes per year by 2100.

Excess nitrogen to soil slows the tree growth, interferring with chlorophyll synthesis and lowering their resistance to frost. Excess nitrogen also affects diversity of species in ecosystems. Excess nitrogen is now considered the biggest polluion problem in world's coastal ecosystems, where it can cause eutrophication, hypoxia, anoxia, and harm aquatic life.

Nitrogen is seeping into the waters and lands of industrialised nations. In the next decade Asia is expected to use more nitrogen fertiliser (it was 14.2 million metric tonnes in 2000-01). India in 1959 used 23,470 tonnes of urea, but by 2000-01 it had consumed over 11 million tonnes of N-furtilisers. Eighty per cent of nitrogen unused by a human is found in urine, which goes to water. The disruption of the nitrogen cycle—from land to water and not from land to land–is as serious as the disruption of global carbon cycle, the cause of global warming.

Due to addition of domestic waste (sewage), phosphates, nitrates etc. from wastes or their decomposition products in water bodies, they become rich in nutrients, especially phosphates and nitrate ions. Thus with the passage of these nutrients through such organic wastes, the water bodies become highly productive or eutrophic resulting into the phenomenon of **eutrophication**. It must be remembered that ponds, lakes etc. during their early stages of formation are relatively barren and nutrient-deficient, thus supporting no or very poor aquatic life. This state of these bodies is known as **oligotrophic**. With the addition of nutrients , there is stimulated luxuriant growth of algae in water. There is also generally a shift in algal flora and blue-green algae begin to predominate. These start forming algal blooms, floating scums or blankets of algae. Blooms of algae are generally not utilised by zooplanktons. The algal blooms compete with other aquatic plants for light for photosynthesis. Thus oxygen level is depleted. Moreover, these blooms also release some toxic chemicals which kill fish, birds and other animals, thus water begins to stink. Decomposition of blooms also leads to oxygen depletion in water. Thus in a poorly oxygenated water with higher CO_2 levels, fish and other animals begin to die and clean water body is turned into a stinking drain.

In U.S.A., Lake Erie is an excellent example of eutrophication due to man-made problems. In 1965, there were added more than 80 tons of phosphates daily in the lake. Each 400 g of PO_4 encourages about 350 tonnes of algal slime. Due to this algal growth these appeared as big mounds on the shores of lake producing unpleasant odor, clogging pipes and interfering with fishing and navigation.

Eutrophication is thus limiting factor in supply of clean water for drinking, fishing and navigation etc. Following are the methods to stop or reverse eutrophication:

(1) To treat the waste water before its discharge into lake or river. This would limit its nutrient input.

(2) To stimulate bacterial multiplication in order to reduce the amount of nutrients solubilised in water. This would help disruption of algal food-web.

(3) To check recycle of nutrients into the water through harvest, and removal of algal blooms upon their death and decomposition.

(4) To remove dissolved nutrients from water by physical or chemical methods. For instance, phosphorous can be removed by precipitation; nitrogen by biological nitrification and denitrification or by air stripping of NH_3 from an alkalised waste water, or by ion exchange, electrodialysis or reverseosmosis.

According to H.H. Koepf, an eminent soil chemist, modern agriculture can honestly claim only two notable crops - "**disease and pests**." To this we can now add a third - **poison** (as nitrites, nitrates). Nitrate fertilisers used on soil enter our wells and ponds. These waters thus are very rich in nitrates. It not only makes water unfit for drinking but also causes diseases. After drinking this water by us, the nitrates are converted to nitrites by microbial flora of intestine. These nitrites then combine with the hemoglobin of blood to form methaemoglobin, which interferes with the O_2-carrying capacity of the blood. The disease thus produced is called **blue baby syndrome** or **methaemoglobinaemia**. Since red blood cells can not carry oxygen, which subsequently turns an infant's lips blue, the disease is called blue baby syndrome. This leads to various ailments as damage to respiratory and vascular system, blue coloration of skin and even cancer. A healthy person contains 0.8% of methaemoglobin, whereas in methaemoglobinaemia this level reaches to 10% in the blood. At above 20% there occurs headache, giddiness, and above 60% there begin consciousness, stiffness, ocular problem etc. At 80% death occurs. Nitrate poisoning is frequent in Rajasthan due to hard and saline water. Several children have died due to this problem. In 1976, there were cases of NO_3-poisoning of cattle in Nagpur. In Rajasthan NO_3 levels in water are very high, 800 mg/litre, which is much beyond the permissible conc. of 45 mg/l by WHO. Consumption of vegetables grown in NO_3-rich soil may also lead to this disease, especially in sick persons and children.

ACID MINE DRAINAGE

Pyrite (FeS_2) is a major source of sulphur in the lithosphere. During mining of ore deposits and of bituminous coal, pyrite is exposed to oxygen and moisture and becomes the source of an acidic, iron-rich leachate known as **acid mine drainage**. The drainage can have a pH of less than 2 and can seriously impair the quality of receiving waters, the surface streams or rivers. The process is initiated spontaneously, but as the local pH drops, a sulphur- and iron-oxidising bacterium, *Thiobacillus ferrooxidans*, also begins to participate in the reaction. It is chemoautotroph deriving energy for carbon fixation and growth from the oxidation of inorganic sulphur- and iron-containing compounds, such as pyrite. Other microbes, such as the thermophilic archaebacteria, *Sulfolobus* and *Acidianus* are also associated with acid mine drainage.

These are acidophilic chemolithotrophs capable of maximal growth around pH 1.5 to 2.5.

Oxidation of elemental sulphur to sulphate leads to environmental accumulation of sulphuric acid. Acid mine drainage is the result of the metabolism of sulphur and iron-oxidising bacteria, which occur in hot sulphur springs and smouldering coal wastes-at high temperatures. Since there exist extreme acidic conditions also in such areas, the bacteria are acidophiles too. Hence the term thermoacidophiles is used for this group of bacteria. Coal in geological deposits is often associated with pyrite (FeS_2) and when coal mining activities expose pyrite ores to atmospheric oxygen, the combination of autooxidation and microbial sulphur-and iron-oxidation produces large amounts of acid. The acid kills aquatic life and renders water unfit for drinking. Strip mining of coal recovery causes acid mine drainage. This method removes the overlying soil and rock, leaving a porous rubble of tailings exposed to O_2 and percolating water. Oxidation of reduced iron and sulphur in the tailings produces acidic products, causing the pH to drop rapidly. A strip-mined piece of land continues to produce acid mine drainage until most of the sulphide is oxidised and leached out; recovery of this land may take 50-150 years.

The overall reaction of the oxidation of pyrite can be summarised as: $2FeS_2 + 7.5 O_2 + 7H_2O \rightarrow 2Fe(OH)_3 + 4H_2SO_4$. The acid produced accounts for high acidity and the precipitated ferric hydroxide. The mechanism of pyrite oxidation in acid mine drainage is quite complex. At neutral pH, oxidation by atmospheric O_2 occurs rapidly and spontaneously but below pH 4.5, autooxidation is slowed drastically. The stalked iron bacterium, *Metallogenium* in the pH range of 4.5-3.5 catalyses the oxidation of iron. As the pH drops below 3.5, the acidophilic bacteria (spp. of *Thiobacillus*) oxidise the reduced iron sulphide in the pyrite. The rate of microbial oxidation of FeS is several hundred times greater than the rate of spontaneous oxidation. Microbial oxidation of sulphur and iron is infact responsible for the continued production of high levels of acid mine drainage. *Thiobacillus spp.* are also used in bioleaching processes for mineral recovery.

METAL AND CONCRETE CORROSION

Microbially-mediated corrosion damage can occur on metal structures submerged in water, such as the hulls of ships, and in pipelines carrying water, chemical, gas and oil products. It is a major cause of pipeline failures in water treatment and chemical industries and also associated with corrosion failures and souring in gas and oil production and storage. Sulphate-reducing bacteria (SRB) have been shown to be involved in metal corrosion mechanism. They are active under strictly anaerobic conditions, utilising sulphate instead of O_2 in utilisation of organic compounds (they are chemoorganotrophs). SRB-mediated corrosion requires anaerobic conditions. Though oxygen is present in water around the metal surface, biofilm formation creates anaerobic microsites within them. Such sites develop due to rapid utilisation of O_2 by bacteria present in the outer layers of biofilms. The anaerobic biofilm microsites now support the growth and activity of SRB which cause metal surface corrosion.

Metal corrosion is initiated by two spontaneous electrochemical reactions: (i) set up of a differential aeration cell in which the metal surface acts as the anode to produce metal ions (anodic reaction ie. $Fe^0 \rightarrow Fe^{2+} + 2e^-$), and electrons thus produced from the elemental metal iron, accepted by oxygen (cathodic reaction ie. $\frac{1}{2} O_2 + H_2O + 2e^- \rightarrow 2(OH)^-$). (ii) creation of a concentrated cell under anaerobic conditions in which the anodic reaction remains the same but the cathodic one produces H_2 as follows: $Fe^0 \rightarrow Fe^{2+} + 2e^-$ (anodic reaction) and $2H^+ + 2e^- \rightarrow 2H \rightarrow H_2$ (cathodic reaction). Microbially-mediated corrosion refers to the involvent of biofilms in stimulating these electrochemical reactions. As already mentioned, a biofilm is formed on metal surface that facilitates the establishment of these separate electrochemical reactions by establishing an anaerobic environment at metal surface (Fig. 10.1). Then SRB utilise H_2 as electron donor, therby removing it form the environment and providing a driving force for the anodic reaction. Finally, the end product of sulphate reduction is sulphide (S_2^-), which reacts with Fe^{2+} to from metal sulphide precipitate as follows:

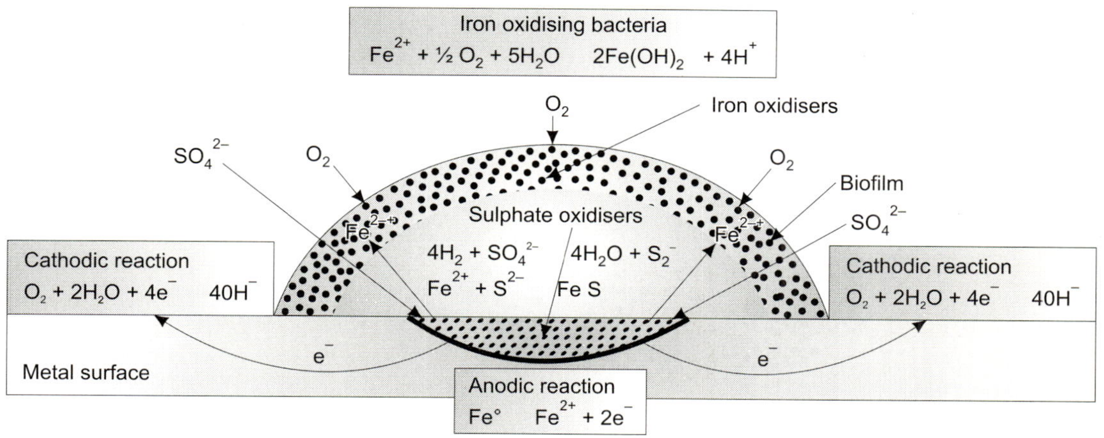

FIG. 10.1

General schematic representation of microbially-mediated corrosion of a metal surface

Step 1. $Fe^{2+} + 2H_2O \rightarrow Fe\,(OH)_2 + H_2$ (spontaneous)

Step 2. $4H_2 + SO_4^{2-} \rightarrow H_2S + 2OH^- + 2H_2O$

(SRB- e.g. *Desulphovibrio desulphuricans*)

Step 3. $H_2S + Fe^{2+} \rightarrow FeS \downarrow + H_2$ (spontaneous)

Overall. $2Fe^{2+} + SO_4^{2-} + 2H_2 \rightarrow FeS \downarrow + Fe\,(OH)_2 \downarrow + 2OH^-$

Microorganisms have also been shown to participate in the corrosion of concrete and stone structures through the production of acidic metabolites. Concrete, ceramic and natural stone objects are sensitive to acids and are easily corroded by acid-producing microbes. Chemoautotrophic sulphur-oxidising and nitrifying bacteria produce sulphuric acid and nitric acid respectively. These and other microbes are found to colonise the surface of concrete structures. Concrete sanitary sewer pipes are susceptible to microbial corrosion. Sewage carrying pipes are corroded by collective action of SRB under anaerobic conditions (generating sulphide, converted to volatile H_2S) and sulphide-oxidising bacteria (oxidising H_2S to H_2SO_4). The common sulphur-oxidiser, *Thiobacillus thiooxidans* grows abundantly in concrete (Fig. 10.2).

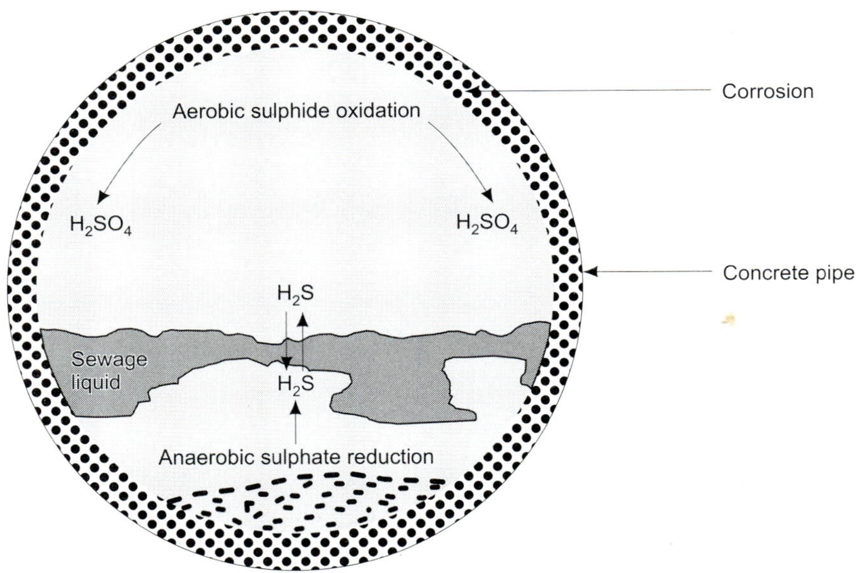

FIG. 10.2

A cross section of sewage pipe showing microbially-mediated concrete corrosion

11

MICROBIAL BIODEGRADATION OF ORGANIC POLLUTANTS

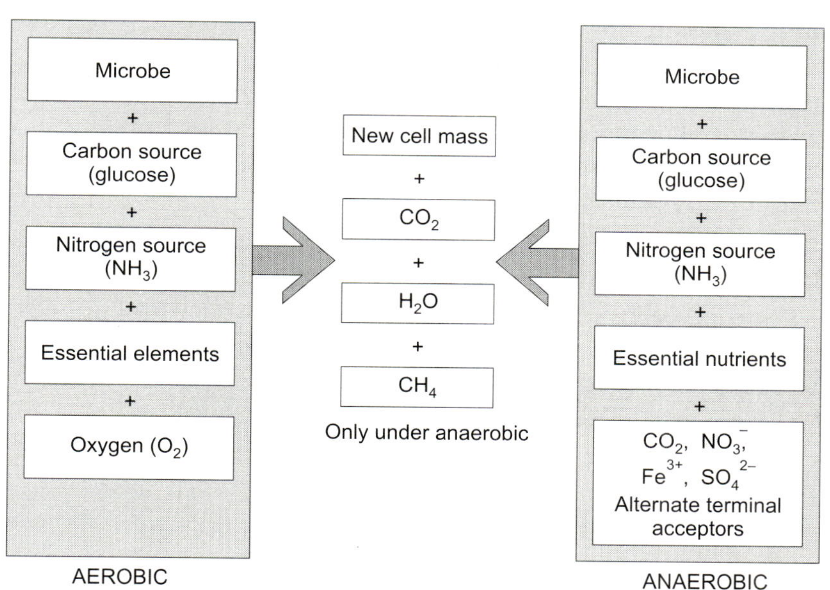

Chapter Outline

Not only environmental scientists but even general public is aware of the severity of pollution problems resulting from careless waste disposal. Agriculture and industry have been the main sources of waste products. Energy generating industries produce large amounts of waste during processing of coal and oil and also nuclear energy production. During past these wastes have been buried without proper care, and so migrating through the soil into groundwater supplies. This is **point source pollution**. Agriculture sector is largest user of pesticides and fertilisers and application of these chemicals over vast land areas resulted into **nonpoint source pollution**. These agro-chemicals are now frequently present in our surface water and groundwater supplies.

I do not intend to write a detailed account of the physical, chemical and biological processes involved in the organic pollutant-microbial interactions. Rather, a brief account of microbial interactions with organic pollutants that can be harnessed to help prevent pollution and clean up the polluted environments will be presented in this chapter.

Biodegradation should not be confused with biodeterioration. **Biodeterioration** is the chemical or physical alteration of a product that decreases its usefulness for its intended purpose, caused by microorganisms or their enzymes. Though both bacteria and fungi are involved in biodeterioration of items of daily use, such as textiles, leather, plastic, paints etc. the latter are relatively better known and more studied for mechanism of the process and factors which favour their growth in substrate.

Besides the items of daily life, microorganisms are involved in deterioration of agricultural produce also. Bacteria and fungi are largely responsible for contamination of grains, fruits and vegetables during storage, transport and marketing spoiling their commercial value. Not only this, microbes produce poisonous toxins in cereal and oil seed crops. Mycotoxins produced by fungi are deadly poisonous. These aspects of biodeterioration and contamination of agricultural produce will be presented later in separate chapter on "Microbes and Agriculture" (Chapter 14). Microbes also cause spoilage of highly perishable and semiperishable foods, such as bread, meat, egg, butter and other milk and dairy products. This is a subject of microbiology of foods.

MICROBIAL BIODEGRADATION

Microbial biodegradation is the process of breakdown of generally complex, organic pollutants (contaminants) to smaller, simpler products by the activities of microorganisms. These organic substances serve as the microbial food source or substrate. Biodegradation in fact involves a series of biological degradation steps (pathways) that finally result into the oxidation of the parent compound, and often into energy generation. Complete biodegradation or mineralisation involves oxidation of parent compound to form carbon dioxide and water, a process providing both carbon and energy for growth and reproduction of microbial cells. Mineralisation of an organic compound (taking glucose, as an example) under aerobic or anaerobic conditions is shown in Fig. 11.1. Irrespective of the structure of the carbon source (a simple sugar like glucose or a complex plant polymer such as cellulose, or a pollutant molecule), each degradation step in the pathway is catalysed by a specific enzyme made by the degrading microbial cell. Lack of appropriate biodegrading enzyme is one common reason for the persistence of organic pollutants in environment. Thus, pollutants that have structures similar to those of natural substrates are normally easily degraded. Those that are different from natural substrates tend to be degraded slowly or not at all. As mentioned earlier, some organic pollutants, due to lack of appropriate enzymes are degraded partially but not completely. A second type of incomplete degradation is **cometabolism**, in which partial oxidation of the substrate occurs but the energy derived from such oxidation is not used for microbial growth. An example of cometabolism is oxidation of the industrial solvent, trichloroethylene (TCE) by methanotrophic bacteria growing on methane as a sole carbon source. TCE, a most frequent pollutant at hazardous waste sites is generally resistant to biodegradation. The first step in oxidation of methane and TCE is catalysed by nonspecific methane monooxygenase (acting on both, CH_4 and TCE), receiving no energy from cometabolic degradation step (Fig. 11.2).

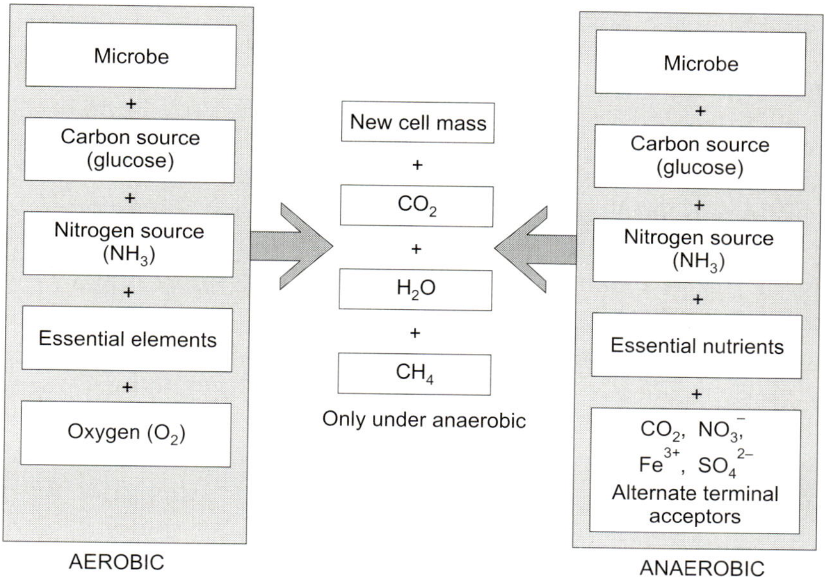

FIG. 11.1
Aerobic or anaerobic biodegradation (mineralisation) of an organic compound

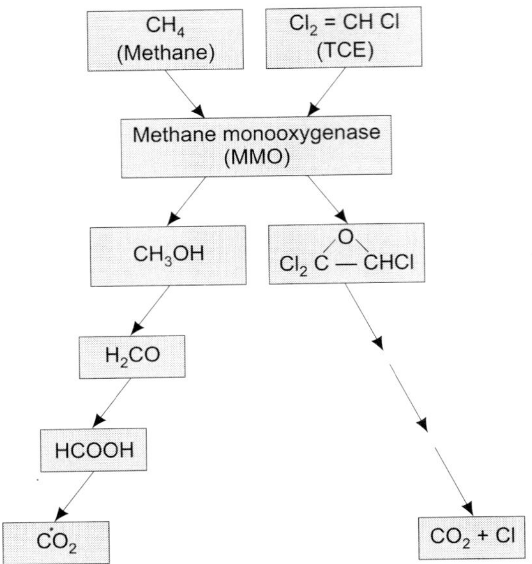

FIG. 11.2
Oxidation of methane by methanotrophic bacteria

Subsequent steps may be catalysed spontaneously by other bacteria, including methanotrophs in some cases.

Factors Affecting Biodegradation

Though large amount of the organic carbon available for microbial biodegradation in the environment is plant material, an environmental microbiologist's major concern are the environments receiving large inputs of carbon from industry and agriculture (petroleum products, organic solvents, pesticides, fertilisers etc.). Activities of microorganisms may be limited by several physico-chemical and biological factors. These include, environmental factors, microbial characteristics and the physico-chemical properties of the pollutant. **Environmental factors** affecting biodegradation include, oxygen, organic matter content, nitrogen level, temperature, pH, salinity and water activity. **Microbial characteristics** chiefly include the genetic potential, or the presence and expression of appropriate degrading genes by the indigenous microbial community. The various physico-chemical properties of the organic pollutant are, bioavailability (or the effect of limited water solubility and sorption on the rate at which a pollutant is taken up by a microbial cell), its structure (including both steric and electronic effects), and toxicity (the inhibitory effect of the pollutant on metabolism of microbial cell).

BIODEGRADATION OF ORGANIC POLLUTANTS

Large quantities of oil are consumed for heating, electricity generation and gasoline. Coal, natural gas and nuclear power are other sources of energy. Textile, transport, electronics, defense and metal industries also produce large amount of waste, including solvents, acids, bases, and metals. In this way industry and agriculture produce large amounts of chemical pollutants. Various such pollutants added to our environment in large quantities by anthropogenic activities belong to three structural classes: aliphatic, alicyclic, and aromatic. A variety of complex molecules can be formed, for use in agriculture and industry by combining or additions to these three basic structures of organics. The major sources and uses of these pollutants are shown in Table 11.1.

TABLE 11.1 Sources and uses of representative pollutants

Pollutant (hydrocarbon type)	*Name*	*Sources and uses*
Aliphatics	Propone (gas), Hexane (liquid), Hexatriacontane (solid)	Petroleum
Substituted aliphatics	Chloroform (liquid), Trichloroethylene (liquid)	Anthropogenically manufactured and used as solvents, degreasing agents and in organic syntheses
Alicyclics	Cyclopentane (liquid), Cyclohexane (liquid)	Petroleum
Aromatics	Benzene (liquid), Naphthalene (solid) Phenanthrene (solid)	Petroleum
Substituted aromatics	Phenol (liquid)	Coal tar or manufactured for use as disinfectant, resin, dyes
	Pentachlorophenol (liquid)	Manufactured for use as insecticide and wood preservative
	Toluene (liquid)	Tar oil, used in manufacture of organics, explosives and dyes, and also as solvent
Heterocyclics	Dibenzodioxin (solid), Chlorinated dioxins (solids)	Created during incineration, and are contaminants in manufacture of herbicides, including 2, 4-D and 2, 4, 5-T
	Pyridine (liquid)	Coal tar, used as solvent
	Thiophene (liquid)	Coal tar, coal gas and crude oil; used as solvent and in manufacture of resins, dyes and pharmaceuticals
Pesticides		
Organic acids	2-4-dichlorophenoxy-acetic acid (solid)	Herbicide
Organophosphates	Chlorpyrifos (solid)	Insecticide, acaricide
Triazenes	Atrazine (solid)	Herbicide
Carbamates	Carbaryl (solid)	Insecticide
Dithiocarbamates	Ferbam (solid), Ziram (solid), Maneb (solid)	Fungicides

Aliphatics enter the environment from common sources as petroleum, detergents and solvents, and include alkanes, alkenes and chlorinated aliphatics. Alkanes are usually the most readily biodegradable type of organic pollutants. Monooxygenases as well as dioxygenases are involved in their breakdown. Like alkanes, alkenes are also easily degraded. Chlorinated aliphatics, such as TCE are used extensively as solvent in industries. These have been detected in groundwaters. Biodegradation of halogenated aliphatics occurs by three basic reactions: substitution, oxidation and, reductive dehalogenation. Oxidative enzymes required for their oxidation are produced by methanotrophic bacteria, *Nitrosomonas europaea* (growing on ammonia, and producing ammonia monooxygenase) and *Mycobacterium vaccae* JOB5 (producing propane monooxygenase to oxidise propane).

Alicyclics, the major components of crude oil are found elsewhere also in nature as components of plant oils and paraffins, microbial lipids, and pesticides. Microbes have been detected that are able to degrade alicyclics completely. However, their biodegradation is thought to occur primarily by commensalistic and cometabolic reactions.

Unsubstituted aromatics are natural products as lignin components in plant products. They are also formed during burning of organic matter as in forest fires. At the same time fossil fuel processing and utilisation and burning of wood and coal have added enormous quantities of such compounds to the environment. These compounds have toxic effects on microorganisms. Due to adverse effects of aromatics on human health, their biodegradation has been studied extensively. A wide variety of bacteria and fungi can biodegrade such organics, partially or completely under a variety of environmental conditions. Eukaryotic microorganisms initially attack aromatics with a cytochrome P-450 monooxygenase.

The white rot fungus, *Phanerochaete chrysosporium*, is unique in the sense that under certain conditions (N_2-limiting conditions, which induce production of lignolytic enzymes), the aromatics are completely mineralised. This fungus has been shown to degrade a variety of aromatics including the pesticide DDT, the herbicides 2, 4-D and 2, 4, 5-T, the wood preservative PCP, PCBs, PAHs and chlorinated dioxins.

Among **substituted aromatics**, chlorinated aromatics are of special interest as they have been used extensively as solvents and fumigants, and wood preservatives. They are also the basic compounds for pesticides such as DDT and 2, 4-D. They are degraded with difficulty by the microbes. Chlorinated phenols are toxic to microbes. Methylated aromatics like toluene are major components of gasoline and are commonly used as solvents. These compounds are attacked either on the methyl group or directly on the ring by microbes.

Among aromatics, dioxins and dibenzofurans are created during waste incineration and are also partly present in the released smoke stack effluent. Bacteria and fungi, including *Phanerochaete chrysosporium* could biodegrade only minimal amounts of these aromatics.

PESTICIDES

Pesticides are the biggest nonpoint source of chemical pollutants added to the environment. Most of them used as organic pesticides are easily degraded. Synthetic pesticides of complex structures have been developed from time to time, but most can be traced to relatively simple aliphatic, alicyclic, and aromatic base structures. These basic structures are coupled with required varieties of halogen, amino, nitro, hydroxyl, carboxyl, and phosphorus substitutes. Based on their structure, there are pesticides that can be biodegraded easily, whereas also those whose degradation is much slower.

A range of pesticides are in current use in agriculture to control pests, and diseases as well as to protect the agriculture produce from contamination by microbes responsible for their deterioration in transit and storage. The rising cost of manual labour and increasingly large-scale agricultural operations in developed and developing countries have promoted the growing use of chemicals in pest control. In USA there were only 3 insecticides in 1948 in agricultural use, but in 1975, 1170 pesticides, 425 herbicides, 410 fungicides and 335 insecticides were registered for use in that country. The pesticide used in agriculture should not have any adverse effect on nontarget organisms. Unfortunately some chemicals have created environmental problems, and one of these of much concern to us is biomagnification.

Besides having no adverse effect on nontarget organisms, a pesticide should be biodegradable in order to minimise any possible harmful ecological side effects. Different pesticides exhibit vastly different residence times in the environment, with some persisting even indefinitely (Table 11.2).

If a persistent compound enters the biosphere, it can be detected anywhere, being transferred from one organism to another through food chain. Chlorinated hydrocarbon insecticides have been detected in remote Arctic regions, 1000 miles away from the point of their introduction. Even though distribution tends to dilute the compound, they still cause concern. They do so because of the phenomenon called **biomagnification**, which is the increase in concentration of a chemical in biological organisms compared to its concentration in the environment. Biomagnification occurs when the pollutant is both, persistent and lipophilic. Because of their lipophilic nature, they are partitioned from the surrounding water into the lipids of both, prokaryotes and eukaryotes, and their concentrations in microbial cells may be one to three orders of magnitude higher than those in the surrounding environment (Fig. 11.3). In the food web, the chemical is neither degraded nor excreted to significant

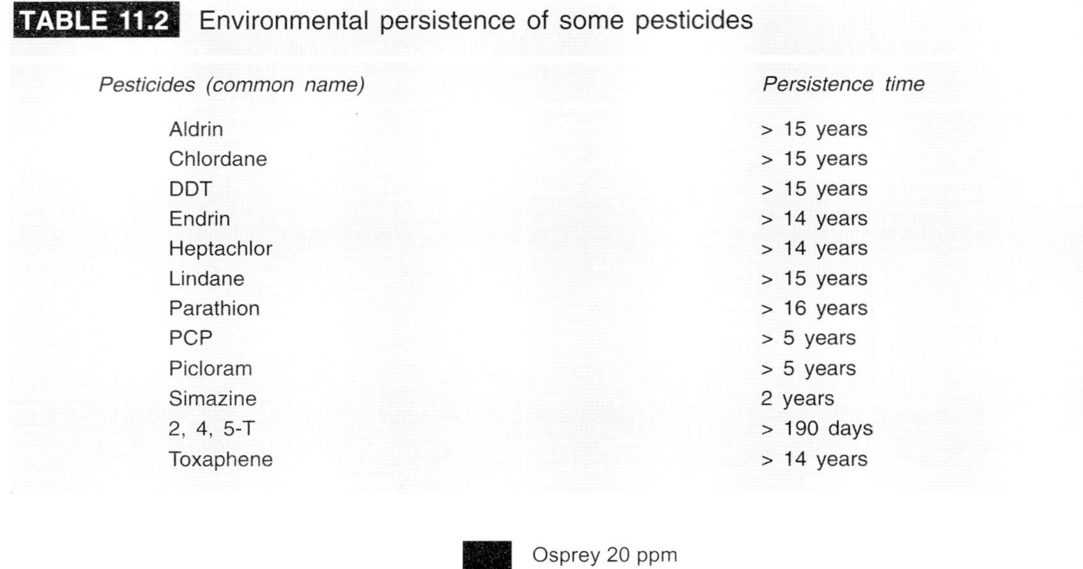

TABLE 11.2 Environmental persistence of some pesticides

Pesticides (common name)	*Persistence time*
Aldrin	> 15 years
Chlordane	> 15 years
DDT	> 15 years
Endrin	> 14 years
Heptachlor	> 14 years
Lindane	> 15 years
Parathion	> 16 years
PCP	> 5 years
Picloram	> 5 years
Simazine	2 years
2, 4, 5-T	> 190 days
Toxaphene	> 14 years

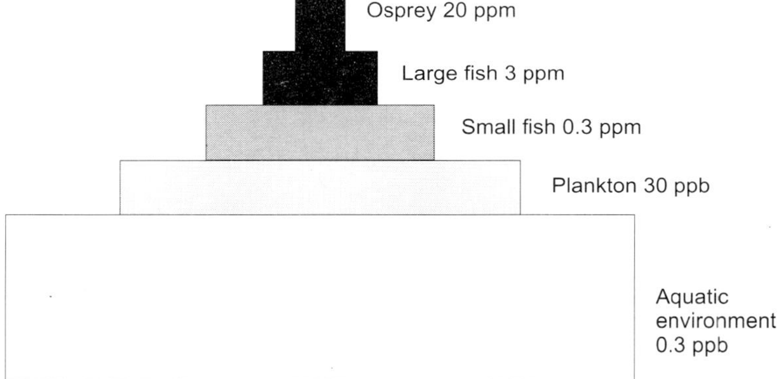

FIG. 11.3
Biomagnification of DDT in food chain

levels. Thus concentrations of such chemicals increase as they are transferred to higher trophic levels. Consequently, the top consumer as birds of prey, carnivores and large fish may carry a body burden of the pollutant that exceeds the environmental concentration by a factor of $10^4–10^6$ or even more. A biomagnified pesticide may cause death or serious problem to the organism. DDT and other such chemicals are known for such effects.

MICROBIAL BIODEGRADATION OF PESTICIDES

An ideal pesticide registered for use in agriculture must be biodegradable, generally by microbes. Most of the currently used organic pesticides are biodegradable. For breakdown of chemicals in the environment there are **three** major natural **mechanisms**: (i) autooxidation, (ii) degradation by sunlight, and (iii) microbial degradation: Most xenobiotics (chemicals foreign to the system) go to soil and water. Since sunlight conditions are very poor in these habitats, microbial degradation is usually the only means for complete mineralisation of xenobiotics in soil and water.

Certain **metabolic processes** which take place only in microbes making them unique tools for biodegradation or detoxification are, (i) fermentation, (ii) anaerobic metabolism, (iii) chemolithotrophic metabolism, (iv) metabolism through exoenzymes, and (v) adaptability to adverse environments through mutation and induction. The microbes biodegrade/detoxify chemicals by enzymatic or non-enzymatic processes. The compounds are biotransformed through oxidation, reduction, hydrolysis and condensation. Synthetic pesticides mostly contain simple hydrocarbon skeletons with a variety of substituents, such as halogens, amino, nitro, hydroxyl and other functional groups. Aliphatic hydrocarbons are oxidised to fatty acids (Fig. 11.4). The fatty acids are then degraded via. β-oxidation sequence, and the resulting C_2 fragments are further metabolised via TCA cycle. Aromatic ring structures are metabolised by dihydroxylation and ring cleavage mechanisms. Prior to these transformations, substituents on the ring may be completely or partially removed. Substituents like halogens, nitro-and sulphonate impede oxygenation and cause recalcitrance. Often a simple change in the substituents of a pesticide may make the difference between recalcitrance (complete resistance to biodegradation) and biodegradability. The chemical structures of some biodegradable and recalcitrant pesticides are compared in Fig. 11.5. The herbicide 2, 4-D is biodegraded within days, but 2, 4, 5-T, which differs only by an additional chlorine substitution in the *meta*-position, persists for many months. Propham is cleaved by microbial amidases rapidly, whereas propachlor with a tertiary amine group is not attacked by these enzymes and persists for longer time. Methoxychlor is less persistent than DDT because the *p*-methoxy groups are subject to dealkylation, and the p-chloro substitution renders DDT greater biological and chemical stability. In some cases, one portion of the pesticide molecule is susceptible to degradation and another is recalcitrant. Some acylamide herbicides are cleaved by microbial amidases, and the aliphatic moiety of the molecule is mineralised (converted to CO_2 and H_2O). The aromatic moiety, stabilised by chlorine substitutions, resists mineralisation. Figure 11.6 shows some of the transformations of the acylanilide herbicide propanil. Here microbial acyl amidases, oxidases and peroxidases are involved.

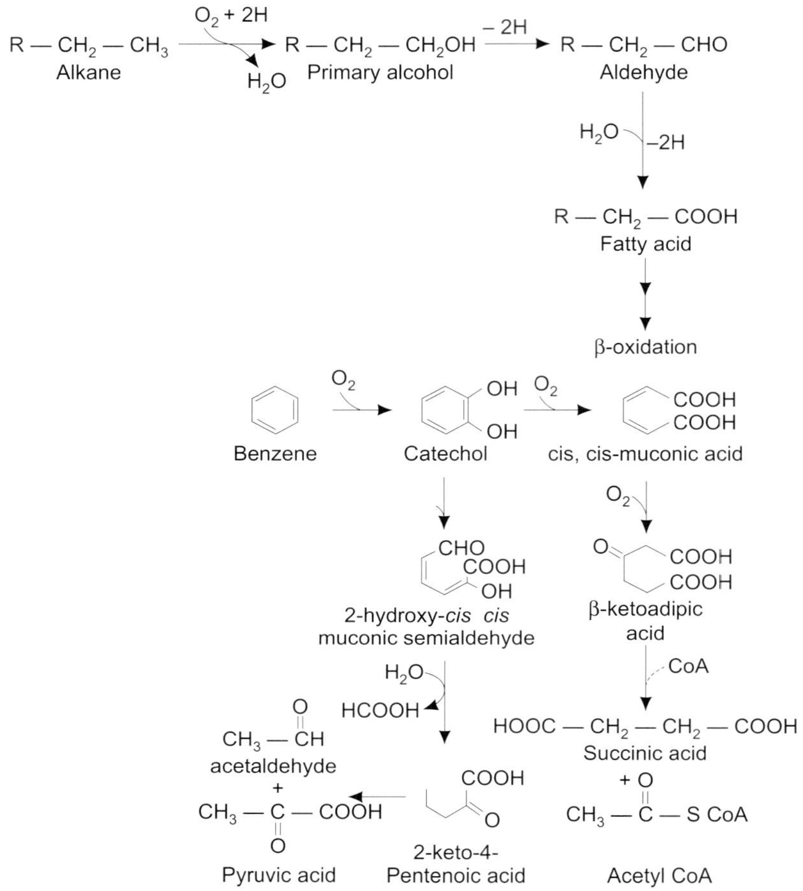

FIG. 11.4

Degradation pathways for aliphatic and aromatic hydrocarbon pesticide moieties

A range of microorganisms have been found to have **environmental use in biodegradation** of different pesticides. Examples of such microorganisms alongwith available information on the enzymes and processes of metabolism are given in Table 11.3.

MICROBIAL BIOREMEDIATION TO CLEAN UP THE ENVIRONMENT

Microorganisms have enough potential to degrade the hazardous chemicals entering all sector of our environment: soil, water, atmosphere. Microbes do occur naturally in

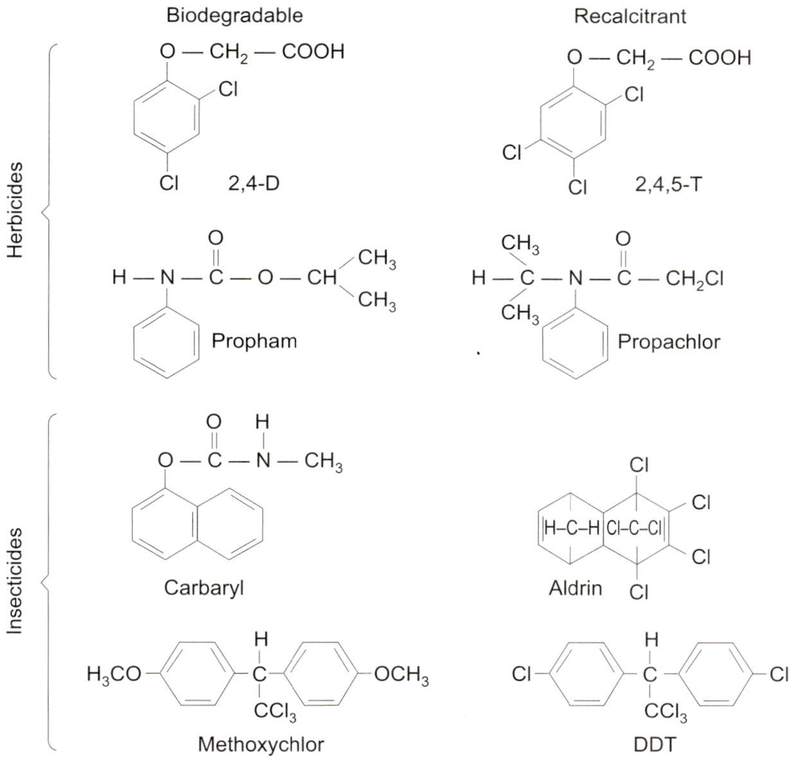

FIG. 11.5

Structures of some biodegradable and recalcitrant pesticides. The molecular features rendering the compounds recalcitrant are the 5-chloro-substitution in 2, 4, 5-T, the N–alkyl substitution in propachlor, the multiple chloro-substitution in aldrin, and the two p-chloro-substitution in DDT

such polluted environments. Traditional waste and water treatment systems using both, anaerobic and aerobic microbes is well known. During recent past abilities of microbes, equipped with adequate enzyme make-up, to degrade pollutants in sites polluted with organic and metal pollutants, have also been well documented. Some of the microbial systems have also become established means of bioremediation of polluted sites on land as well as in water. Some microbial systems are emerging gradually. However, the question is whether these microbial activities will occur rapidly enough naturally, or whether they can and should be enhanced. Such aspects of this expanding area of microbial biotechnology to clean up polluted environments will be examined in detail in Chapter 13, "Microbial Bioremediation of Polluted Environments-Current Status".

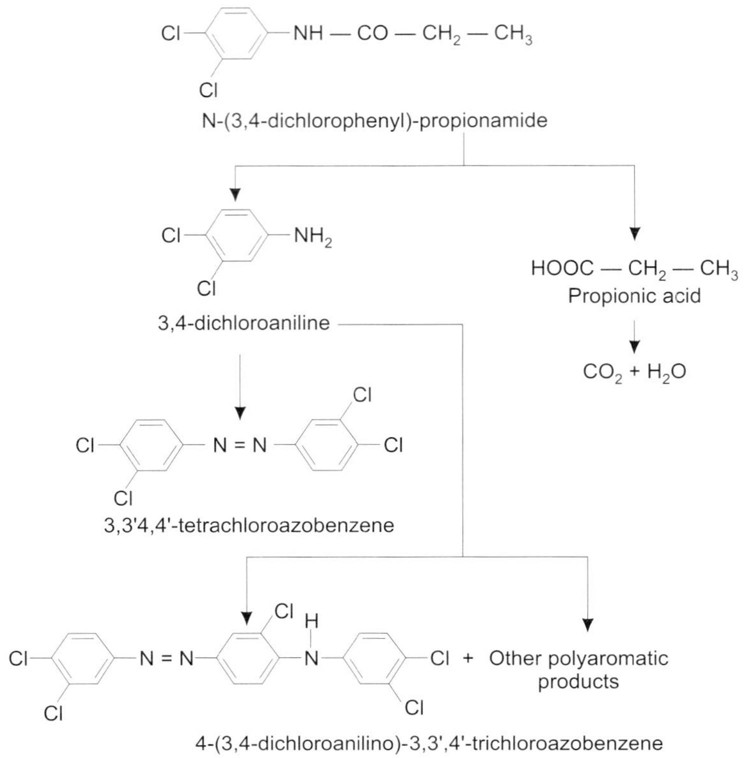
Cl
NH — CO — CH_2 — CH_3
Cl
N-(3,4-dichlorophenyl)-propionamide
Cl
NH_2
Cl
3,4-dichloroaniline
HOOC — CH_2 — CH_3
Propionic acid
CO_2 + H_2O
Cl
N = N
Cl
Cl
Cl
3,3'4,4'-tetrachloroazobenzene
Cl
H
N = N
N
Cl + Other polyaromatic products
Cl
Cl
Cl
4-(3,4-dichloroanilino)-3,3',4'-trichloroazobenzene

FIG. 11.6

The biodegradation pathway for the herbicide propanil, N-(3,4-dichloro-phenyl)-propionamide. The aliphatic portion of the molecule is degraded, but the aromatic one is dimerised and polymerised to persistent residues. The transformation involves microbial synergism

TABLE 11.3 Examples of environmental use of microorganisms in biodegradation/detoxification of pesticides used in agriculture

Pesticides	*Biodegraded by*	*Pathways/Enzymes used*
DDT, Lindane, Heptachlor, Chlordane (chlorinated hydrocarbons)	*Aerobacter aerogenes, E. coli, Klebsiella pneumoniae, Proteus vulgaris, Clostridium, Pseudomonas fluorescens, P. putida, Nocardia, Streptomyces, Aspergillus flavus, Saccharomyces* sp.	Reductive dechlorination; (anaerobic); Dehydrochlorination (aerobic)
Diazinon, Parathion, Chloropyrifan, Paraoxon Fenitrothion, Malathion (organophosphates) Carbaryl, Carbofuran	*Arthrobacter, Streptomyces, Flavobacterium, Pseudomonas, Nocardia, Corynebacterium, Trichoderma viride Pseudomonas striaia, Achromobacter* strain WM III	Hydrolysis of alkyl and aryl bonds, Reductive transformation; Phosphotriesterase; Reductive demethylation, Hydrolases

12

MICROBIAL INTERACTIONS WITH METAL POLLUTANTS

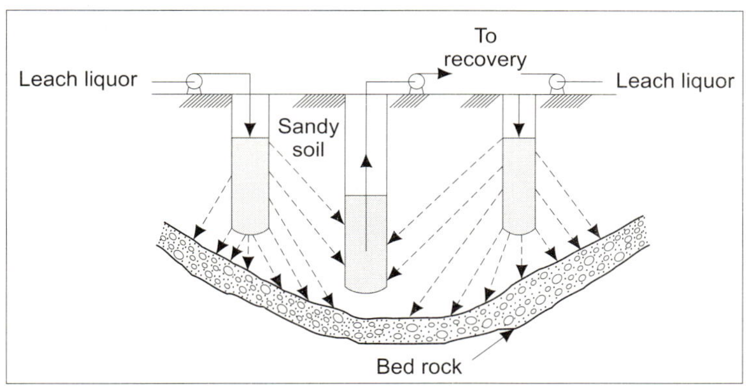

Chapter Outline

Metallic elements are an intrinsic component of the environment. Their presence is considered unique in the sense that it is difficult to remove them completely from the environment once they enter it. Metals constitute an important class of toxic substances which are encountered in numerous occupational and environmental circumstances. The impact of these toxic agents on human health is currently an area of intense interest due to the ubiquity of exposure.

With the increasing use of a wide variety of metals in industry and our daily life, problem arising from toxic metal pollution of the environment have assumed serious dimensions. Metal pollution has become a global concern.

SOURCES AND EMISSIONS

Toxic metals, to a large extent, are dispersed in the environment through industrial effluents, organic wastes, refuse burning, transport, and power generation (Fig. 12.1). They can be carried to places many miles away from the sources by wind, depending upon whether they are in gaseous form or as particulates. Metallic pollutants are ultimately washed out of the air by rain into land or the surface of water ways. Thus air is also a route for the pollution of environment.

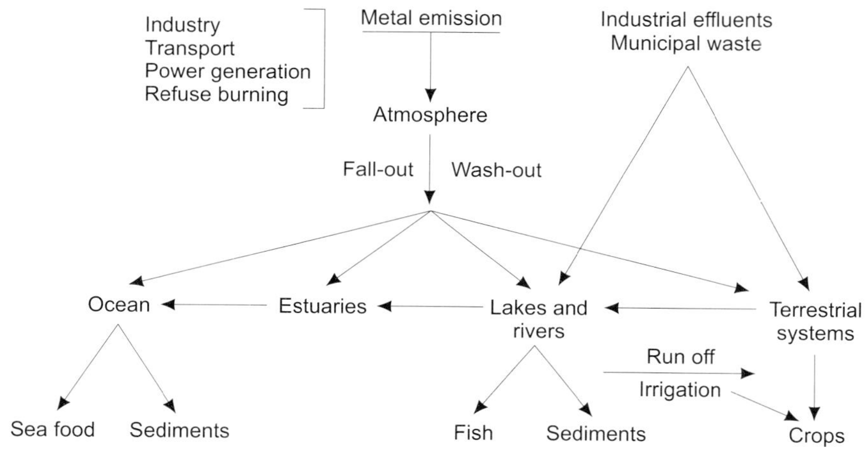

FIG. 12.1
Dispersion of metals in the environment

Metal-polluted environments pose serious health and ecological risks. Metal containing industrial effluents constitute a major source of metallic pollution of hydrosphere. Another means of dispersal is the movement of drainage water from catchment areas which have been contaminated by waste from mining and smelting units. The chief toxic metals in industrial effluents are shown in Table 12.1.

TABLE 12.1 Toxic metals in industrial effluents

Metals	*Manufacturing industries*
Arsenic	Phosphate fertiliser, metal hardening, paints and textiles
Cadmium	Phosphate fertiliser, electroplating, pigments and paints
Chromium	Metal plating, tanning, rubber, photographic
Copper	Plating, rayon, electrical
Lead	Paint, battery
Nickle	Electroplating, iron, steel
Zinc	Galvanising, plating, iron, steel
Mercury	Chlor-alkali, scientific instrument, chemical

Toxic Effects

In general, the toxicity of metal ions to mammalian systems is due to chemical reactivity of the ions with cellular structural proteins, enzymes and membrane systems. The target organs of specific metal toxicities are usually those organs that accumulate the highest concentrations of the metal in vivo. This is often dependent on the route of exposure and the chemical compound of the metal i.e. its valency state, volatility, lipid solubility etc.

The target organs and clinical manifestations of chronic exposures to the metals are given in Table 12.2.

Besides the general toxicities of metals, we are today also concerned with the potential carcinogenicity of metal compounds. Certain metals such as chromium and nickel have been linked with cancers in exposed human populations.

BIOMAGNIFICATION

Because of their toxic nature, metals are not as amenable to bioremediation as organics. Unlike organics, metals are persistent in the environment and can not be degraded through biological, chemical or physical means to an innocuous byproduct.

There are metals like lead, mercury and copper that show biomagnification in a food chain. Mercury is most harmful. Both, inorganic and organic compounds of this element are present in industrial effluents, polluting chiefly the water bodies. Mercury pollution due to methyl mercury is a global problem. Methyl mercury was responsible for the **Minamata** epidemic in Japan and Sweden. The tragedy had

TABLE 12.2 Clinical aspects of chronic metal toxicities

Metal	*Target organs*	*Primary sources*	*Clinical effects*
Arsenic	Pulmonary, nervous system, skin	Industrial dusts, medicinal uses, polluted water	Perforation of nasal septum, respiratory cancer, peripheral neuropathy, dermatoses; skin cancer
Cadmium	Renal, skeletal, pulmonary	Industrial dusts and fumes, polluted water and food	Proteinuria, glucosuria, osteomalacia, aminoaciduria, emphysemia
Chromium	Pulmonary	Industrial dusts and fumes; pollute food	U!cer, perforation of nasal septum, respiratory concer.
Lead	Nervous system, hematopoietic system, renal	Industrial dusts and fumes, polluted food	Encephalopathy, peripheral neuropathy, anemia.
Manganese	Nervous system	Industrial dusts and fumes	Central nervours system disorders
Mercury	Nervous system, renal	Industrial fumes, vapors, polluted water and food	Central and peripheral neuropathies, proteinuria
Nickel	Pulmonary, skin	Industrial dusts, aerosols	Chronic rhinitis and sinusitis, respiratory cancer, dermatitis
Tin	Nervous, pulmonary system	Medicinal uses, industrial dusts	Central nervous system disorders, visual defects and EEG changes pneumoconiosis

occurred due to consumption of heavily mercury-contaminated fish (27 to 102 ppm, average 50 ppm) by the villagers. Methyl mercury had actually accumulted in the tissues of fish. Mercury poisoning, responsible for Minamata disease occurred in Minamata Bay of Kyushu in Japan during 1953-1961, and at Niigata, also in Japan in 1965. The mercury in Minamata Bay had entered from a nearby chloride producing industry using $HgCl_2$, as a catalyst. In Sweden also many rivers and lakes became polluted due to widespread use of mercury compounds as fungicides and algicides in paper and pulp industries, and in agriculture. Chloral alkali plants were the chief source of mercury containing effluents. Methyl mercury is persistent and accumulates in food chain. It accumulated in fatty tissues of animals. Mercury concentration in blood and brain of the affected fetus was found about 20% higher than mother.

From the effluents mercury compounds enter the water body and at their bottom these are metabolically converted into methyl mercury compounds by anaerobic microbes. Methyl mercury is highly persistent and thus accumulates in food chain. Methyl mercury is soluble in lipids and thus after being taken by animals it accumulates in fatty tissues. Fish may accumulate methyl mercury ions directly. There may be nearly 3000 times more mercury in fish than in water. In Minamata Bay all the mercury in sea food is as organic methyl mercury compounds. The symptoms of Minamata include malaise, numbness, visual disturbance dysphasia, ataxia, mental deterioration, convulsions and final death. Mercury readily penetrated the central nervous system of children born in Minamata causing teratogenic effects, Methyl mercury penetrates through placenta. Swedish fish eaters have high mercury content in blood. In *Drosophila* methyl mercury (0.25 ppm) treatment brought chromosomal dysjunction in gametes. Mercury poisoning is caused due to inactivation of several sulfhydral enzymes by replacement of hydrogen atoms in sulfhydral groups. The antidote, BAL (dimercaprol) is used for mercury poisoning.

METAL TOXICITY ON MICROBES

Metals are toxic because due to being strongly ionic in nature, they bind to many cellular ligands and displace native essential metals from their normal binding sites (Fig. 12.2). For instance, arsenate can replace phosphate in the cell. Metals also disrupt proteins by binding to sulfhydryl groups and nucleic acids by binding to phosphate or hydroxyl groups. Cadmium competes with cellular Zn and nonspecifically binds to DNA. Metals may also affect oxidative phosphorylation and

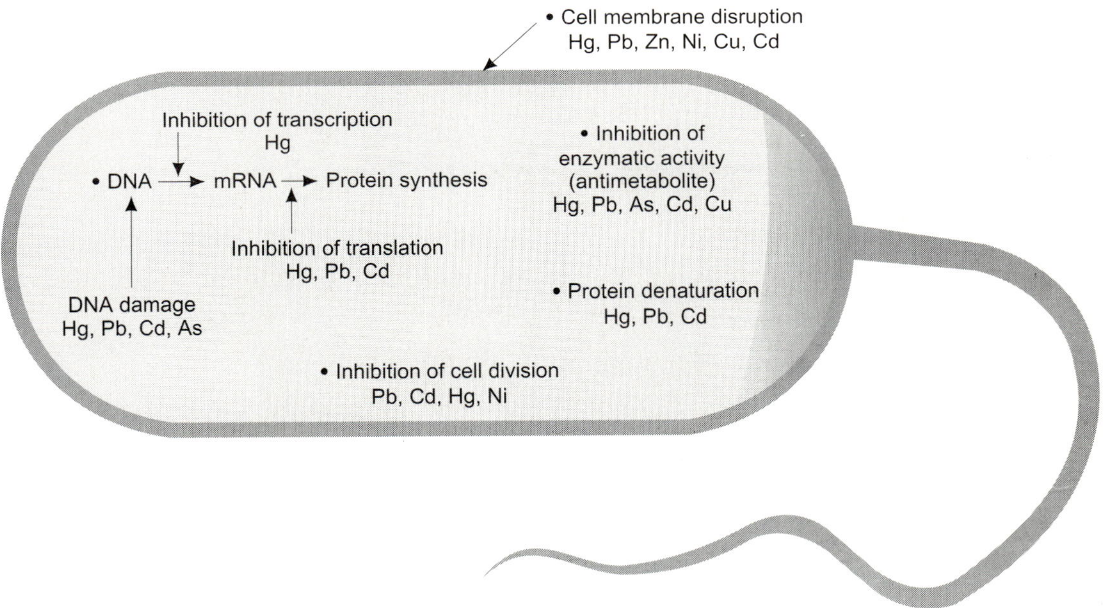

FIG. 12.2

Diagrammatic sketch to show the various toxic influences of metals on the microbial cell demonstrating the widespread metal toxicity in the cell. Metal generally inhibits cell division and metabolism. In response to this microorganisms have to develop "global" mechanisms of resistance that protect the entire cell from metal toxicity

membrane permeability. These metal-microbe interactions result in decreased growth, abnormal morphological changes, and inhibition of biochemical proceses.

MICROBIAL METAL RESISTANCE AND DETOXIFICATION

Some microbes seem to have evolved metal resistance due to their exposure to toxic metals after life began. Others appear to have evolved this mechanism of metal resistance in response to recent exposure to metal pollution over the past few years. In response to metals in the environment, microbes have evolved ingenious mechanisms of metal resistance and detoxification (Fig. 12.3). Resistance mechanisms may be **general** or **specific**, plasmid coded and dependent on a specific metal for activation. **General mechanisms** independent of metal stress often serve several functions. Slime layers or exopolymers provide an effective barrier against metal entry into the cell. Algal surfaces contain carboxylic, amino, thio, hydroxo and hydroxy carboxylic groups that strongly bind metals.

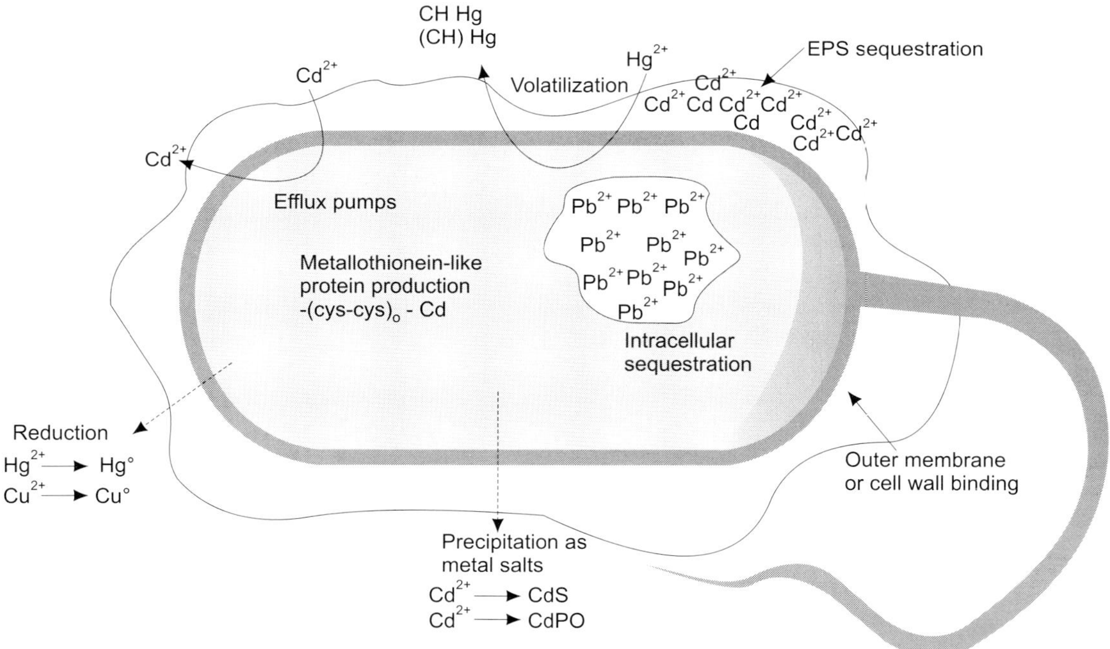

FIG. 12.3

In response to metal toxicity many microorganisms have developed unique mechanisms of resistance and detoxification of harmful metals. These mechanisms of resistance may be intracellular or extracellular and may be specific to a particular metal, or a general mechanism able to interact with a variety of metals

Some extracellular molecules produced under metal stress also provide general resistance to metals. Some microbes produce **siderophores**, which are iron-complexing low molecular weight organic compounds produced in habitats with low (stress) iron content. Siderophores facilitates iron transport into the cell. At the same time, siderophores may interact with other metals (similar to iron in chemistry), such as aluminium, gallium and chromium. By binding such metals, siderophores can reduce metal availability and thereby metal toxicity. Siderophores are shown to reduce copper toxicity in cyanobacteria.

Metal-dependent resistance mechanisms in bacteria are not well understood. The best known mechanism involves metal binding or sequestration by **metallothioneins**. Metallothioneins are low molecular weight, cysteine-rich proteins with a high affinity for Cd, Zn, Cu, Ag and Hg metals. They have been reported in algae, yeast, some fungi, bacteria, cyanobacteria and even plants. They are produced in presence of a particular metal and are involved in detoxification. Metallothionein-like proteins have been isolated from the cyanobacterium *Synechococcus* spp, *E. coli* and *Pseudomonas putida*.

DETRIMENTAL IMPACTS OF METAL-MICROBE INTERACTIONS

Adverse effects of microbial activities on metallic, ceramic and natural stone objects have already been referred to in Chapter 10 "Detrimental Impacts of Diverted Biogeochemical Cycles". These include **acid mine drainage** resulting into impairment of the quality of surface streams, and rivers, and **corrosion** of metal, stone and concrete structures. Metals in geological formations are found in reduced state as iron in form of pyrite (FeS_2). Pyrite is often associated with metal ore deposits. As explained in Chapter 10, when pyrite is exposed to oxygen, this autooxidation coupled with microbial oxidation of iron and sulphur results in the production of large amounts of sulphuric acid. The acid now facilitates metal solubilisation, resulting in metal-rich leachate called acid mine drainage. The leachate is highly acidic and contains high levels of a dark brown ferrous hydroxide precipitate. Ferric iron remaining in the leachate can, in the presence of sulphate and a monovalent cation such as potassium, precipitates as a complex sulphate mineral, $KFe_3\ (SO_4)_2\ (OH)_6$, which is a yellow-brown product characteristic of acid mine drainage. In addition, the leachate can contain other acid-soluble elements including aluminium. The leachate is carried by surface water and groundwater flow into receiving streams and river waters. These waters become highly toxic to plants and animals resulting into widespread fish kills. Acid mine drainage is a problem associated with any type of mining activity including subsurface mining (metal deposits become exposed to atmospheric oxygen), strip mining (large expanses of land exposed to oxygen), and mine tailings (metal deposits brought to the surface).

Besides acid mine drainage, microbial production of acids is also involved in microbially-induced **corrosion** of metal pipes. Fuel and storage tanks are susceptible to microbial corosion, which is the result of acid production by bacteria or fungi and or attack by sulphate-reducing bacteria (SRB). Acid corrosion of ferrous metals is caused by sulphuric acid produced by *Thiobacillus* spp. and other sulphur oxidisers. The fungus, *Cladosporium resinae* produces organic acid which causes corrosion of aluminium fuel tanks in aircraft. In some cases, microbial organic acids destroy protective metal coatings, exposing the metal to attack by other microbes. Sulphate-reducing bacteria (SRB) are primarily responsible for destruction of buried and water-immersed iron or steel structures and also participate in aerobic corrosion by forming biofilms on the metal surface. This has already been explained in Chapter 10. Typical of the SRB genera are *Desulphovibrio* and *Desulphotomaculum* which use sulphate as terminal sulphate-reducing electron acceptor, reducing it to sulphide (a black precipitate visible on metal surface).

Under certain conditions, the microbial transformation of metals results in increased toxicity. Metals are known to be methylated by microorganisms. Biomethylation of nickel, tin, antinomy, mercury, lead, arsenic, selenium and geranium in environment is brought by microbes, with mercury methylation being the most intensively studied. Biologically mediated linking of an alkyl group, e.g. –CH_3 to a metal results into formation of organometal, which is more toxic than the parent metal. Methylation increases the toxicity because methylated form of metal is more volatile and more soluble in lipids. A consequence of increased lipid solubility is that methylated metals are less easily excreted and thus accumulate in live organisms causing toxicity. Biomethylation of mercury occurs in sediments of lakes, rivers, estuaries and other water bodies. The primary generators of methyl mercury in the environment are believed to be the sulphate-reducing bacteria, although a variety of other microbes are capable of methylating mercury. Several outbreaks of mercury poisoning have occurred throughout history, including the Minimata Bay, Japan. Arsenic is reported to be methylated by some fungi, such as *Scopulariopsis brevicaulis*, to mono-, di-, and trimethylarsenes, highly toxic forms of arsenic. Fungi growing on damp wall papers convert and volatilise arsenate (colouring agent) in the wall paper, causing illness due to inhalation of methylarsenes.

BENEFICIAL METAL-MICROBE INTERACTIONS

Metal-microbe interactions like those involved in acid mine drainage are environmentally damaging. Similar reactions can be beneficial provided that these are harnessed in a controlled way. Controlled microbe-metal interactions proved useful in metal recovery and desulphurisation of coal. Such interactions are two way beneficial to us. One, it reduces metal waste (waste containing toxic metals) and bioavailability, and second, such oxidations of minerals can be used in the commercial recovery of copper, uranium and gold from low-grade ores (< 0.4% metal content).

ENHANCED RECOVERY OF METALS (BIOLEACHING)

Since high-grade ore deposits are easily accessible, these become rapidly depleted. It thus becomes necessary to recover mineral resources from low-grade ore deposits. However, no appropriate technology is still available for recovery of metals from low-grade deposits. It is encouraging to find some microorganisms who could do it efficiently. This potential of microbes could only be realised recently and efforts are being made to use them for enhanced recovery of mineral resources from natural deposits.

It was in 1957 that a relationship between the presence of *Thiobacillus ferrooxidans* and the dissolution of metals in copper-leaching operation was recognised by American microbiologists. *T. ferrooxidans* and *T. thiooxidans* are thermoacidophilic archaebacteria. They are autotrophs and grow in acidic and hot environments. It has been demonstrated that these *Thiobacillus* spp. can be used for extraction of copper and uranium from insoluble minerals. This implication of microbial activity in weathering, leaching and deposition of mineral ores could develop into a recent field of biotechnology—**biohydrometallurgy**. **Biomineralisation** is the deposition of metals as insoluble oxides and sulphides due to microbial activity.

Microbial mining is the process of **bioleaching** which recovers metals from ores that are not suitable for direct smelting due to their low metal content. Bioleaching uses microbes to alter the physical or chemical properties of a metallic ore so that the metal can be extracted. Metals can be extracted economically from low-grade sulphide or sulphide-containing ore by exploiting metabolic activities of thiobacilli, particularly *T. ferrooxidans*. Under optimal conditions in the laboratory, as much as 97% of the copper in low-grade ores has been recovered by bioleaching, but such high yields are not achieved in actual mining operations. The process is being commercially used for recovery of copper and uranium from low-grade ores. Laboratory experiments could show that recovery of other metals such as Ni, Zn, Co, Sn, Cd, Mb, Pb, Sb, As and Se from their low-grade sulphide-containing ores is also possible through bioleaching. The leaching process can also be used to separate the insoluble lead sulphate ($PbSO_4$) from other metals that occur in the same ore.

Different approaches have been used for metal recovery by controlled bioleaching. These include, heap leaching, dump leaching, vat leaching, and *in situ* or in-place leaching (Fig. 12.4). **Heap leaching** usually involves excavating a long narrow ditch about 30 cm deep and lining it with polyethylene. Air lines are evenly placed along the ditch to provide aeration. The ditch is filled with crushed ore to a height of 2 to 3 m and then saturated with acidic water (p^H 2 to 3) containing inoculnm of the bacterium, *Thiobacillus ferroxidans*.

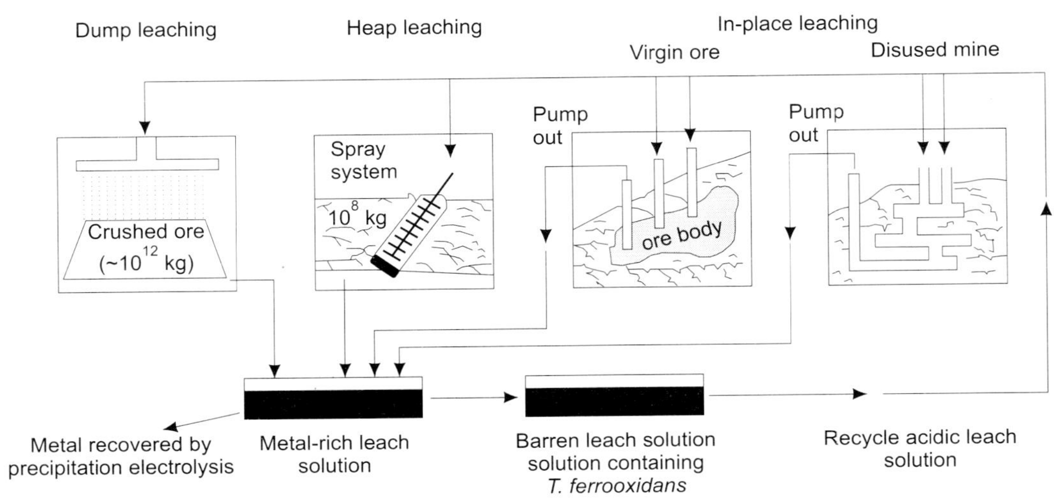

FIG. 12.4

Various applications of micrometallurgy in the recovery of metals from crushed ores. Metals can be recovered from ores either in place, in heaps, or in dump leaching. In each method, an acidic leach solution containing *Thiobacillus ferrooxidans* is flushed through the ore, leaching out the metal to be recovered via precipitation or electrolysis

The leachate is collected at the bottom and recycled through the heap. Like heap leaching, **dump leaching** occurs on an impermeable surface. However, there is no artificial aeration so that set up costs are lower. But extraction efficiency decreases.

Vat leaching involves the continuous stirring of an ore slurry in large tanks. The ore is mixed with acidified water and *T. ferrooxidans*. In this case advantage is that leaching conditions can be controlled efficiently, which is not possible in heap leaching. **In-place leaching** is carried out underground in spent mines. Up to 39% of the metal can remain in these mines. The mine is flooded with acid-bacterial solution, resulting in acid-rich, metal-rich leachate several months later. Metal is extracted and recovered from the leachate. This method has lower associated energy costs and has a less negative environmental impact.

The general process carried out by *T. ferrooxidans* (T.f.) and related species can be shown by the following equation.

$MS + 2O_2 \rightarrow MSO_4$ where M is divalent metal. Because metal sulphide is insoluble and metal sulphate usually water-soluble, this transformation produces a readily leachable form of the metal. T.f., a chemolithotroph derives energy through oxidation of either a reduced sulphur compound or ferrous iron. It exerts its bioleaching action by oxidising the metal sulphide being recovered either **directly** converting S^{2-} to SO_4^{2-} and/or **indirectly** by oxidising the ferrous iron content of the ore to ferric ion. The ferric ion, in turn, chemically oxidises the metal to be recovered to a soluble form that can be leached from the ore.

It is possible to leach the ore *in situ* without first mining it, if the ore formation is porous and overlays a water-impermeable stratum. A pattern of boreholes is established with some of the holes used for injecting the leaching liquour and others for the recovery of leachate (Fig. 12.5). More frequently, however, this bioleaching process is used after the ore is mined, broken up and piled in heaps on a water-impermeable formation or on a specially constructed apron (Fig. 12.6). Water is then pumped to the top of ore heap and trickles down through the ore to the apron. A continuous reactor leaching operation (vat leaching) for recovery of cooper from its low-grade sulphide ore is shown in Fig. 12.7. The leaching water and ore usually supply enough dissolved mineral nutrients required by T.f., but in some cases NH_3 and PO_4 may be added. The leached metal is extracted with an organic solvent and then removed from solvent by stripping. Both the leaching liquor and the solvent are recycled.

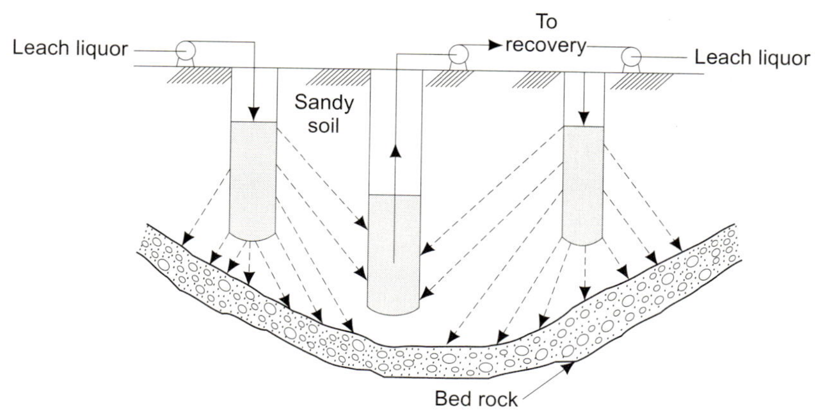

FIG. 12.5

Hole-to-hole bioleaching process used for low-grade ores employing *Thiobacillus* spp. Leaching is from multiple injection wells to a central collection well

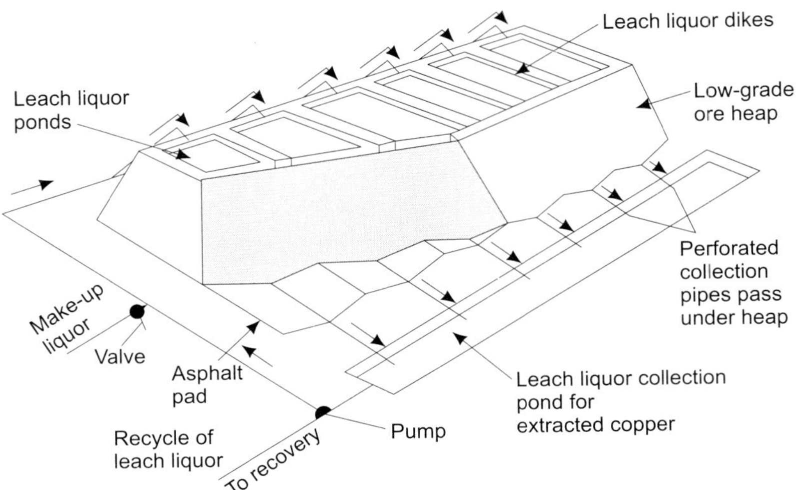

FIG. 12.6

The heap bioleaching process also used for low-grade ores. Ore is mined, crushed and heaped into the shape of a cone with its point cut off on an impervious asphalt pad. Leaching liquor is pumped to the top of the heap and percolates through the ore. The leachate is collected, processed and recycled

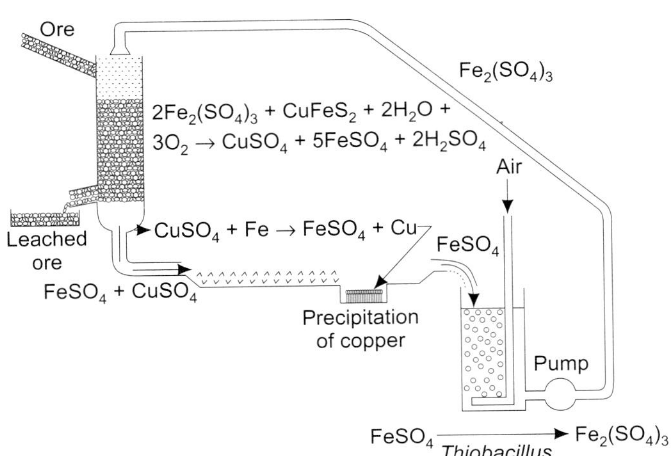

FIG. 12.7

A continuous reactor leaching operation (vat leaching) for extraction of copper from low-grade ore. *Thiobacillus ferrooxidans* oxidises sulphide and ferrous iron generating the acid for leaching. Copper is precipitated by exchange, using scrap iron

The bacterium, *T. ferroxidans* is used for recovery of copper from the ores, such as chalcopyrite ($CuFeS_2$), chalcocite (Cu_2S), and covellite (CuS). It can participate in both, the **direct** and **indirect leaching** of metals as copper.

Copper is generally in short supply. Low-grade copper ore contains 0.1-0.4% Cu. The pregnant leaching solution may contain 1 to 3 g of Cu/l. In copper leaching operations, the bacterium participates in both, direct oxidation of CuS and indirect oxidation of CuS via generation of ferric ions from ferrous sulphide, present in most of the important copper ores such as chalcopyrite ($CuFeS_2$). In the latter case, copper replaces iron i.e. $CuSO_4 + Fe^x \rightarrow Cu^x + FeSO_4$. In 1980s various firms began to utilise bioleaching for extracting copper.

Direct leaching in recovery of copper from chalcocite occurs as follows:

$$\underset{\text{(chalcocite)}}{2Cu_2S} + O_2 + 4H^+ \rightarrow \underset{\text{(covellite)}}{2CuS} + \underset{\text{(leached copper)}}{2Cu^{2+}} + 2H_2O$$

More important overall are the indirect leaching reactions:

$$Cu_2S + 2Fe_2(SO_4)_3 \rightarrow 2CuSO_4 + 4FeSO_4 + S^0 \quad \text{...(1)}$$

$$2FeSO_4 + \tfrac{1}{2}O_2 + H_2SO_4 \rightarrow Fe_2(SO_4)_3 + H_2O \quad \text{...(2)}$$

where copper is spontaneously oxidised by the presence of the ferric ion (Fe^{3+}) and acid (eq. 1) and then the resulting ferrous ion (Fe^{2+}) is reoxidised biologically by *T. ferrooxidans* (eq. 2). Finally, the copper ions (Cu^{2+}) can be recovered from solution by spontaneous precipitation in the presence of scrap ion

$$Cu^{2+} + Fe^0 \rightarrow Fe^{2+} + Cu^0$$

The recovery of uranium, a nuclear fuel, can also be enhanced by microbial activities, which should help overcoming global energy crisis. Moreover, the current controversies about nuclear plants may also be diluted/solved, atleast from economics point of view, if not safety. Insoluble tetravalent uranium oxide (UO_2) occurs in low-grade ores. There is no evidence for direct oxidation. But UO_2 can be indirectly converted to leachable hexavalent form (UO_2SO_4) by T.f., which oxidises ferrous iron in pyrite (FeS), that often accompanies uranium ores. The oxidised iron as an oxidant converts UO_2 to UO_2. SO_4 chemically, which can be recovered by leaching. In Canada, bioleaching was first employed in 1970 for extraction of uranium.

In recovery of uranium, in-place leaching has received considerable attention as it involves less risk of human exposure. Uranium leaching relies on the bacterially facilitated oxidation of insoluble UO_2 to the acid-soluble U^{6+} (eq. 1). *T. ferrooxidans* reoxidises Fe^{2+} to Fe^{3+} facilitating the uranium oxidation using the indirect method of leaching as explained for copper (eq. 2).

$$UO_2 + 2Fe^{3+} \rightarrow 6\ UO_2^{2+} + 2Fe^{2+} \quad ...(1)$$

$$4\ Fe^{2+} + O_2 + 4H^{+} \rightarrow 6Fe^{3+} + 2H_2O \quad ...(2)$$

Recovery of copper and uranium through bioleaching depends on several factors, such as type of the geological formations, ore characteristics and prevailing conditions under which the concerned microbe is to grow. Also, oxidative activity of *Thiobacillus* results into high temperatures, and other bacteria like *Sulpholobus* (obligate thermophile and acid tolerant) can be useful that can oxidise ferrous iron and sulphur in a manner similar to thiobacilli. *Sulpholobus* has been used for bioleaching of molybdenite (molybdenum sulphide), whereas *Thiobacillus* is intolerant of high concentrations of molybdenum, mercury and silver.

Each of these approaches produces a metal containing leachate. A number of methods can be used to recover the metal from the leachate, including solvent extraction, increasing the p^H to precipitate the metals, and electrolysis. In gold recovery, cyanidation is used, where free gold (Au^{2+}) is precipitated as a cyanogold (I) species as follows:

$$2Au + 4CN^{-} + O_2 \rightarrow 2Au(CN)_2^{-} + O_2^{-}$$

Besides bioleaching, some microbes including fungi are able to accumulate metals in their cells at concentrations higher than in the surrounding media. Such bioconcentration has the potential for extracting rare metal ores from dilute solutions and for recovery of metals (gold, silver) from industrial effluents. *Rhizopus* binds uranium from low-grade ores and nuclear wastes. Theoretically, microbes could be used to recover gold from the sea. In South Africa efforts are being made for extraction of gold through bioleaching.

ENHANCED RECOVERY OF PETROLEUM

Besides metals, microbes can also be used to enhance recovery of petroleum hydrocarbons. The **tertiary recovery of petroleum** (the use of biological and chemical means to enhance oil recovery), and the enhanced recovery of hydrocarbons from oil shales are important due to depletion of recoverable oil resources. Tertiory recovery of oil uses solvents, surfactants and polymers to dislodge oil from geological formations. Xanthan gums produced by some bacteria, such as *Xanthomonas campestris*, are useful compounds in oil recovery. These polymers have high viscosity and flow characteristics that allow them to pass through small pores in rock layers containing oil deposits. Xanthan gums are added during water flooding operations (water is pumped into oil reservoirs to force put oil). These help push the oil toward the production wells. The polymers are produced by conventional fermentation in which *X. compestris* is grown and the gums are recovered. Many oil shales contain large amounts of carbonates and pyrites and their removal increases the porosity of shale, enhancing recovery of oil. Acid dissolves the carbonates and these can bc produced by *Thiobacillus* spp. growing on sulphur and iron in the pyrite. Thus bioleaching of oil shales by microbes has also the potential for enhancing the recovery of hydrocarbons.

Microorganisms with potential of bioremediation to clean up the metal-polluted soils and sediments have now been identified. Though physical and chemical remedial methods of metal-polluted soils as well as aquatic systems are well established and straight forward, these are costly. Hence there is felt a need of possible biological means of remediation of such environments. This subject of microbial bioremediation of metal-polluted sites has been examined in the next chapter.

13

MICROBIAL BIOREMEDIATION OF POLLUTED ENVIRONMENTS—CURRENT STATUS

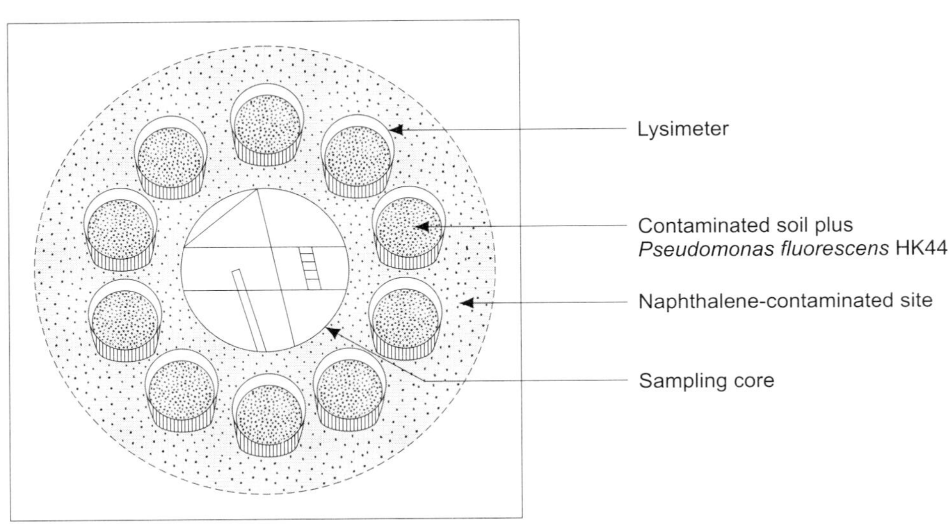

Chapter Outline

- *What is Bioremediation?*
- *Environmental Applications of Bioremediation*
- *Foci of R&D for Bioremediation*
- *Current Status*

It can be very aptly concluded from what has been presented in Chapters 11 and 12 that naturally-occurring microorganisms have abilities to biodegrade (partially or completely), detoxify as well as accumulate variety of organic and metallic pollutants entering into our environments: land, water, and atmosphere. A range of microorganisms are able to degrade complex pesticide wastes used in agriculture and those generated from pesticide industries as well as toxic heavy metals present in wastes from thermal power stations, metal and refining industries, electroplating units, sewage sludges and tanneries etc. Some microbes are also able to degrade petrol and petroleum products, entering as oil pollutants due to oil spills in the marine waters. During past few years enough progress has been made to evolve technologies for use of microorganisms in prevention of environmental pollution caused by accumulation of normally non-biodegradable pollutants, wastes, heavy metals etc. released in atmosphere, water and land.

The objective of this chapter is to examine various microbial activities that can be harnessed to help prevent pollution and clean up polluted environments. Such microbial activities occur naturally in such environments. Is it possible to enhance such activities of microbes through amendments in such habitats (making them more favourable for microbial growth) or by use of tailored microbes by appropriate genetic engineering technologies? It is now beyond any doubt that although microbes may not remove pollutants entirely, but they certainly have capabilities to reduce pollution significantly.

WHAT IS BIOREMEDIATION?

Bioremediation is "the use of living organisms (primarily microorganisms) to degrade environmental pollutants or to prevent pollution through waste treatment." Bioremediation is emerging as most ideal alternative technology for removing pollutants from the environment, restoring contaminated sites, and preventing further pollution. This environment friendly technology is expanding the range of organisms to be used to clean up pollution, and forms a vital component of the so-called **green movement** of maintaining the nature's overall ecological balance, an issue at present being top priority of environmental awareness and public policy.

The term "bioremediation" implies the alliance of a biological system and an engineering system for the remediation of undesirable chemical pollution. The objective of bioremediation is to exploit naturally occurring biodegradative processes to clean up polluted environments.

Need and Scope of Bioremediation

Application of **bioremediation**–the use of living organisms to degrade environmental pollutants or to treat waste streams to control pollution- is expanding world over. The OECD (Organisation for Economic Cooperation and Development) in a report entitled **Biotechnology For a Clean Environment**, issued in 1994 estimated that world market potential for all environmental biotechnologies would nearly double in the 1990s, rising from $40 billion in 1990 to $75 billion by 2000. Bioremediation is needed because certain chemicals accumulate in the environment to levels that threaten human health or environmental quality. It was in fact in mid-1960s, when DDT accumulation changed the old belief (that microbes were able to degrade all compounds entering the environment) that serious efforts were begun to exploit the natural capacity of microorganisms to degrade complex compounds. OECD has been sponsoring meetings of scientists and government representatives from U.S.A., Canada, Japan and West European countries since 1991 to consider the environmental applications of biotechnology. At such meetings, besides other areas, representatives have examined the state of the art of bioremediation and how governments could extend aid to its research and development. At a meeting of OECD workshop, held in November 1994 in Tokyo, participants recognised that bioremediation can have local, regional, and global applications and that both indigenous and gentically- engineered microbes may play important roles.

ENVIRONMENTAL APPLICATIONS OF BIOREMEDIATION

Research projects are being modified to expand the range of microorganisms used for bioremediation. There is search for naturally occurring microbes that have better pollutant degradation kinetics, attack a wider range of pollutant compounds, and do so over a wider range of microbial growth conditions. There is also search for microbes that could grow under extreme environmental conditions, such as tolerance to organic solvents, growth under extremely alkaline substrates, or high temperatures. This information would widen the scope of bioremediation even to nonaqueous pollutants and those found in the environmental conditions unfavourable to growth of most microorganisms.

Researchers have also been using genetic engineering to develop new microbial strains with novel biodegradative capabilities. For instance, modified microbes may be produced through genetic coding for the attack of complex chlorinated hydrocarbons such as dioxins which are non-degradable by naturally occurring microorganisms. Adding genes that code for enzymes that breakdown toxic chemicals to microbes able to survive and grow in much disturbed and harsh environments would greatly extend the range of compounds that might be treated with bioremediation. For example, in Japan a research team has already isolated a species of *Pseudomonas* that can grow in solvents containing more than 50% toluene, a condition that kills most organisms through disruption of cell membrane. Adding appropriate genes for catabolic enzymes to this strain has great potential for expanding the range of bioconversions into nonaqueous solvents.

There are several types of bioremediation. **In situ bioremediation** is the in-place treatment of a polluted site. **Ex situ bioremediation** may be implemented to treat polluted soil or water that is removed from a polluted site. **Intrinsic bioremediation** or **natural attenuation** is the indigenous level of pollutant biodegradation that occurs without any stimulant or treatment.

FOCI OF R&D FOR BIOREMEDIATION

Three different foci of R&D for bioremediation research have emerged worldwide. While **Europeans** are expanding their traditional waste and water treatment systems to cope with specific chemical pollutants, **Americans** focus on site-specific cleanup of soil and water contaminated with petroleum and xenobiotics, and the **Japanese** take aim at global environmental problems. Each of the three technologies of bioremediation in environmental cleanup will be considered in somewhat detail below.

European Focus (Upgrading Traditional Waste and Water Treatment Systems)

Several European countries, particularly Germany, the Netherlands, Belgium, Austria and Italy have programmes to foster R&D in environmental biotechnology. German government agencies promote technology transfer and co-operative research between government-sponsored research institutes and private industry. In the Netherlands, the most developed country in environmental biotechnology, the government supports an innovative programme in bioremediation R&D and subsidises industries participating in the programme. For example, long back in 1978, the Dutch Nuisance & Air Pollution Act took advantage of the development of biofilters, a modification of traditional trickling filters, permitting installation of these biofilters to meet air quality standards. Following are examples of the developments made in the waste and water systems in some European countries.

Biogas (energy) from solid wastes and refuse

It could be recognised by some newly developed European waste treatment systems that some microbial degradation or transformation occur only in the absence of oxygen, and such a process could be important in environmental biotechnology. Under anaerobic conditions, microorganisms growing on wastes can sometimes produce valuable fuels. The dry anaerobic composting (Dranco) process converts the organic fraction of biodegradable organic solid waste and refuse into energy in the form of **biogas** (methane and carbon dioxide) and a humus-like material. The biogas is produced by a consortium of anaerobic bacteria that includes methanogens (methane-producing archaebacteria). Full-scale installations using Dranco process are in operation at Brecht, Belgium and Salzburg, Austria. The systems treat source-separated plant residue and paper from municipal waste. In Salzburg, Austria both anaerobic and aerobic microbes are used. The first phase of treatment, anaerobic digestion takes place in a large tank lasting for about three weeks. The digested material then moves for aerobic maturation which takes about two weeks. The process is shown in Fig. 13.1. The facility built by Organic Waste System of Ghent, Belgium can treat 20,000 tons of biowaste every year. Such contained aerobic and anaerobic treatment systems replace the traditional landfills for disposal of solid wastes. These new systems produce biogas that is about 55% methane, which is burnt to produce electricity.

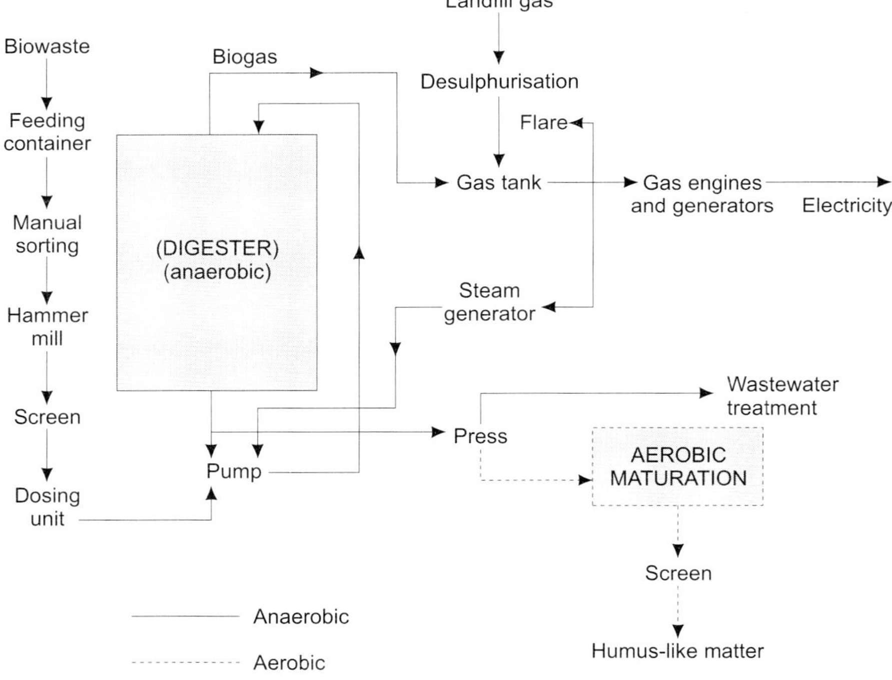

FIG. 13.1

Various steps of treatment facility of municipal waste by using both, anaerobic and aerobic microorganisms

Removal of inorganic compounds

Some wastewater systems, originally designed only to remove organic compounds from water aerobically to reduce BOD, have been modified to include anaerobic zones to remove inorganic compounds as well. One such system, a pilot-scaled fluidized bed reactor that removes nitrate from water, has been tested at the municipal water treatment facility in Blankaart, Belgium. The removal of nitrate from wastewater helps **prevent eutrophication** of the waterways receiving the treated water. At Blankaart, nitrate concentration was reduced from 75.0 mg per l to 0.1 mg per l when water was passed through bioreactor. The reactor contains the methylotrophic bacteria, such as *Methylophilus methylotrophus* that carry out denitrification. Methanol is first added to bioreactor to support the growth of methylotrophs and later removed by trickling filters and granular activated carbon filters before the water is discharged for use. The bacteria convert nitrate to nitrite and then to molecular nitrogen, which is released into the atmosphere.

Removal of toxic chemicals from industrial wastewater

New waste treatment systems can also maintain populations of desired specific microorganisms that can biodegrade toxic compounds like hydrocarbons and chlorinated solvents found in industrial plant wastewater. These compounds generally escape biodegradation in traditional wastewater treatment systems. Bacteria like *Pseudomonas cepacia* are able to biodegrade chlorinated hydrocarbons present in the effluents of pesticide industries manufacturing DDT, heptachlor, chlordane etc. In Matera, Italy, the wastewater is treated anaerobically in a bioreactor in a factory that produces epoxy resins from epichlorohydrin and phenolics. Besides alkaline hydrolysis and subsequent chloride removal, effluents containing unreacted toxic compounds are also treated with specific microbes. These acclimatised anaerobic microorganisms in the waste treatment digester are able to tolerate the concentration of epichlorohydrin and glycidol in the waste stream, and degrade these and other chlorinated organics so that the detoxified water can be released safely into the environment.

Textile and dye industries in Hong Kong are using the bacterium, *Acetobacter liquefaciens* S-1 to treat their wastewater. This bacterium is able to consume bright coloured azo dyes (agar plate containing 100 ppm methyl red) in culture.

Biological gas treatment systems (biofilters, biofilms, bioscrubbers, biotrickling filters)

Traditional water treatment systems, particularly aerobic trickling filters, have been modified to treat air pollutants. In **bioscrubbers** and **biotrickling filters**, multiple microbial communities grow on solid surfaces to produce multilayered complexes called **biofilms**. When gas streams (coming out of water treatment plant) containing organic pollutants are passed through these systems, the pollutants are degraded. Several companies in the Netherlands and Germany have taken the lead in developing such biological gas treatment systems. Since 1978 when Nuisance & Air Pollution Act was enacted in the Netherlands more than 200 **biofilters** have been installed and this biotechnology is now in widespread use in that country to remove organic contaminants from air. Biofiltration has been used since 1989 to treat gases given off by soybean toasters in Hengelo, the Netherlands. The gases pass through a column packed with a proprietary solid support. A biofilm growing on this support biodegrades more than 95% of the organic compounds in the gases. This biofilter, covering an area of 240 sq. meters can process 300 cubic meters of gas per cu m of packing per hour. A cattle feed extrusion plant in Zwolle, the Netherlands, also uses biofiltration to remove more than 99% of the foul gases from its emissions. The facility handles 600 cu m of gas per cu m of packing per hour.

Air treatment bioreactors are in use in the Netherlands to remove formaldehyde from air released from plywood production facilities and phenols from resin producers. Similar biofiltration systems are being tested to remove solvents from indoor air at point production facilities. Some fungi are being exploited in biofilters for the treatment of volatile organic compounds in air. Some fungi like *Candida tropicalis* are able to assimilate styrene - a fragrant liquid unsaturated hydrocarbon used chiefly in making synthetic rubber, resins and plastics and in improving drying oils. In laboratory tests, such styrene - assimilating fungi are grown on a ceramic support. The mycelium of the fungi gives the biofilters a large surface area and greater capacity to eliminate pollutant than the conventional compost biofilters. In addition to styrene the biofilters can remove toluene, xylenes, α-methylstyrene and propene from test gases. The technique is being scaled up and marketed by TNO Institute for Experimental Sciences in Delft, the Netherlands in collaboration with four Dutch companies.

A German animal rendering plant treats 214,000 cu m of gas per hour using a biofilter filled with peat and heather that has a volume of 3,240 cu m. Odours are reduced 94 to 99%. A ceramics factory in Southern Germany uses biofilters to remove more than 99% of the ethanol and isopropyl alcohol released into the air from drying ceramics. The alcohols (90% ethanol and 10% isopropyl alocohol) are released at a concentration of 230 ppm organic carbon and a flow rate of 30,000 cu m per hour. A simple biofilter with a volume of 200 cu m removes them.

American Focus (Site-Specific Cleanup)

In the United States, bioremediation is mostly being used to clean up sites contaminated by toxic chemical spills or polluted from disposal of chemical wastes (primarily mixtures of nonaqueous chemicals). Federal and State Governments have enforced legally the cleanup of several highly contaminated sites that pose threats to human health, particularly where the pollutants are gradually seeping through soil into aquifers used for drinking water. The so-called **superfund sites** - the sites of highest priority in the list of EPA (The Environmental Protection Agency) - include more than 1200 locations. Superfund sites contaminated with multiple pollutants and leaking underground storage tanks, numbering in the hundreds of thousands are being cleaned up through bioremediation with federal funding of about $20 million since 1990. The federal government supported research on the potential for bioremediation of sites contaminated with heavy fossil fuels, creosote, munitions (such as TNT and other nitroaromatics), and chlorinated compounds, among others. This funding comes particularly from the Departments of Energy and Defense to clean up contaminated sites on federal lands. Several companies have also taken up bioremediation as cost-effective measure of restoring environmental quality.

Bioremediation appears to be an attactive alternative to the physical removal and subsequent disposal/destruction of pollutants. The cost of moving and incinerating of polluted matters is at least 10 times that of biological treatment, especially if the latter is carried out in situ (in situ bioremediation). Several large chemical producers like DuPont, and other manufacturers like General Electric and General Mortors are seeking ways to clean up trichloroethylene (TCE) and polychlorinated biphenyls (PCBs). TCE, once a widely used cleaning solvent, is also the most prevalent ground water pollutant of interest to the Departments of Energy and Defense.

Complex organic pollutants including petroleum products in oil spills

Heavy fossil fuels, related compounds as well as chlorinated compounds pollute sites both, in water and land.

Bioremediation of sites in water. Pollutants are often mixtures of complex chemicals. Crude oil, for example contains thousands of hydrocarbons with different structures; refined oils have hundreds of different components; PCBs have dozens of congeners; and some pollutants are undefined combinations of oils, pesticides, other organic compounds and inorganics such as heavy metals.

Oil spills on ocean waters caused alarming threats to biodiversity and human health. The world first woke up to the disaster of an oil spill when on 18 March 1967 a Liberian tanker, Terry Canyon, ran ground on the southwest coast of Great Britain, near the entrance to the English Channel, spilling 60,000 tons of crude oil into the sea. Oil splattered on to 160 km of coastline killing fish and birds. In January 1969 occurred the Second major oil spill off the coast of Santa Barbara in U.S.A. discharging oil at the rate of 1000 gallons per hour. In 1978, Amoco Cadiz disaster dumped 68 million gallons of oil along the French coast. On March 24, 1989 the Supertanker Exxon Valdez belonging to the Exxon Corporation, plied on to a reef off the coasts of Alaska, U.S.A. spilling over 11 million gallons of oil into the clean waters of Alaska's Prince William Sound. As the oil hit 1930 km of shoreline, 100,000 seabirds died including 150 rare species of bald eagles. Many dead seals sank to the bottom of the ocean and at least 1000 sea otters perished. Indian coasts have also suffered from tanker disasters. In July, 1973, 3000 tons of oil washed on to the Gujarat coast when an oil tanker, Cosmos Pioneer ran aground. In 1974, an American oil tanker, Transhuron collided with one of the atolls of the Laccadives, spilling 5000 tons of special furance oil. In June 1989, a Maltese tanker, M.T. Puppy collided with a British vessel, spilling over 5500 tons of furnace oil into the open seas off Bombay. The massive oil slick in the Gulf in 1991 has been the largest so far, spreading over 700 sq km. This oil slick made history, spilling more than 330 million gallons of oil, roughly 30 times more than the quantity spilled by the Exxon Valdez on the shores of Alaska. The oil slick stretched over an area of more than 80 km long and 20 km wide moving south at a speed of 20 km a day. It hit Sandi, Bahrain, Qatar and the U.A.E. shores.

Spills can be dealt with more easily if they are confined to a small area on the water surface. For this mechanical booms or barriers are spread around an oil slick to check its progress and prevent it from hitting the shoreline. The spilled oil can also be treated with dispersants, which are sprayed from aircrafts or ships. Dispersants cause the oil to spread farther and disperse in a way similar to the manner soap removes oil from our hands, allowing the oil to be emulsified and washed away with water. A dispersant contains a surfactant, a solvent, and a stabiliser. Absorbents are also used to facilitate the cleanup of oil spills. Natural meterials like peat moss, straw, sawdust and pine bark can be used. Synthetic absorbents include polyethylene, polystyrene, polypropylene, and polyurethane. Of all these, polyurethane is the most promising.

By far the safest way of treating an oil slick is bioremediation- the use of biological agents for degrading oil. These microbial surfactants are sprayed from the air. They mix with the oil, emulsify it and disperse it throughout the water body so thinly that it no longer remains hazardous. Bacteria and yeasts can grow on several fractions of hydrocarbons as heptane, decane, hexadecane etc. However, not every clean-up method can clear up all oil spills. Clean-up methods are chosen on a case-by-case basis. Professor Ananda M. Chakrabarty a hydrocarbon biotechnologist, working at the University of Illinois Medical Centre, Chicago, U.S.A has developed many a new strain of oil-eating bacteria. He could develop a very efficient oil-eating "superbug" using species of *Pseudomonas* through recombinant DNA technology.

Several hundred U.S. companies sell microoganisms for environmental cleanup. But many of these microbial products are however, of dubious value and are little more than snake oil. Often, naturally occurring (indigenous) microbes at a contaminated site are already biodegrading the pollutant, and addition of introduced microbial products is of no help. Most cultures sold to biodegrade hydrocarbons in contaminated soil and water do not enhance the rates of biodegradation above those carried by indigenous microbes. Efforts have, therefore been made to modify the environment of the given site to enhance the activity of naturally occurring microorganisms, present already at the contaminated site.

It is necessary to maintain specific, complex microbial populations having necessary biodegradative capacities in order to treat the pollutants. It is difficult to ensure the persistence of these specific populations. A novel method has been developed using **reporter genes**. These genes produce an easily monitored effect when microbial activities are occurring at the site. The method has been developed and demostrated by Gary S. Sayler at the University of Tennessee, Knoxville, U.S.A. Sayler uses expression of the **lux** gene, which codes for bioluminescence, to monitor degradation. The fermenter vessel glows from light emitted by the aerobic bacterium, *Pseudomonas fluorescens* strain HK44 when naphthalene is present as the carbon and energy source and oxygen as the electron acceptor, expressing biodegradative activity of the bacterium (naphthalene degradation). Strain HK44 was engineered to contain both, the **nah** gene, which codes for an enzyme that catabolises napthalene, and the **lux** gene for bioluminescence; both genes being under the control of the same promoter.

The largest bioremediation project in the U.S. so for - the treatment of the Exxon Valdez Alaskan oil spill - highlights the benefits of **environmental modification approach** to bioremedation. Here site cleanup is achieved by enhancing the activity of naturally occurring oil-eating microbes through modification of the environment. A massive cleanup involving 11000 workers attempted to remove oil from more than 1000 miles of shoreline. Physical cleanup of the oil, using high-pressure water to wash the rocks, cost Exxon over $1 million per day and the treatment was slow and also left subsurface oil that recontaminated shorelines. Bioremediation through simple addition of nitrogen-containing fertilisers to the contaminated shorelines, stimulated the metabolism of indigenous hydrogen-degrading microorganisms and degraded both surface and subsurface oil three to five times faster than occurred at untreated test sites. Natural hydrocarbon degrading microbes were already abundant in the water of Prince William Sound and tidal flux aerated the shorelines making bioremediation a viable treatment. The cost of remediating hundreds of miles of contaminated shoreline was less than $1 million and could complete within a couple of year contrary to more than a decade in un-fertilised condition. Though rates of hydrocarbaon degradation were stimulated, it did not lead to complete removal of the hydrocarbons. Residual hydrocarbons were contained in asphalt-like materials that are insoluble in water and should affect biodiversity.

Bioremediation of contaminated sites on land. Petroleum hydrocarbons can also be successfully removed from land-based sites using environment moderation, as demonstrated more than 20 years ago in tests on the cleanup of a gasoline-contaminated aquifer in Amber, Philadelphia. In 1972 in this test, hydrocarbons were removed by adding fertilisers and using forced aeration to stimulate the growth of indigenous hydrocarbon-utilising microorganisms in the groundwater.

Soil microbes in aquifers are responsible for a significant portion of the degradation of aromatic compounds when land sites are contaminated. These microbes degrade benzene, toluene, ethyl benzene, and xylenes. Most subsoil indigenous microbes biodegrade low levels of these compounds if there is enough dissolved oxygen in the groundwater.

Leaking underground storage tanks in the U.S. have contaminated soil and groundwater with low molecular weight aromatics that can threaten human health, when these hydrocarbons get into drinking water supplies. Bioremediation can treat both soil and groundwater in one step. The cleanup of pollutants from leaking underground storage tanks at a bus maintenance and fueling facility in Denver is an example of such a treatment. Gasoline, diesel fuel, and lubricating oil contaminated the soil and underlying groundwater near the leaking tanks. Beginning in 1993, the polluting hydrocarbons were removed by **air sparging** (physical transfer to the atmosphere) and **bioventing** in which the contaminated water was pumped to the surface and reinjected. Oxygen introduced into the water in this way supported the biodegradative metabolism of indigenous microorganisms. The dissolved concentration in the petroleum-contaminated groundwater was as high as 2.8 mg per l and natural bacteria are able to consume fuel residues at a rate of about 4.2 g of hydrocarbon degradation per l of water per year. In 1993, the. Air Force completed a full-scale soil bioventing project to clean up a 27000-gal jet-fuel spill at Hill Air Force Base in Utah. During this 18 month project, jet-fuel residues in soil were reduced from an average total petroleum hydrocarbon concentration of about 900 mg per kg to less than 10 mg per kg. In this process 60% of the hydrocarbons removed from the soil volatilised directly into the atmosphere and the remaining 40% was biodegraded to carbon dioxide and water.

Bioreactors are being tested by the Department of Energy at its Savannah River site in South Carolina as well as by companies such as Envirogen in Lawrenceville, N.J. to treat volatile compounds. TCE and other organic contaminants can be removed from soil and groundwater by simultaneous use of in situ thermal catalytic oxidation and bioremediation. In such a faster cleanup of contaminants (Fig. 13.2), concentrations of volatile organics can be quickly reduced from the soil above the water table by vacuum extraction. These contaminants are then converted to carbon dioxide and hydrochloric acid by catalytic oxidation treatment at the surface. The Savannah River system uses a proprietary-metal catalyst and operates at about 425°C. Bioremediation, accomplished by encouraging indigenous methane-oxidising microorganisms, removes contaminants from water-saturated soils as well as much of the residual contamination after volatilisation. Methane, oxygen, and nutrient gases that supply phosphorus and nitrogen are pumped into the contaminated soil to encourage indigenous microbes which convert the contaminants to carbon dioxide and hydrochloric acid through the action of the enzyme, methane monooxygenase. The system can degrade more than 250 organic compounds, reducing TCE concentration to less than 5 ppb-low enough to meet drinking water standards.

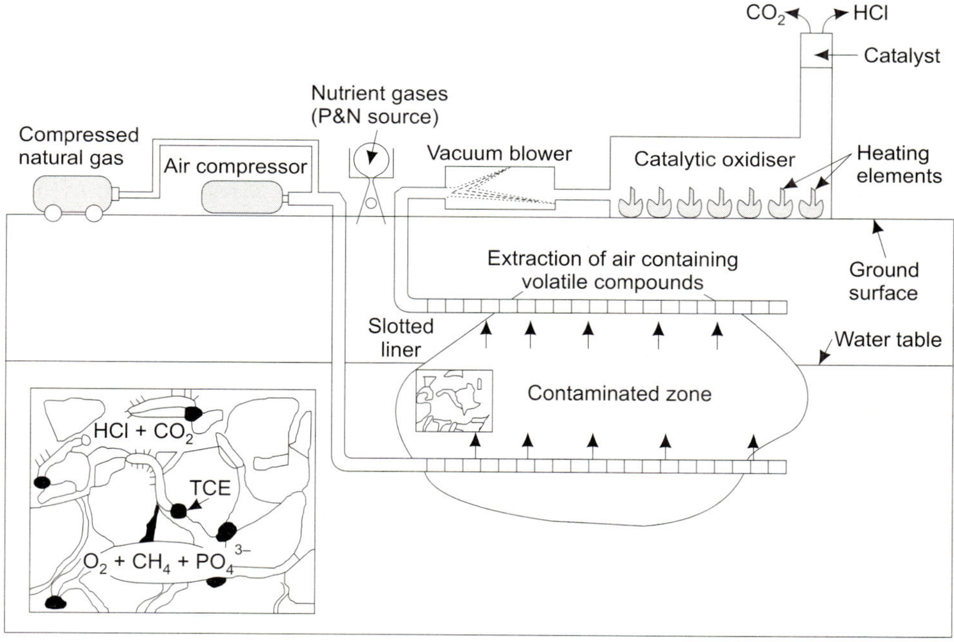

FIG. 13.2

A schematic of a full-scale demonstration facility developed for simultaneous use of in situ bioremediation and thermal catalytic oxidation for a faster cleanup of trichloroethylene (TCE) and other organic pollutants of soil and groundwater

Genetically - engineered bacteria like *Pseudomonas fluorescens* strain HK44 have been successfully used to treat naphthalene-contaminated soil in field tests at Qak Ridge National Laboratory, E.Tennessee, U.S.A. In the tests, underground steel tanks, called **lysimeters** (Fig. 13.3) surround a central sampling core. Each lysimeter is about 8 feet in diameter and 10 feet deep. It is fitted with a cover, sampling ports, and sprinklers to stimulate rainfall. The lysimeters are loaded with contaminated soil and *Pseudomonas fluorescens* HK44 and monitored for the parameters such as colonisation, gene transfer, bioluminescence, and naphthalene degradation. Bioremediation has been only occasionally used in Europe to clean up sites contaminated with petroleum spills. The situation may change over times as Eastern European countries are beginning to face such problems severely soon. Many sites in Czech Republic, Lithuania, Latvia, Ukraine and Russia are at least as polluted as the super fund sites in U.S.A.

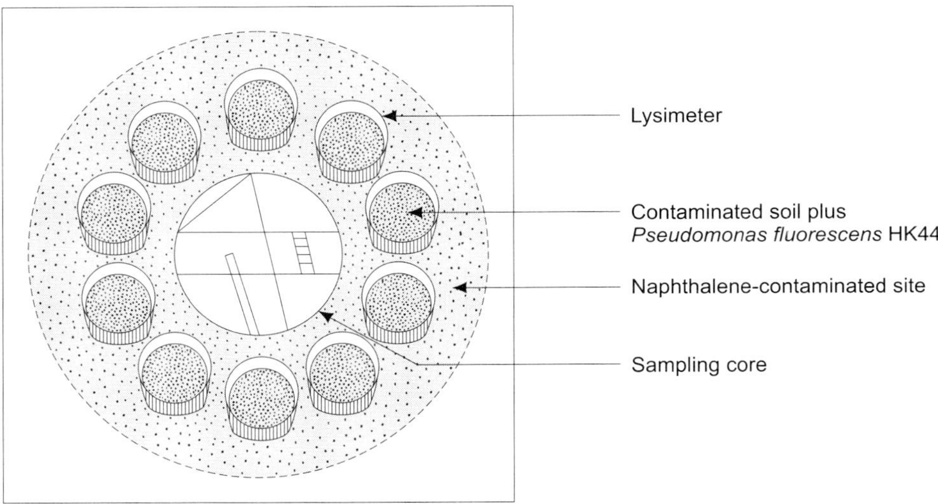

FIG. 13.3
A schematic of an experimental site showing field tests to monitor the activity of naphthalene-degrading bacterium

Integrated bioremediation of polluted groundwater as well as soil zone. Most of the developed bioremediation technologies are based on two standard practices: addition of oxygen and addition of other nutrients. In case appropriate biodegrading microbes are not present in soil, or their population reduced due to any toxic effects of the pollutant, microbes can also be added as introduced organisms to enhance the existing populations.

A typical bioremediation system used to treat a contaminated aquifer as well as the contaminated zone of soil (vadose zone) above the water table is shown in Fig. 13.4a. Several technologies have been developed to overcome a lack of oxygen. The system contains a series of injection wells or galleries and a series of recovery wells. First, the recovery wells remove contaminated groundwater, which is treated above ground, using a bioreactor, containing microbes that are acclimated to the contaminant. This would be considered *ex situ* treatment. After bioreactor treatment, the clean water is supplied with oxygen and nutrients, and then it is reinjected into the site. The reinjected water provides oxygen and nutrients to stimulate *in situ* biodegradation. The reinjected water also flushes the vadose zone to aid in removal of the contaminant for aboveground bioreactor treatment. Oxygen to the site of contamination in the vadose zone is added by **bioventing** (Fig. 13.4b) which is a technique of merging of soil vapour extraction technology and bioremediation. A series of wells are constructed around the zone of contamination (just below the water table). A vacuum is drawn on these wells to force accelerated air movement through the contamination zone. This increases O_2 supply throughout the site. Oxygen to the saturated zone (Fig. 13.4c) is added by **air sparging**, in which an air sparger well is used to inject air under pressure below the water table. Methane can also be added with oxygen in extracted groundwater and reinjected into the saturated zone. Methane is used specifically for methanotrophic microbes.

Since many contaminated sites contain carbon-rich organic matter but minimal amount of nitrogen and phosphorus, these two **nutrients** are added to the contaminated site. Nutrient solutions are injected from an aboveground batch feed system to optimise C : N : P ratios to approximately 100 : 10 : 1. Bioremediation under anaerobic conditions can also be carried by adding alternative electron acceptors, which include nitrate, sulphate, iron (Fe^{3+}), and CO_2. Bioavailability of contaminants can also be increased by addition of surfactants (e.g. detergents). If appropriate biodegrading microbes are not present in soil or in case their populations are very low, specific microorganisms can also be added as introduced organisms to enhance the population. This is known as **bioaugmentation**. "Superbugs", the genetically-engineered microbes, capable of degrading the pollutants at faster rates have also been developed by scientists. Superbugs can be used in bioreactor systems under controlled conditions. Bioaugmentation can also be achieved by addition of specific gene sequences (degrading genes), as microbial cells or as "naked" DNA.

Metal-polluted soils and sediments

Microbial remediation of metal-contaminated soils and sediments can be achieved by (i) immobilisation of metal *in situ* to reduce metal bioavailability and mobility or (ii) removal of the metal from soil (Fig. 13.5). Several methods have been developed for microbial remediation of contaminated soils. These include microbial leaching,

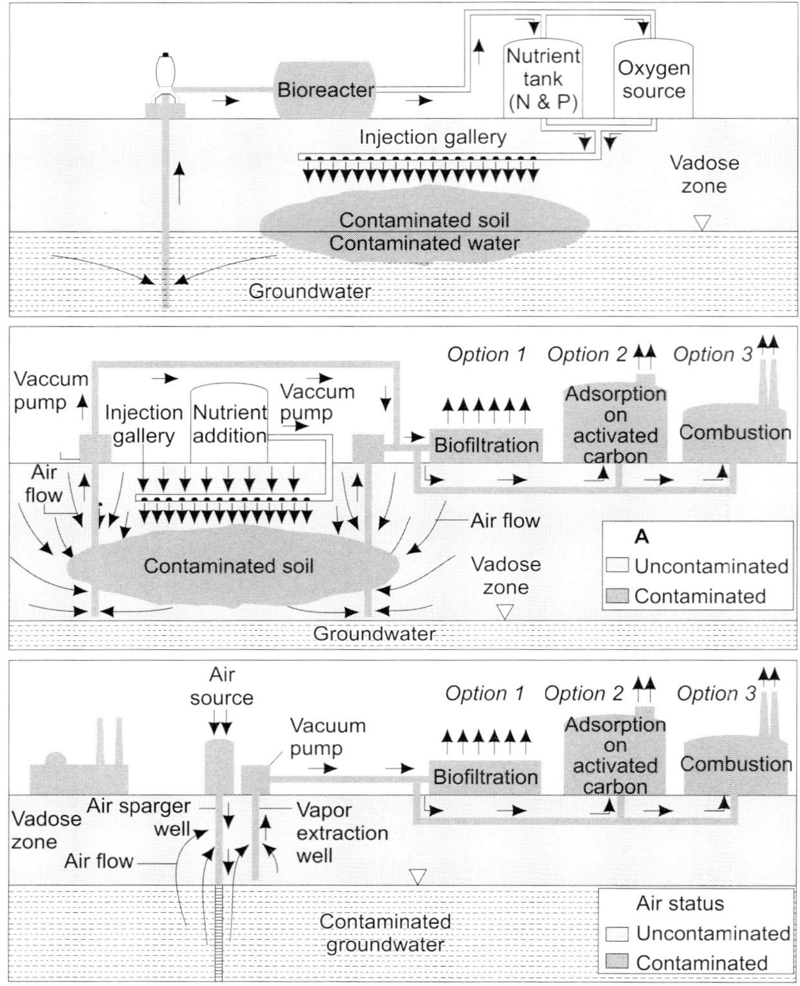

FIG. 13.4

(a) *In situ* bioremediation in the vadose zone and groundwater. Nutrients and oxygen are pumped into the contaminated area to promote *in situ* processes. This figure also shows *ex situ* treatment. *Ex situ* treatment is for water pumped to the surface and uses an aboveground bioreactor, as shown, or other methods, e.g., air stripping, activated carbon, oil-water separation, or oxidation. An injection well returns treated water to the aquifer. (b) Bioventing and biofiltration in the vadose zone. Air drawn through the contaminated site (bioventing) stimulates *in situ* aerobic degradation. Volatile contaminants removed with the air are treated in a biofilter, by adsorption on activated carbon, or by combustion. (c) Bioremediation in the groundwater by air sparging. Air pumped into the contaminated site stimulates aerobic degradation in the saturated zone. Volatile contaminants brought to the surface are treated by biofiltration, activated carbon, or combustion

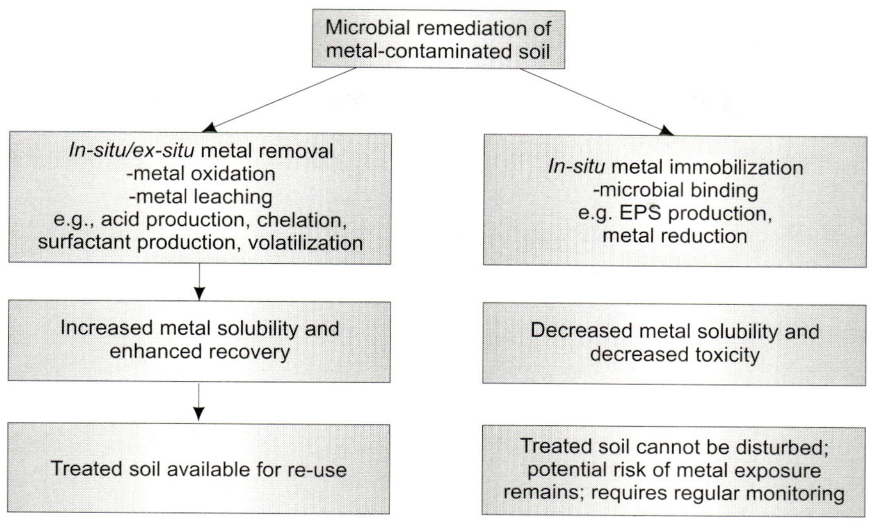

FIG. 13.5

Microbial remediation in metal-contaminated soils relies on either metal removal or, more commonly, metal immobilization

microbial surfactants, microbially induced volatilisation, and microbial immobilisation and complexation.

Microbes like *Thiobacillus ferrooxidans* are known to remove metals from soils through **metal solubilisation** or **leaching**. Bioleaching is used to recover copper, lead, zinc and uranium. Bioleaching can also be used in removal of metals from contaminated sites. Unfortunately this could not receive much attention so far. Microbes can also increase leaching for recovery through **surfactants**. Some surfactants, such as the rhamnolipid produced by *Pseudomonas aeruginosa* shows specificity for cadmium and lead. Emulsan produced by *Acinetobacter calcoaceticus* can also help in metal removal.

Like leaching, **volatilisation** of metals, specifically methylation, increases metal bioavailability. Methylated metals are more lipophilic. Inspite of their being toxic, many microbes volatilise metals to facilitate their removal from environment. In this way methylation of some metals has been used as bioremediation tactic: the most famous example is the removal of selenium from contaminated soil in San Joaquin, California in U.S.A. Though mercury is also commonly methylated, it is susceptible to bioaccumulation in the food chain. **Metal sequestration** is another approach, which exploits the ability of some microbes to produce metal-complexing polymers. It has been shown effective in laboratory studies only, and yet to be taken to field trials.

Microbial bioremediation can also be used to treat sites contaminated with heavy metals or radionuclides. Microbes-algae, bacteria. and fungi as well as higher plants have capabilities to uptake these pollutants. After uptake, these either accumulate or are assimilated by them. Accumulated heavy metals are recovered for recycling or disposal. For example, *Zooglea ramigera* adsorbs copper and cadmium up to the levels of 300 and 100 mg of metal per g dry wt. respectively. *Pseudomonas putida, Arthrobacter viscous* and *Citrobacter* spp remove several toxic heavy metals from industrial effluents. Radioactive metals as uranium and thorium are removed by *Rhizopus arrhizus,* and *Penicillium chrysogenum* can accumulate radium. The yeast, *Saccharomyces cerevisiae* accumulates uranium from dilute solution. The bacteira, like *Thiobacillus thiooxidans* bring about bioleaching of zinc, cobalt and nickel from sulphide rocks. Fungi belonging to the genera, *Trichoderma, Aspergillus, Aureobasidium, Ophiostoma* and *Rhodotorula* are shown to have biosorption ability of heavy metals and these seem to play important role in detoxification of industrial effluents. The Royal Dutch Shell uses sulphate-reducing bacteria to immobilise metals at an old zinc refining site at Budelco, the Netherlands. The groundwater at this site was heavily contaminated with zinc and cadmium. For bioremediation, groundwater is pumped through a bioreactor to which ethanol, ammonia and phosphate are added to support the growth of sulphate-reducing bacteria. These bacteria convert the sulphate in the groundwater to hydrogen sulphide, which reacts with heavy metals to form insoluble metal sulphides. A flocculent is used to retain the precipitated sulphides.

Some of the limitations of bioremediation can be overcome through genetic engineering. One can produce strains of microbes that are resistant to heavy metals, have greater degradative capabilities and able to grow in more than one kind of environments. The first patent for a genetically engineered organism was granted in the U.S. in 1981, for a bacterium that degrades hydrocarbons. This has never been a candidate for cleaning up oil spills, as it was engineered to degrade low molecular weight hydrocarbons such as toluene and octane that evaporate rapidly. Efforts to engineer genetically modified microbes that can degrade specific pollutants in situ have yet to meet with great success. In most countries environmental regulations greatly restrict the release of these microbes into the environment for in situ bioremediation. The real benefits of genetically modified microbes for bioremediation are likely to be in contained bioreactors.

Japanese Focus (Global Applications)

In Japan, academic, industrial and government research is coordinated by the Ministry of International Trade & Industry (MITI). Developing long-term global applications of environmental biotechnology is one of the R&D priorities of this agency. Japanese bioremediation research focuses on ways to reverse environmental degradation in the 21st century. The research agenda of MITI places great emphasis on global problems such as **climate change** and **desertification** rather than on site-specific cleanup. Japanese are therefore concentrating on much broader issues. Following are the ways in which they plan to handle environmental problems through bioremediation at global scale.

Replacement for petrochemicals (Hydrogen fuel)

There has been a search for the microbes that produce some substances that can act as replacement for petrochemicals. Biotechnology research for producing hydrogen for use as automotive fuel that will not contribute to global climate change is of particular interest. One thrust of the work of Japanese researchers is use of microbes to produce such fuels and materials (as hydrogen) that do not emit carbon dioxide and, thus reducing air pollution. Researchers at several institutions including the National Institute of Bioscience and Human Technology and the National Institute for Resources and the Environment, both in Tsukuba, are seeking hydrogen-producing algae as sources of this clean-burning fuel.

Reversal of global warming

Even more far-reaching is the Japanese effort to reverse global warming. Besides looking for hydrogen and other fuels that would not contribute to global climate change there are also simultaneous efforts to develop bioremediation system to remove carbon dioxide already emitted into the atmosphere from the burning of carbon-based fossil fuels. This removal of the "extra" carbon dioxide polluting the atmosphere due to fossil fuel burning would be an extremely dramatic demonstration of bioremediation. The "extra" carbon dioxide contributes to global warming. Microbes can remove enough of this greenhouse gas and some of them convert carbon dioxide to various organic compounds. The recombinant DNA technology should be of much use in such efforts to produce new strains of microbes that would theoretically do so more efficiently. At the Marine Biotechnology Laboratory at Kamaishi, Japan algae have been isolated that convert carbon dioxide to carbohydrates 10 times faster at 40°C than terrestrial green plants. Such algae could be grown in bioreactors near power plants where they would remove carbon dioxide from the air as it is released by fossil fuel burning at power plant. However, the organic compounds produced by assimilation of this greenhouse gas are likely to be subsequently biodegraded to carbon dioxide and water releasing this gas back to air. A successful bioremediation programme using such algae and other microbes should therefore depend on finding such microbes that produce lignin or other polymeric compounds resistant to biodegradation as the original fossil fuels were. Such conversions would immobilise carbon, reducing the build up of carbon dioxide in the atmosphere.

Japanese researchers have been growing some algae in bioreactors at the Marine Biotechnology Laboratory, Kamaishi. The alga, *Chlorococcum littorale,* isolated from Kamaishi Bay and grown successfully in laboratory could tolerate high levels of carbon dioxide, converting it to polysaccharides. Another alga being studied in the same laboratory, *Prasinococcus capsulatus* produces large amounts of extracellular mucilaginous polysaccharide. This alga is being studied for its potential to remove carbon dioxide from the atmosphere and for commercial applications of the polysaccharide.

Another group of microbes, the marine foraminifera, can also convert carbon from carbon dioxide into calcium carbonate. This process, occurring in coral reef formation, removes carbon from the usual biogeochemical carbon cycle that circulates carbon between organic compounds and carbon dioxide. Such a treatment would consume enormous quantities of calcium, creating its own impact on the environment. Despite these limitations the efforts may succeed in attacking global pollution problems. Such Japanese strategies if developed would raise the need for international understanding in the uses of biotechnology.

Biodegradable plastics

Biodegradable plastics made of microbially produced polyhydroxybutyrate have already been developed and being marketed by ICI in the U.K. These ICI's biodegradable plastics are used worldwide in trash bags that biodegrade in landfills, In Japan and elsewhere studies are being made on microorganisms that produce such biodegradable polymers that form solid wastes. Researchers under the guidance of Yoshiharu Doi in the polymer chemistry division of the Institue of Physical and Chemical Research (the Riken Institute) in Saitama have isolated a strain of the bacterium, *Alcaligenes eutrophus* that produces a copolyester of 3- and 4-hydroxybutyrate. The cells of this bacterium produce 80% of their biomass as this polymer, when gorwn on 1, 4-butanediol or γ-butyrolactone.

Biodesulphurisation

Japanese have also developed microbial systems to remove inorganic sulphur from industrial air emissions. The Bio-SR (system developed by Dowa Mining has been put to commerical use. Industrial exhaust gas containing hydrogen sulphide passes through an acidic ferric sulphate solution which oxidises the sulphide to elemental sulphur while reducing the ferric ion to ferrous ion.

An iron- and sulphur-oxidising bacterium, *Thiobacillus ferrooxidans* regenerates the ferric sulphate solution. The bacterium is grown in a bioreactor at a pH between 2 and 3. At a commercial facility of NKK Corporation, Tokyo, concentrations of hydrogen sulphide in an effluent gas are reduced from a range of 400 to 2000 ppm to less than 10 ppm. Thus 99% of the sulphide that would have otherwise gone to the atmosphere is captured as sulphur.

In the U.S. also, the Energy Biosystems in Woodlands, Texas have been developing similar systems to remove organic sulphur-containing compounds from fossil fuels. The bacterium *Rhodococcus* biodegrades dibenzothiophenes and other organosulphur compounds in the fuel. Biodesulphurisation produces a cleaner fuel that does not produce sulphur oxide emissions when it burns. The process may be used to produce low-sulphur diesel fuel that can meet the sulphur dioxide air emission standards of the Clean Air Act.

Japan's official policy for the development of its biochemical industry since 1990 has required the industry to work to preserve the global environment, treat wastes, and conserve natural resources. The policy calls for the biochemical industry to use the **sophisticated functions of microorganisms that make them friendly to the environment**. Under this policy MITI has initiated research projects at several national laboratories to work in collaboration with industry and Universities. These projects include work to develop biotechnology systems to combat global environmental changes.

Reversal of desert formation

At present, 4.5 million sq km (35% of the total land area of the planet) is threatened by desertification due to loss of water from various soils. Efforts are being made to develop microbes that could help reverse desert formation. Such biological systems produce products like biopolymers that retain water reversing the phenomenon on large landscapes. In Japan, a MITI - supported programme initiated the use of the bacterium, *Alcaligenes latus* to produce a **superbioabsorbent**, a polysaccharide composed of glucose and glucuronic acid that can absorb and hold more than a thousand times its own weight in water. This ability is yet to be tested in the field.

CURRENT STATUS

Although it is often not thought of as bioremediation, domestic sewage waste has been treated biologically for many years. The first successful application of bioremediation outside sewage treatment was the cleanup of oil spills, and success in this area is now well documented. In the past few years many new bioemediation technologies have emerged that are being used to address other types of pollutants. The current status of bioremediation (Table 13.1) reflects satisfactory progress made in the area during last few years providing impetus for future research.

The above account shows that bioremediation has promising future with several potential applications to clean up the polluted environments and treat wastes. Compared to physical cleanup methods, in situ bioremediation is much less expensive and causes less environmental perturbation. It may alter manufacturing practices,

TABLE 13.1 Current status of bioremediation

Chemical class	*State of occurrence*	*Status of bioremediation*
Hydrocarbons and derivatives		
Gasoline, fuel oil	Very frequent	Established
PAHs	Common	Emerging
Alcohols, ketones, esters	Common	Established
Ethers	Common	Emerging
Halogenated aliphatics		
Highly chlorinated	Very frequent	Emerging
Less chlorinated	Very frequent	Emerging
Halogenated aromatics		
Highly chlorinated	Common	Emerging
Less chlorinated	Common	Emerging
Polychlorinated biphenyls		
Highly chlorinated	Infrequent	Emerging
Less chlorinated	Infrequent	Emerging
Nitroaromatics	Common	Emerging
Metals (Cr, Cu, Ni, Pb, Hg, Cd, Zn, etc.)	Common	Possible

fuel production and composition, the way of individual's disposal of wases, cost of products, and environmental quality.

However, employing bioremediation effectively in many application requires both further R&D and clarification of government policies, particularly in release of genetically-engineered organisms, a controversial issue, scientifically as well as politically. Most European countries as well as the U.S. and Japan have regulations that ban the deliberate release of genetically engineered microorganisms. However, it may be possible to use them in contained bioreactors. Most successful applications of bioremediation, such as the treatment of the shorelines contaminated by the Exxon Valdez oil spill, have relied on indigenous microorganisms and simple environmental modifications such as nutrient applications and aeration. The current and emerging applications of bioremediation are different in Europe and Japan than they are in U.S.A. Research in U.S.A. promises to contribute to the development of cost-effective solutions for bioremediation of contaminated sites. This new biotechnology will enable contaminated soils and waters to be cleaned up for reuse. In several European countries existing biological waste treatment technologies are to be extended to new areas for removal of harmful toxicants from air \and industrial waste streams. If current trends continue, the Netherlands and Germany are likely to become worldwide leaders in biotreatment of wastes by the end of the century.

Japanese research could aptly combine a quest for improved global environmental quality with a drive for energy self-sufficiency. Their efforts to reduce global warming, if successful, will have far-reaching effects. Japanese programmes to replace petroleum with biologically produced hydrogen and to replace gasoline-powered automobiles with cars that burn hydrogen would greatly reduce future buildup of carbon dioxide in the atmosphere. A change to hydrogen fuel produced by microbes would be a major new use of bioremediation for pollution prevention that would revolutionise not only the petrochemical and automotive industries but the worldwide economy.

14

MICROBES AND AGRICULTURE

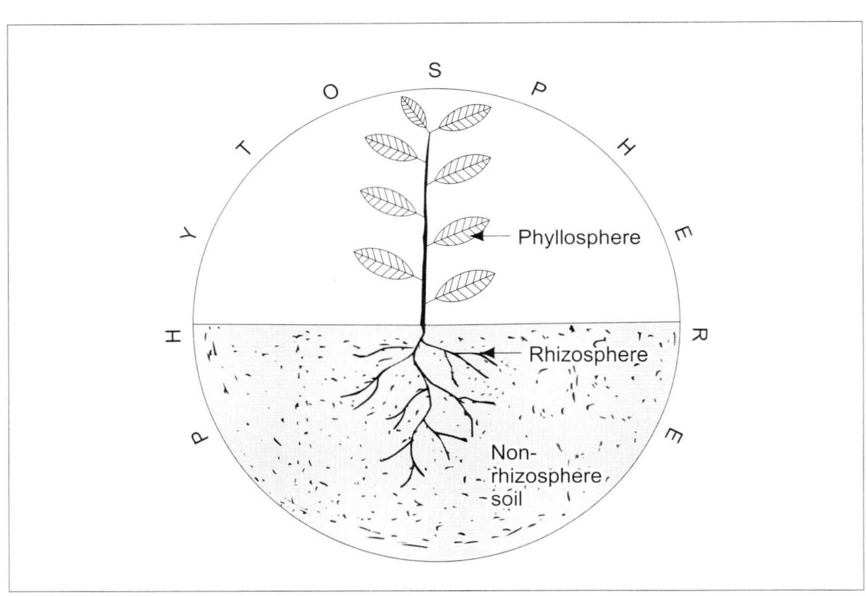

Chapter Outline

- *Biological Nitrogen Fixation*
- *Mycorrhizal Associations*
- *Biological Control of Plant Diseases*
- *Microbial Pesticides to Control Insects, Nematodes and Weeds*
- *Microbial Biodeterioration*

Plant and microbe are so intimately associated with each other that they are designated as **plant-microorganism system**. Plant surfaces, both above–as well as below-ground remain under the influence of activities of microorganisms. Surfaces of aerial plant part (shoot) and underground part (root) provide characteristic microhabitats for the growth of microbial communities. Plants too modify these two microhabitats in different ways. Plant-microbe interactions occur in these microhabitats influencing each other's life in different ways. Below the ground such interactions occur between plant roots and the microbes growing in the soil adjacent to living roots. The term **rhizosphere** was introduced in literature by Hiltner in 1904 for "the zone of soil around plant roots characterised with immense microbial activity". Microhabitat of the surface proper of the root is designated as **rhizoplane** (Fig. 14.2). Microbes also inhabit the root itself and are known as **endophytes**. Interactions between aerial plant surfaces (leaves, branches, flowers) and the microbes landing on them from the atmosphere have also been reported. In analogy to rhizosphere, the term **phyllosphere** was more or less simultaneously introduced in literature by Last (1955) in U.K. and Ruinen (1956) in Indonesia for the microhabitat around the living leaving. Since the microbes associated with leaves in fact are the forms isolated from their surfaces (landing from the surrounding air), a more appropriate term **phylloplane** was suggested for this microhabitat by Last and Deighton (1965), agreeing with a view expressed earlier by Kerling (1958, 1964). Some authors designate the two microhabitats, rhizosphere and phyllosphere together, with a collective term, **phytosphere** for the composite communities of microbes around the surface of underground and aboveground parts of the plant (Fig. 14.1).

Microbial-plant interactions, though do occur in both, rhizosphere as well as phyllosphere, historically more studies remained centered around the former. Nutrients for microbial growth are available in the substrates released by plant roots in rhizosphere and shoots in the phyllosphere. However, when compared with rhizosphere, phyllosphere is a harsh and stressed niche for microbial growth and multiplication. This has been the chief limiting factor in experimenting with the phyllosphere microbes in relation to improvement of plant growth, and most such studies have been made on rhizosphere microorganisms.

Substrates Released From Plant

Roots as well as shoots are surrounded by microorganisms which include viruses, bacteria, actinomycetes, fungi, algae, protozoa and nematodes. Microhabitats along with the plant and associated microorganisms thus constitute interacting systems in both, soil and the air, Accordingly, there can be recognised rhizosphere (soil) - plant (root) - microorganism system, below the ground, and phyllosphere (air) - plant

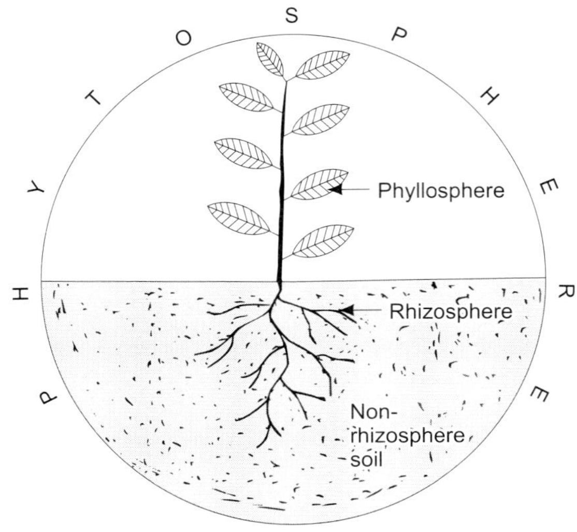

FIG. 14.1
A schematic representation of phytosphere

(shoot) - microorganism system, above ground. The components of each system interact with each other, which distinguishes the rhizosphere from the bulk of soil. Such microbial populations within rhizosphere and phyllosphere affect the plant growth in beneficial or detrimental ways. Synergistic interactions between plant and microbes are important in providing the nutritional requirements of both. Within the

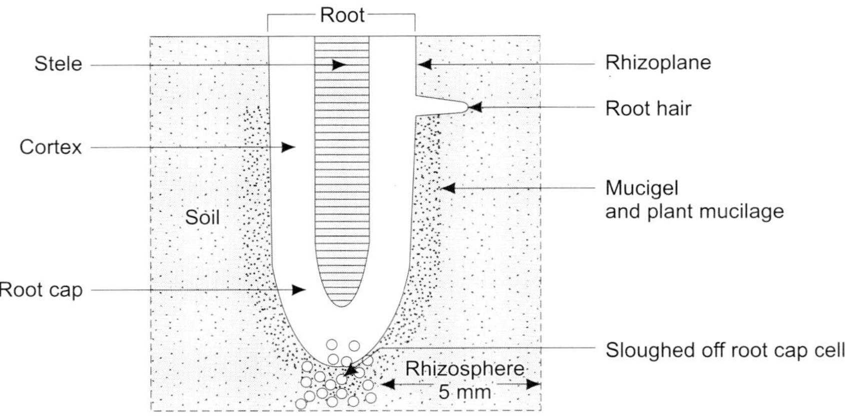

FIG. 14.2
A schematic representation of root structure and rhizosphere

rhizosphere, the plant roots exert a direct influence on the soil bacteria, known as **rhizosphere effect**. Likewise, the microbial populations in the rhizosphere have pronounced effect on the growth of the plant. As a result of these interactions there is **qualitative effect** and microbial populations reach much higher densities in the rhizosphere than in the free (non-rhizosphere) soil. There is also a **quantitative effect**. Short Gram-negative rods predominate in the rhizosphere, while Gram-positive rods and coccoid forms are less numerous in the rhizosphere than elsewhere in soil. The interactions of plant roots and rhizosphere microbes are based largely on interactive modifications of the soil chemical environment by processes such as water uptake by the plant; release of organic chemicals in soil by roots; microbial production of plant growth factors; and microbially mediated availability of mineral nutrients.

Within the rhizosphere, the microbes obtain required growth factors, such as sugars, vitamins and aminoacids from the plant roots. In turn, rhizosphere microbes have a marked influence on the growth of plant. Microbial populations in the rhizosphere benefit the plant by (i) removing H_2S, which is toxic to roots, (ii) increasing solubilisation of mineral nutrients, (iii) synthesising vitamins, amino ,acids, auxins, and gibberellins that stimulate plant growth and (iv) antagonising potential root pathogens through competition and antibiosis. **Allelopathic substances** produced by some microbes, may not allow the invasion of that habitat by other plants, thus help rendered by rhizosphere microbes to their host plants.

Substrates released from roots have many origins and have been originally classified by Rovira *et al* (1979) as follows:

Exudates. Compounds of low molecular weight that leak nonmetabolically from intact plant cells.

Secretions. Compounds metabolically released from active plant cells.

Lysates. Compounds released by the autolysis of older cells.

Plant mucilages. Polysaccharides from root cap, root cap cells, primary cell wall, and other cells.

Mucigel. Gelatinous material of plant and microbial origin.

Though, nature and chemistry may differ, various organic compounds are released from aerial surfaces of plant also in the phyllosphere. When compared with rhizosphere, the subject of phyllosphere began growing relatively quite late, and more information thus accumulated on rhizosphere-plant-microorganism system.

Bacteria including actinomycetes are the most abundant inhabitants of rhizosphere. Pseudomonads and gram-negative bacteria are especially competitive. Fungal plate counts are generally less than bacterial counts. But they are prevalent and equally important. Besides these protozoa, nematodes and microarthropods received much attention in rhizosphere. Many plant parasitic nematodes are also present.

In this chapter I will concentrate over the activities of microbial populations affecting plant growth in beneficial or detrimental ways in context with agriculture systems. I do not intend to write a detailed text on this subject. Rather I will restrict to the recent developments in the field of agricultural microbiology.

Several microbial activities in the rhizosphere are beneficial for plant growth. Three of such beneficial activities, biological nitrogen fixation by prokaryotic bacteria; enhanced plant phosphorus uptake through mycorrhizal associations of eukaryotic fungi; and biological control of root diseases by different microorganisms predominate the interacting system below soil.

BIOLOGICAL NITROGEN FIXATION

Biological nitrogen fixation (conversion of nitrogen gas to ammonia) is mediated by prokaryotes, including bacteria, cyanobacteria and the actinomycete *Frankia*. These nitrogen-fixers can exist as independent, free-living organisms or as part of complex interactions with other microbes, plants, and animals. Organisms that can utilise atmospheric nitrogen gas as their sole source of nitrogen for growth are known as **diazotrophs**.

There can be recognised **three** systems of biological nitrogen fixation: Free-living, associative symbiotic (involving free-living), and symbiotic systems based on the levels of association of diazotrophs with the plant. In terms of benefits to agriculture, the systems involving symbiotic relationships have been major nitrogen supply systems to plants. They are the major players in terms of importance to agriculture.

Free-Living Nitrogen Fixers

Nitrogen fixation occurs in a diverse array of prokaryotes including bacteria, actinomycetes and cyanobacteria. They include both autotrophic as well as heterotrophic forms utilising different terminal electron acceptors for energy generation (Table 14.1) Relatively, they fix small amounts of nitrogen (2-25 kg/ha/yr).

Associative Nitrogen Fixers

Some of the free-living diazotrophs have developed the ability to form associations with plants. They are found established on or in plant cells, where carbon becomes available to them. In return they fix nitrogen providing it to plant. This casual or associative symbiosis do not appear to require genetic interaction between the plant and the microbe and no morphological modifications develop. Examples of associative symbiosis include tropical grasses such as *Paspalum notatum* with *Azotobacter paspali*

TABLE 14.1 Representative genera of free-living nitrogen fixers

Oxygen relation	*Energy generation mode*	*Genera*
Aerobe	Heterotrophic	*Azotobacter, Beijerinckia, Acetobacter, Pseudomonas*
Facultative anaerobe	Heterotrophic	*Bacillus, Klebsiella*
Microaerophile	Heterotrophic	*Xanthobacter, Azospirillum*
Strict anaerobe	Autotrophic	*Thiobacillus*
	Heterotrophic	*Desulphovibrio, Clostridium*
Aerobe	Phototrophic	*Anabaena, Nostoc*
Facultative anaerobe	Phototrophic	*Rhodospirillum*
Strict anaerobe	Phototrophic	*Chlorobium, Chromatium*

where the microbe exists in grass rhizosphere. *Azospirillum* spp. are also found associated with a range of plants including sugarcane, rye, and sorghum. Besides rhizosphere, some of these microbes are also present in outer layers of root or even in internal tissues. *Acetobacter diazotrophicus* has been shown to fix nitrogen from within the internal cells of sugarcane. Attempts have been made to enhance crop production by inoculation with free-living *Azospirillum* and *Acetobacter*. They are able to fix upto 20 kgN/ha/yr.

Symbiotic Nitrogen Fixers

The major systems in this category are the following symbiotic relationships.

Rhizobia (bacterium)–Legumes

Frankia (actinomycete)–Nonlegume

Anabaena (cyanobacterium)–*Azolla* (fern)

Among these, rhizobia-legume symbioses are the major players. This is a formal symbiosis in which both parteners benefit. Gram-negative heterotrophic bacteria originally classified within the genus *Rhizobium* interacts with roots of legumes, forming root nodules. Nodules develop in response to the presence of specific soil borne rhizobia. The rhizobia themselves undergo physiological changes and are known as **bacteroids**, which conduct the process of nitrogen fixation. Grain legumes such as pea, bean and soybean can fix about 50% of their total nitrogen requirement, with rates of fixation up to 100 kg/ha/yr.

Originally all rhizobia were classified within the genus *Rhizobium*, and the species were identified on legume host basis. However, it became evident that some rhizobia could nodulate more than one host and different classification schemes evolved. The original genus *Rhizobium* is now divided into four genera and 16 species as follows:

(1) ***Rhizobium:*** *R. leguminosarum* - Three biovars: *trifolü* (*Trifolium*), *viciae* (*Pisum, vicia, Lathyrus, Lens*) and *phaseoli* (*Phaseolus*).
R. loti (*Lotus*), *R. tropici* (*Phaseolus, Leucaena, Macroptilium*), *R. etli* (*Phaseolus*), *R. galegae* (*Galega, Leucaena*),
R. haukuü (*Astragalus*), *R. ciceri* (*Cicer*),
R. mediterraneum (*Cicer*).

(2) ***Simorhizobium:*** *S. meliloti* (*Melilotus, Medicago, Trigonella*), *S. fredü* (*Glycine*), *S. Saheli* (*Sesbania*), *S. teranga* (*Sesbania, Acacia*)

(3) ***Bradyrhizobium:*** *B. japonicum* (*Glycine*), *B. elkanü* (*Glycine*), *B. liaoningense* (*Glycine*)

(4) ***Azorhizobium:*** *A. caulinodans* (*Sesbania*)

Root nodules develop as a result of interaction between the host and the rhizobium. Overall, rhizobia infect the growing nodule and ultimately inhabit it as nitrogen fixing bacteroids. The success of symbiosis is due to both plant and bacterial genes that are turned on sequentially during nodule initiation and development. These interactions are highly specific and only specific genetic combinations lead to successful **effective nodule** formation that can fix nitrogen. Two different groups of microbial genes are required for infection. For *Rhizobium* spp. most of these genes are plasmid borne, whereas for *Bradyrhizobium* spp, *Azorhizobium* spp. and *R. loti* they are chromosomal. The common nodulation genes (*nod ABC*) are found in all rhizobia. A fourth gene (*nod D*) is also common to all rhizobia but is the only *nod* gene expressed in absence of suitable host. It is believed that the product of *nod D* interacts with the appropriate plant host and initiates nodule formation (Fig. 14.3). Specifically, flavenoids are secreted by plant in the rhizosphere. If the correct symbiont is present, the flavenoids interact with the *nod D* product and cause expression of the *nod ABC* genes. Host-specific nodulation genes differ depending on the type of rhizobia, and more than 50 genes have been defined in different rhizobia.

Rhizobia travel along an infection thread produced by plant. As rhizobia are released from the infection thread, cell division occurs and a visible nodule can be seen (1-2 weeks after infection). Within the nodule, the rhizobia enlarge and elongate and change physiologically to forms known as **bacteroids**. The nodule also contains leghemoglobin, which protects the nitrogenase enzyme from the presence of oxygen.

Commercial legume crops are often aided in terms of nitrogen fixation through the application of rhizobial inoculants. Rhizobia of **desired characteristics** (Table 14.2) are introduced into the soil using suitable inoculants. They are impregnated into a peat-based carrier with about 10^9 rhizobia/g peat.

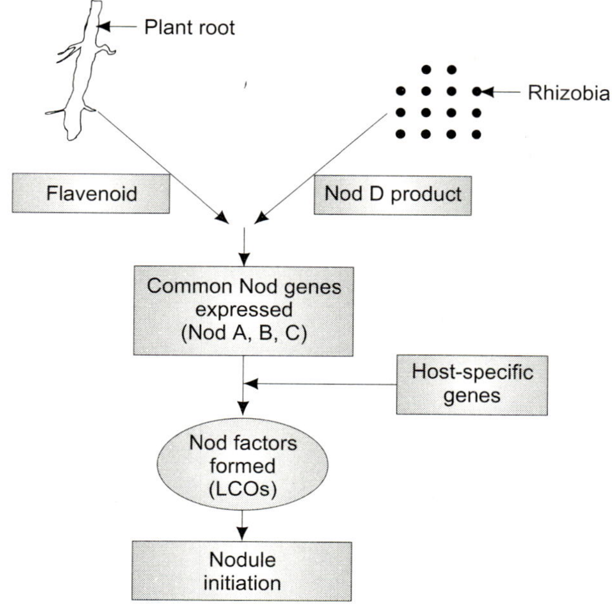

FIG. 14.3
Plant-rhizobia genetic interactions that initiate nodule formation

TABLE 14.2 Desired characteristics for rhizobia used as commercial legume inoculants

Characteristic	*Definition*
Infective	Capable of causing nodule initiation and development
Effective	Capable of efficient nitrogen fixation
Competitive	Capable of causing nodule initiation in presence of other rhizobia
Persistent	Capable of surviving in soil between crops in successive years

MYCORRHIZAL ASSOCIATIONS

Soil microbial communities besides uptake of nitrogen have also evolved mechanisms to enhance plant uptake of phosphates. The establishment and growth of most plants are enhanced by the presence of special group of fungi in soil that form close association with their roots. These fungi are called **mycorrhizal fungi**. In fact they are said to act as an extension of the plant root system, which helps in uptake of all nutrients, particularly phosphates. Mycorrhizal fungi naturally infect most plants but in some commercial crop systems, as establishment of forest tree seedlings (as pine), plants can be infected with known sensitive strains of fungi. Since phosphates have low solubility in soil solution, mycorrhizal fungi help in their solubilisation and subsequent uptake by roots. There are several types of mycorrhizal fungi. Basically **three** types of mycorrhiza are found in association with plants: (i) ectomycorrhiza, (ii) endomycorrhiza or VAM, and (iii) ericaceous-orchidaceous mycorrhiza. All the three types are shown in Fig. 14.4.

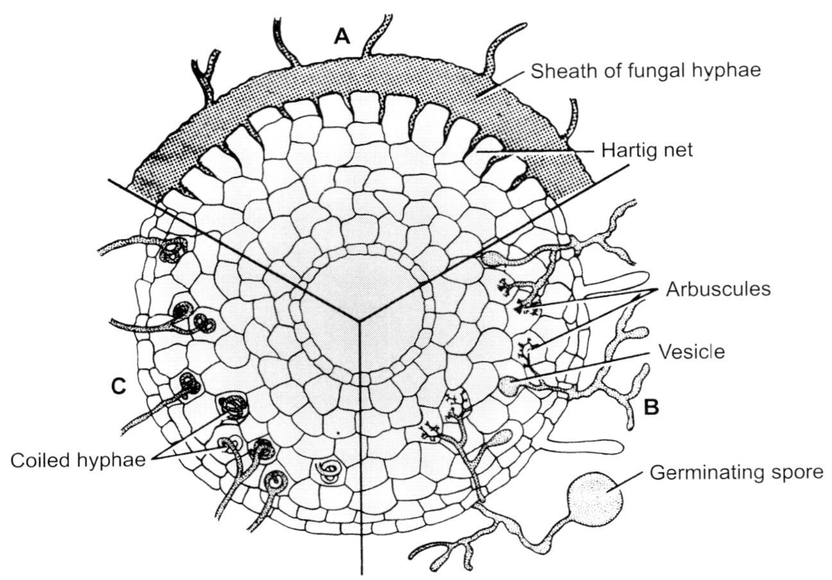

FIG. 14.4

Diagrammatic representation of three types of mycorrhiza shown in T.S. of a root. A-ectomycorrhiza of forest trees, B-endomycorrhiza or vesicular-arbuscular mycorrhiza (VAM) of many herbaceous plants, C-ericaceous-orchidaceous mycorrhiza (some hyphae digested intracellularly)

Ectomycorrhiza

Here the ultimate absorbing rootlets of the root system are completely surrounded by a distinct mantle or sheath of fungal tissue (thus aptly called sheathing mycorrhiza) from which hyphae penetrate between the outermost cell layer or layers of the root (Fig. 14.4A). The mantle is connected to the network of intercellular hyphae found in the root contex known as the Hartig net. About 3% of all seed plants have ectomycorrhiza, being particularly common in northern temperate forest trees, especially in Fagacae and Pinaceae but also occur in Betulaceae and Tiliaceae. In tropics, they are well-represented in Myrtaceae and Dipterocarpaceae.

Majority of the fungi involved are Basidiomycotina, especially Hymenomycete Agaricales. Over 100 species are shown to form ectomycorrhiza, majority of which are in the genera *Amanita, Boletus* and *Tricholoma,* but there are also many species in the genera *Cortinarius, Lactarius* and *Russula.* Most work on these myorrhiza is done by J.L. Harley. The fungus depends on the host for carbon source. These mycorrhiza are responsible for enhanced mineral ion uptake (N, P, K) and hydrolysis of organic phosphorus in soil. Their sheath serves as a reservoir of mineral ions.

Endomycorrhiza or Vesicular-Arbuscular Mycorrhiza (VAM)

Though a wide variety of endomycorrhizal associations exist in nature, we would consider here only **vesicular-arbuscular mycorrhiza (VAM)**. VAMs are produced by aseptate mycelial fungi which produce vesicles and arbuscules in plant roots after infection. They are found in bryophytes, pteridophytes, gymnosperms, excluding the Pinaceae and in virtually all families of angiosperms. They are of general occurrence in Gramineae, Palmae, Rosaceae and Leguminosae. But Chenopodiaceae, Cruciferae, Cyperaceae and Resedaceae are odd exceptions. In VAM fungi, hyphae produce very thick-walled, brown to black, balloon-like chlamydospores, 100-250 μm in diameter. They are produced singly or in clusters as the blown-out ends of hyphae. Loose aggregations of spores may also be formed within a poorly differentiated reproductive structure called a **sporocarp**. Chlamydospores are most common in soil, though zygospores may also be formed. The hyphae infect epidermal cells of root.

Arbuscules are formed inside the parenchymatous cells of cortex (Figs. 14.4B, 14.5). Arbuscules are infact complex haustoria. The hyphae may also form vesicles, arising as swellings on terminating hyphae or in intercalary position along a hypha. They are thin -or thick-walled, ovoid or spherical. Though there has been much controversy over the taxonomic position of VAM fungi, all are believed to belong to Endogonaceae (Mucorales, Zygomycotina). Most fungi belong to the genus *Glomus,* although some also to *Gigaspora* and *Sclerocystis.* Spores extracted from soil have been used to infect plants and have produced VAM mycorrhiza but so far these fungi have not been grown in pure culture. Chlamydospores will germinate on nutrient agar but the hyphae stop growing when the food supply in the spore is used up so they can not be sub-cultured. They are thus obligate biotrophs but are unusual in showing lack of host specificity. Spores from infected maize plant may be used to produce typical VAM in strawberry, onion, red clover and soybean.

VAM infection results enhanced growth of host plants and an increased uptake of phosphate from Soil. Since virtually all crop plants possess VAM and because they benefit so much from these in their mineral nutrition, research has been concentrated on these microbes. Much emphasis is being placed on their use as biological fertilizers.

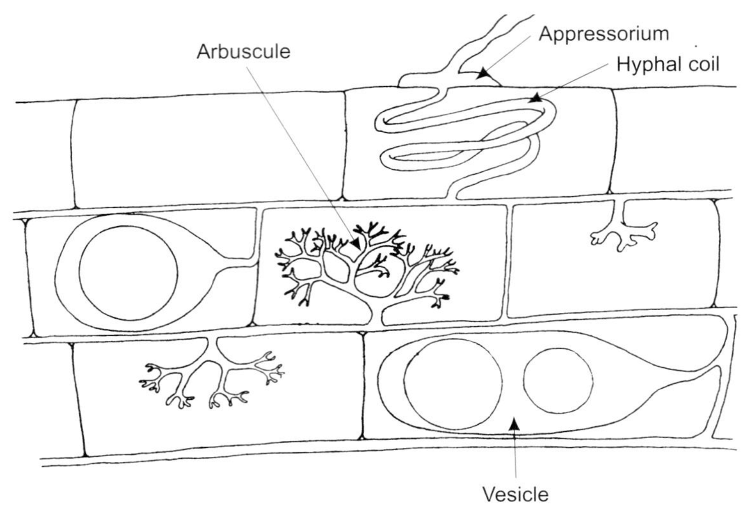

FIG. 14.5

Vesicular-arbuscular mycorrhiza, showing irregular external hyphae, appressorium, hyphal coil in the first cell penetrated, intercellular hyphae, vesicles and arbuscules in various stages of lysis

The VAM spores may be incorporated into pellets around seeds. Inoculation of seeds with VAM fungi may make it possible to grow plants in soils with very low available phosphate.

Since anyone fungus may infect a number of different plant species, a series of plants may become connected via the hyphae of their common VAM fungus. This could lead to **inter-plant transfer of carbon and mineral nutrients** via fungus. The fungus in VAM depends on its host for its carbon supply.

Ericaceous Mycorrhiza

They are another type of endomycorrhiza, which involve different kind of fungi. In plants of Ericaceae, the rootlets are covered by a very sparse, loose weft of dark brown, septate hyphae. From this weft some branches penetrate the cortical cells forming compact intercellular hyphal coils invaginating and enclosed by host plasmalemma (Fig. 14.6). After some time cell contents of host degenerate and these coils collapse. These mycorrhiza help in phosphorus and nitrogen uptake by plants. Most of the fungi involved are membres of Ascomycotina and Deuteromycotina

A number of genera in Ericaceae and plants of other families in Ericales possess mycorrhiza, different from the above. They are somewhat intermediate between ecto-

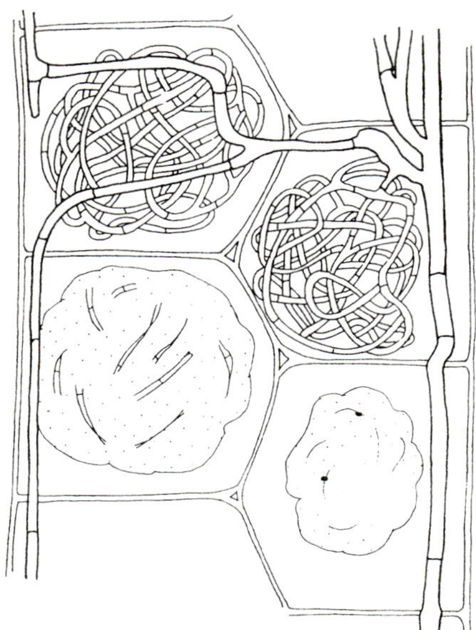

FIG. 14.6

Ericoid mycorrhiza, showing four cells of a rootlet with a loose weft of septate hyphae and hyphal coils, some undergoing lysis in the cortical cells

and endomycorrhiza and have been called **ectendomycorrhiza**. *Arbutus* and *Arctostaphylos* (Ericaceae) possess these mycorrhiza. Their short roots are swollen and are invested with a hyphal sheath. There is no Harting net as in sheathing mycorrhiza but intercellular coils develop in the outer cortical cells that are eventually lysed.

Orchidaceous Mycorrhiza

These fungi are much more specific than typical VAM fungi and infect only plants of the orchid family, most members of which are tropical. The hyphae form coils (like ericoid mycorrhiza) in cortical cells of plant roots (Fig. 14.4C). In this association this is the fungus supplying carbon source to plant. This is remarkable association in this respect. It is also of interest that many orchids are associated with *Rhizoctonia* spp., including *R. solani*, which are common plant pathogens.

Undoubtedly, mycorrhiza have evolved independently in several groups of fungi and in different types of plants. Ectomycorrhiza of forest trees are important in the plant nutrition. They play role in the uptake of mineral nutrients from soil. Most work on the roles of mycorrhiza in nutrient uptake has been focused on phosphorus, since this element normally exists as insoluble calcium phosphate or organic phosphates (as inositol phosphate) in soil and is, largely unavailble to plant roots not in immediate contact with these insoluble materials. In addition to increase in mineral uptake, mycorrhiza have other consequences also. For example, mycorrhizal infection is known to increase the availability of water to plants growing in dry habitats; to increase the degree of establishment and the efficiency of root nodules in legumes; and to offset or reduce plant disease. Foresters have paid much attention to the role of ectomycorrhiza in protecting trees from infection by the root pathogen, *Phytophthora cinnamomi.* Mycorrhiza are, therefore, significant for the plant pathologist, as well as for the botanist or microbiologist at large.

BIOLOGICAL CONTROL OF PLANT DISEASES

Microorganisms have been experimented for their ability to control soil-borne as well as air-borne diseases caused by various pathogenic microorganisms as well as insect and nematodal pests. During last few years, some success could be achieved in this area and some microbial preparations are even in use as viable and effective components of integrated disease and pest management programmes in agriculture and forestry.

Birth of Biological Control

Interest in biological control in fact first arose in 1920s and 1930s when some plant pathogens could be suppressed by introducing some antibiotic- producing microbes to the natural habitats. A turning point for research on biological control of plant pathogens came after a gap of more than 30 years when in 1963 an International Symposium on "Ecology of Soilborne Plant Pathogens-Prelude to Biological Control" was held at Berkeley, U.S.A. The outcome of this symposium was the book entitled "Ecology of Soil-Borne Plant Pathogens", edited by K.F. Baker and W.C. Snyder (1965). The first book devoted wholly to biological control, entitled "Biological Control of Plant Pathogens" by K.F. Baker and R.J. Cook appeared in 1974. The first Standing Committee on biological control was established within the American Phytopathological Society in 1976. The Beltsville Agricultural Research Centre (B.A.R.C.) Symposium held in 1981 was a landmark in the history of biological control where plant pathologists, entomologists, weed scientists and nematologists were brought together for the first time.

Biological control proved to be more successful todate in rhizosphere than in phyllosphere as not even one of the commercial biocides is so far directed against leaf or other aerial plant part pathogen. The limited economic success in phyllosphere appears to be due to the fact that as compared to rhizosphere, phyllosphere is a highly stressed niche, and there are practical problems in using antagonists in this niche. For instance, pronounced fluctuations in temperature, free moisture and relative humidity, intense radiation, relatively low and variable nutrient availability, presence of antimicrobial components of host origin (e.g. gallic acid) and scouring action of wind and rain in the phyllosphere make this microhabitat inhospitable for microbial growth and multiplicition (as compared to rhizosphere). Biological control, therefore, has always been more challenging in phyllosphere than in the rhizosphere. This seems to be the chief reason of slow progress in biological control in air infection courts.

Tactics of Biological Control

Eradication/Reduction of inoculum

Under some situations biological methods can be used to eradicate or reduce the pathogen inoculum. This is achieved through suppressive soils and hyperparasites. Several soil-borne pathogens, such as *Fusarium oxysporum* (vascular wilts), *Gaeumannomyces graminis* var. *tritici* (take-all of wheat), *Phytophthora cinnamomi* (root rot of several fruit and forest trees), *Pythium* spp. (damping-off) and *Heterodera avenae* (oat cyst nematode) thrive and cause severe diseases in some soils, called **conducive soils**, whereas they develop much less and cause much milder diseases in other soils, known as **suppressive soils**. In most cases, this failure to establish in suppressive soils has been attributed to the presence of several microorganisms, antagonistic to the pathogens in these soils. These antagonists do not allow the pathogen to reach high enough populations to cause severe disease by producing antibiotics, lytic enzymes, by competition for food or through direct prasitism of the pathogen.

Numerous kinds of antagonistic microbes have been found to increase in suppressive soils; most commonly, however, pathogen and disease suppression has been demonstrated for fungal antagonists, such as *Trichoderma, Penicillium* and *Sporidesmium,* or bacterial antagonists of the genera *Pseudomonas, Bacillus* and *Streptomyces.* Suppressive soil added to conducive soil can reduce the extent of disease by introducing microbes antagonistic to the pathogen. For example, soil amended with soil containing a strain of *Streptomyces* sp. antagonistic to *Streptomyces scabies* (potato scab), resulted in potato tubers significantly free from the disease. Suppressive, virgin soil has been used to control *Phytophthora* root rot of papaya by planting papaya seedlings in suppressive soil placed in holes in the orchard soil which was infested with the root rot fungus, *Phytophthora palmivora.* Continuous monoculture of the same crop in a conducive soil, after some years of severe disease, eventually leads to suppression of the disease, through increased population of antagonistic microbes. For example, continuous cultivation of wheat or cucumber leads to reduction of take-all of wheat and *Rhizoctonia* damping-off of cucumber, respectively. Pasteurisation of the soil at 60°C for 30 minutes completely eliminates the suppressiveness of the soil. Some type of soil suppressiveness also develops after crop rotation of appropriate crops for sufficient duration.

Hyperparasites proved ideal organisms for several soilborne as well as aerial plant pathogens. **Soilborne pathogens** could be eliminated/reduced by using appropriate hyperparasites, both fungi and bacteria. The mycelium and resting spores (oospores) or sclerotia of several plant pathogenic soil fungi such as *Pythium, Phytophthora, Rhizoctonia, Sclerotinia* and *Sclerotium,* are invaded and parasitised (mycoparasitism) or are lysed (mycolysis) by several fungi (non-pathogens). These fungi, including some oomycetes, chytridiomycetes, and hyphomycetes, and some pseudomonads and actinomycetous bacteria infect the resting spores of several pathogenic fungi. Among the most common mycoparasitic fungi are *Trichoderma* spp., mainly *T. harzianum,* which parasitises mycelia of *Rhizoctonia* and *Sclerotium,* inhibits the growth of many other fungi like *Pythium, Phytophthora, Fusarium* and reduces the diseases caused by most of these pathogens. Similar other mycoparasitic fungi are listed below:

Hyperparasite(s)	*Target pathogen(s)*
Laetisaria arvalis (*Corticium* sp.)	*Rhizoctonia, Pythium*
Sporidesmium sclerotivorum, Glicoladium virens and *Coniothyrium minitans*	*Sclerotinia sclerotiorum* and several other *Sclerotinia* spp.
Talaromyces flavus	*Verticillium* (wilt)
Pythium spp.	*Phytophthora* spp. and other *Pythium* spp.
Pichia gulliermondii	*Botrytis, Penicillium*

Besides fungi, bacteria of the genera *Bacillus, Enterobacter* and *Pseudomonas* are known to parasitise and/or inhibit the soilborne pathogenic fungi like *Sclerotium ceptivorum, Phytophthora* sp., *Pythium* sp. and *Gaeumannomyces graminis tritici,* whereas the mycophagous nematode, *Aphelenchus avenae* parasitises *Rhizoctonia* and *Fusarium.*

Several fungi are known to parasitise the fungal pathogens of **aerial plant parts** also. Some examples of hyperparasites of pathogens of aerial plant parts are:

Hyperparaslte(s)	*Target pathogen(s)*
Tuberculina maxima	*Cronartium ribicola* (white pine blister rust)
Darluca filum, Verticillium lecanii	Several rusts
Ampelomyces quisqualis	Several powdery mildews
Tilletiopsis spp.	*Sphaerotheca fuliginea* ((cucumber powdery mildew))
Nectria inventa, Gonatobotrys simplex	*Alternaria* spp.

Protection by introduced antagonists

As with chemicals, strategies have also been evolved during recent past for direct protection of plants through the use of biologicals, mainly microbial antagonists (fungal as well as bacterial). There are several cases of successful biological control, only for soil borne pathogens so far, and it is likely that success could also be achieved for aerial plant pathogens in near future. Biological control practices for direct protection involve the deployment of antagonistic micoorganisms at the infection court before or after infection. Although hundreds, possibly thousands of microbes are shown to inhibit the growth of plant pathogens in the laboratory, greenhouse, or field and to provide some protection from diseases, so far only a few, hardly six or seven are registered for commercial use. These include **fungal** as well as **bacterial** antagonists. Several fungi have been used in the biocontrol of both, soilborne as well as airborne disease of plants, and even postharvest diseases.

Though in several instances, fungal antagonists have been shown to suppress soilborne diseases, there are only few cases where success has been achieved. One such discase is biocontrol of root and butt rot of conifers. This is the classic, most successful story of biological control of a soilborne disease by using a fungal antagonist. *Heterobasidion annosum* (= *Fomes annosus*), the cause of root and butt rot of conifers infects freshly-cut pine stumps and then spreads into the roots. Through these infected roots, it then spreads into the roots of standing trees, which are killed by it. To control this disease, the stump surface of pine is inoculated with oidia of the antagonistic fungus, *Phlebiopsis (= Peniophora) gigantea* immediately after the felling of the tree. *P. gigantea* occupies the cut surface and then spreads through the stump into the lateral roots. There it successfully competes with and excludes the pathogen, *Heterobasidion* in the stump, thereby protecting the trees. The oidia are applied to the cut surface either as a water suspension or as a powder, or they are added to the lubricating oil placed on the chain saw, and thus deposited on the surface as it is cut. This method is in use at commercial scale particularly in U.K. and Europe.

Though, no commercial success has so far been achieved in the biological control in phyllosphere, there are instances of effective biocontrols of aerial plant pathogens using fungal antagonists. One such example is biocontrol of chestnut blight. Chestnut blight, caused by the fungus *Cryphonectria* (*Endothia*) *parasitica* is controlled naturally in Italy, and artificially in France through inoculation of cankers, caused by the normal pathogenic strains of the fungus, with hypovirulent strains of the same fungus. The hypovirulent strains carrying virus like double-stranded RNAs (dsRNAs) apparently limit the pathogenicity of the virulent strains. The dsRNAs apparently pass through mycelial anastamoses from the hypovirulent to the virulent strains, the latter are rendered hypovirulent, and the canker development slows down or stops completely. In USA also the control is limited so far to experimental trees, and the work is in progress.

Like fungi, several antagonistic bacteria have also proved effective biocontrol agents for soilborne, airborne as well as postharvest diseases.

Though a large number of bacteria have been demonstrated to effectively reduce the soil-borne diseases, only a few of them could so far become available on commercial scale (three/four preparations). These, and some other instances of successful biocontrol using bacterial antagonists are given below:

Crown gall of pome, stone, and several small fruits (grapes, strawberries) and ornamentals, caused by the bacterium, *Agrobacterium tumefaciens* is being controlled commercially by treating the seeds, seedlings, and cuttings with **Galltrol**, a suspension of strain K84 of the related but nonpathogenic bacterium, *Agrobacterium radiobacter*. This antagonist produces an antibiotic, bacteriocin, called **agrocin 84**. The bacteriocin, specific against related bacteria, selectively inhibits most pathogenic agrobacteria that arrive at surfaces occupied by strain K84.

Damping-off and rots of several crops, including cotton has been successfully controlled at commercial scale by using preparations of bacterial antagonists. For example, rhizobacteria, primarily of the *Pseudomonas fluorescens, P. putida, P. cepacia* and *P. aureofaciens* groups, applied to seeds, seed pieces, and roots of plants resulted in the reduction of damping-off and soft rot, and increased yield in several crops.

While handling research project on biological control of bacterial rots of vegetable crops including soft rot and black leg of potato (caused by *Erwinia carotovora* pv. *atroseptica*), researchers in USA incidentally came across with a distinct group of bacteria, the fluorescent pseudomonads. The plant roots, instead of any conventional antagonist consistently yielded such fluorescent pseudomonads. While presenting their findings at the Fourth International Conference on Plant Pathogenic Bacteria, held at Angers in France, Suslow *et al* (1978) and Kloeper and Scroth (1978) more or less simultaneously proposed the term **rhizobacteria** for such bacteria in the rhizosphere that aggressively colonised the root system. These bacteria are shown to stimulate the plant growth in greenhouse and field trials. Such rhizobacteria were designated as plant-growth-promoting-rhizobacteria (PGPR) by Suslow *et al* (1978) for potato and by Kloeper and Scroth (1978) for radish. These bacteria suppressed the disease by competing with pathogenic bacteria through production of siderophores. The characteristic of **aggressive colonisation** distinguishes rhizobacteria from traditional rhizosphere and rhizoplane bacteria. Rhizobacteria are said to have co-evolved with the plants. Formulations of *Pseudomonas fluorescens,* sold in the market as **Dagger G** is used to control *Rhizoctonia* and *Pythium* damping-off of cotton. Preparations of *Bacillus subtilis,* is being sold as **Kodiak**, which is used as a seed-treatment for the control of damping-off and rot of several crops including potato, sugarbeet and wheat.

Besides the above three commercial formulations of bacterial antagonists to control crown gall, damping-off and rot of different plants, several other soilborne diseases could also be successfully controlled by using bacterial antagonists. The most common, from among many, soilborne diseases controlled by soil borne bacteria are damping-off and root rot diseases caused by *Pythium, Phytophthora, Rhizoctonia, Fusarium* and *Gaeumannomyces. Bacillus cereus,* especially its strain UW 85 provides effective control of damping-off of legumes.

Though, none of the biocontrol in phyllosphere could so far be developed on commercial scale, there are several antagonistic bacteria which significantly suppressed the diseases of aerial plant parts. One such successful case is biological control of bacteria-mediated frost injury of plants. Normally frost-sensitive plants are injured when temperatures drop below 0°C due to ice formation within their tissues. Small volumes of pure water can be supercooled even to –10°C or below without ice formation, provided that no catalyst centres or nuclei are present to influence ice formation. However, certain strains of at least three species of epiphytic bacteria (*Pseudomonas syringae, P. fluorescens* and *Erwinia herbicola*), common in the phylloplane of so many plants, serve as ice-nucleation-active (INA) catalysts for ice formation at temperatures as high as – 1°C. Such bacteria make up a small proportion (0.1-10%) of the phylloplane (leaf surface) bacteria. By isolating, culturing, mass producing and applying non-ice-nucleation-active (non-INA) strains of these bacteria on the plant surfaces, it has been possible to reduce and replace large numbers of INA bacteria. This protects frost-sensitive plants from injury at temperatures at which untreated plants may be severly injured.

MICROBIAL PESTICIDES TO CONTROL INSECTS, NEMATODES AND WEEDS

A range of microorganisms could be identified which have potential of their use as microbial pesticides to protect plants from insect and nematode pests as well as control weeds. We would focus on possible use of various microbes i.e. viruses bacteria, and fungi being used in pest control including weeds. Products of these microbes are patented and registered for commercial production and use in field.

Microbial populations can be used directly for controlling plant and animal pests. Populations of pathogenic or predatory microbes that are antagonistic toward a particular pest population provide a natural means of pest control. Preparations of such antagonistic microbial populations are called **microbial pesticides**. Microbial control methods have been developed for management of arthropod pests, especially insects. A microbial pesticide should be harmless to man and other valued plant and animal populations.

Microbial Insecticides

The greatest commercial impacts of biocontrol agents have been made in the insecticide markets. The most successful biocontrol agent has so far been the insecticidal bacterium, *Bacillus thuringiensis,* whose sales in forestry, agriculture and public health went much higher. Viruses, bacteria, and fungi have been used as microbial insecticides.

Viral insecticides

Pathogenic viruses possess the potential for use as pesticidal agents. Viruses attack insects and other arthropods. The specificity of virus-host relationship makes them ideal for this purpose. They have been reported from approximately 600 insect species. The most thoroughly studied viruses are, (i) **nuclear polyhedrosis viruses (NPV)**, that develop in the host-cell nuclei; the virions are occluded singly or in groups in polyhedral inclusion bodies, (ii) **cytoplasmic polyhedrosis viruses (CPV)**, that develop only in the cytoplasm of host midgut epithelial cells; the virions are occluded singly in polyhedral inclusion bodies, and (iii) **granulosis viruses (GV)**, developing in either the nucleus or cytoplasm of host fat, tracheal or epidermal cells; the virions are occluded singly or rarely in pairs in small occlusion bodies called capsules. Baculoviruses are perhaps the most carefully studied insect viruses and they include NPV and GV. Pathogenic baculoviruses have been found principally for Lepidoptera, Hymenoptera and Diptera. Viruses have been used in attempts to control outbreaks of several insect pests including gypsy moths, Douglas fir tussock moths, pine caterpillars, red-banded leaf rollers (pest of apples), spruce budworms, codling moths (a pest of apple, walnut and other deciduous fruits), alfalfa caterpillars, cabbage white butterflies, cabbage loopers, cotton bollworms, corn earworms, tobacco budworms, tomato worms, etc. Table 14.3 shows some viral insecticides registered for use against some such insect pests.

TABLE 14.3 Viral insecticides

Insect pests	*Viruses*	*Registered trade names*	*Countries*
Tobacco budworm (*Heliothis virescens*), Cotton bollworm (*H. zea*)	NPV	ELCAR	USA
Gypsy moth	NPV	GYPCHEK	USA
Douglas fir	NPV	TM-BIOCONTROL-1	USA
tussock moth		VIRTUSS	Canada
European sawfly	NPV	NEOCHECK-S	USA
		VIROX	U.K.
Red-headed sawfly	NPV	LECONTVIRUS	Canada
Pine caterpillar	CPV	MATSUKEMIN	Japan
Insects of food crops as codling moths	GV	MATEX GRANUPOM	Switzerland Germany

An interesting example of the use of viral pesticides is the attempt to control rabbit populations of Australia with myxoma virus.

Bacterial insecticides

There are several bacterial pathogens of insects that are being used at present or have potential for use as insecticides. They include endospore-forming *Bacillus* and *Clostridium* spp. as well as non-endospore forming species of *Pseudomonas, Enterobacter, Proteus, Serratia* and *Xenorhabdus*. Of these *Bacillus thuringiensis* has been most extensively used. Commercial preparations of *B. thuringiensis* are registered by more than 12 manufacturers in five countries for use in several agricultural crops, forest trees, and ornamentals for control of various insect pests (Table 14.4). This bacterium has been tested successfully against more than 150 insect species. Four separate toxins are produced by *B.t.* Several vegetable insect pests have been successfully managed by this bacterium. About 16 formulations based on exo- and endotoxins are used in USA, France, Germany, U.S.S.R. and Czechoslovakia. More than 30 subspecies of *B.t.* have been identified and called serotypes. Some registered products of *B.t.* are, Thuricide, Sporeine, Condor, Cutlass, Foil and Invade, mostly in U.S.A. In India also, trials have been made for thuricide against insect pests of lac, cruciferous crops and white grubs of sugarcane and 0.4% thuricide has been found superior over DDT, malathion and endrin.

TABLE 14.4 Some crop pests being controlled by registered *Bacillus thuringiensis* products in U.S.A.

Pest	*Crop*
	Vegetable and field crops
Alfalfa caterpillar	Alfalfa
Artichoke plume moth	Artichokes
Bollworm	Cotton
Cabbage looper	Beans, broccoli, cabbage, cauliflower, celery, cotton, cucumber, lettuce, melons, potato, peach, tobacco
European corn borer	Sweet corn
Cabbage worm	Broccoli, cabbage, cauliflower
Tobacco budworm	Tobacco
Tomato hornworm	Tomatoes
	Fruit crops
Fruit tree leaf roller	Oranges
Orange dog	Oranges
Grape leaf folder	Grapes
	Shade trees, Ornamentals
California oakworm	
Gypsy moth	Several trees and ornamentals
Winter moth	

Of particular interest is the potential of *B. thuringiensis israelensis* (BTI) to control mosquito vectors of malaria. Unlike DDT, it is environmentally safe and there is no resistance shown in mosquito for this bacterium. Testing of BTI for malaria-carrying mosquito has been successful. According to WHO the results of tests in Africa against blackfly, the carrier of widespread river blindness have been excellent.

In India, trials were made for *B.t., B. popilliae* and *Serratia mariscens* to control insect pests of cotton and sugarcane. Success is achieved in control of cotton bollworm pest.

Fungal insecticides

Fungal insecticides could become most common and effective means of control of insect pests in some countries, chiefly in ex-USSR. Products of the entomogenous fungi have been used for insects of field crops, forest trees as well as horticultural and vegetable crops grown in greenhouses. Several preparations have been produced, formulated and used commercially, chiefly in ex-USSR and developing countries like Brazil, Cuba and Israel. In western world fungal products have not yet been as successful as viral and *B.t.* insecticides. Different kinds of formulations have been developed and applied in different ways to insect pests. Though insect mycoses are caused by members of every class of fungi, most studies on entomogenous fungi have been concerned with species of the genera, *Aschersonia, Beauveria, Metarhizium, Verticillium, Hirsutella, Coelomomyces* and *Entomophthora*. Table 14.5 shows examples of fungi used as insecticides.

TABLE 14.5 Entomogenous fungi used as insecticides

Insect pest	*Fungus*	*Registered product, if any*	*Country*
Glasshouse whitefly (many crops)	*Aschersonia aleyrodis*	Aseronija	The Netherlands ex-U.S.S.R.
Colorado potato beetle, Codling moth, European corn borer, Pine caterpillars	*Beauveria bassiana*	Boverin (Biotrol in U.S.A. but could not be registered)	ex-U.S.S.R., China, France,
Green leafhoppers, Chrysomelid beetles, Rice back bug		Boverol	Czechoslovakia, Japan, Brazil
European cock-chafer, Black vine weevil	*B. brongniartii*		France (integrated programme)
Lucerne aphid	*Entomophthora sphaerosperma (Zoophthora radicans)*		Australia, U.S.A., Israel
Citrus rust mite	*Hirsutella thompsonii*	Mycar (but production ceased in 1985)	U.S.A. (used in integrated programmes)
Spittle bug of sugar cane, Pasture cockchafer, Black wine weevil, Coconut pests	*Metarhizium anisopliae*	Metaquino Metabiol	Brazil Australia Japan
Several Lepidoptera on different crops, including velvet bean caterpillar of soybean	*Nomuraea rileyi*		Southern U.S.A.
Aphids of glasshouse crops, Whitefly of glasshouse crops, Soft green scale (on coffee)	*Verticillium lecanii*	Vertalec, Mycotal (both withdrawn in 1986)	Britain (mostly integrated) programme) India

Microbial Nematicides

Most studies have been made with fungal nematicides. Classical nematode-trapping fungi belonging to the genera, *Arthrobotrys, Dactylaria, Dactylella* and *Monacrosporium* have been studied in trials to control nematode genera *Meloidogyne, Heterodera* and *Rotylenchulus,* attacking mostly vegetable crops. These nematodes cause cyst and root knot diseases in these plants. There have been some limitations in the use of classical nematophagous (trap-forming) fungi for the control of these nematodes. It is difficult to manage fungi so that periods of nematode migration and trap formation coincide. Another group of fungi, the soil fungi have been found more ideal nematicides. These are "opportunistic fungi" such as *Verticillium chlamydosporium, Dactylella oviparasitica* and *Paecilomyces lilacinus* that also attack eggs and young females of cyst and root-knot nematodes. Of these *P. lilacinus* has attracted much attention as it is almost ubiquitous in tropical and substropical soils.

Microbial Herbicides

Though viruses, bacteria as well as fungi have been studied for use as herbicides, fungi could be found most suitable for the purpose. In classical strategy fungus parasitic on a particular weed plant is introduced into a new area in which the pathogen is not known to attack that weed. Mostly rust fungi have been used. These include *Puccinia chondrillina* to control rush skeletonweed in Australia, and West U.S., and *Phragmidium violaceum* to control European blackberry in Chile. Recently, however, a new strategy **mycoherbicide** or **inundative strategy**, has been developed. In this method, the weed plant is repeatedly inoculated with inoculum doses of the pathogenic fungus. Several products of different fungi have been developed and are in use on commercial scale in different parts of the world. Table 14.6 shows some fungal herbicides registered for use against different weed plants through mycoherbicide strategy.

TABLE 14.6 Some fungal herbicides

Target weed	*Fungus*	*Registered product*	*Country*
Northern jointvetch	*Colletotrichum gloeosporioides* f.sp. *aeschynomene*	COLLEGO	USA
Milkweed vine	*Phytophthora palmivora*	DE VINE	USA
Velvel leaf	*Colletotrichum coccodes*	VELGO	USA, Canada
Dodders	*Colletotrichum gloeosporioides* f.sp. *cuscutae*	LUBOA	China
Round-leaved mallow	*Colletotrichum gloeosporioides* f.sp. *malvae*	BIOMAL	USA, Canada
Sicklepod	*Alternaria cassiae*	CASST	USA.

In all the above cases of microbial pesticides, it has been found more desirable to use biological control methods in combination with other regulatory, cultural, physical and chemical methods. This led to the development of **Integrated Pest Management (IPM)** programmes, now undertaken in plant protection against pests and weeds.

Commercial Reality

In the scientific literature there are thousands of papers on the isolation and in vitro screening of microorganisms against a plethora of plant pathogens. It is a sad fact that very few of these potential biocontrol agents could actually proceed to commercialisation. It seems it is relatively easier to select for antagonism between two organisms in an in vitro system, but it is much harder to develop a biocontrol agent to be effective in the field. Very rarely in routine academic research papers, any consideration is given to the suitability of a microbe to mass production, formulation and persistence in the field.

Like chemical pesticide, a biocontrol agent can not be considered for development unless its projected market could cover costs of discovery, development and registration plus make a reasonable profit to recycle into the system. According to Lethbride, in 1989, the total global market for biocontrol agents and other microbial based products amounted to about $ 48 million per annum, of which about 50% was accounted for by *Bacillus thuringiensis*, used as bioinsecticide and 18% by *Rhizobium*. The bulk of the remaining section of the market was accounted for by *Agrobacterium radiobacter* (used as biobactericide). Till 1988, only 5% of the deliberate releases of biocontrol agents could actually achieve their aim. The features of successful biocontrol agents and their scientific and commercial requirement have been considered by Powell and Faull (1989). They have included bioinsecticides, bioherbicides as well as biofungicides, with details of fungal insecticides as much is known of their formulation and commercial technology. It has been indicated that biocontrol agents have promising future, and may offer a potential solution to the dilemma from both the environmentalists as well as industrialists. The environmental impact of biocontrol agents is assumed to be less than that of agrochemicals. Their usage is presumed to reduce the global consumption of chemical pesticides, with little resistance problems in target organisms. Biocontrol agents can also be used in integrated pest management (IPM) schemes. On the industrial side, attractions include cheaper discovery and registration of biocontrol agents relative to chemical pesticides, with a consequently shorter lead time.

Biocides have been developed, registered and some even commercialised for use in agriculture and forestry. A number of microbial biocides have so far been developed during recent past originating from a range of microbes - viruses, bacteria and fungi, to control pathogens, insect and nematode pests as well as weeds. These include viral insecticides, bacterial biocides and fungal biocides. However, the commercial reality has so far been disappointing. Desite decades of research (about nine in rhizosphere and four in phyllosphere) only six or seven reasonably effective products could be commercialised for biological control of target pathgens (Table 14.7).

Biological control proved to be more successful todate in rhizosphere than in phyllosphere as not even one of the commercial biocides is so far directed against an aerial pathogen. For soil-borne plant pathogens the formulations containing bacteria, *Pseudomonas fluorescens* (Dagger-G), *Agrobacterium radiobacter* (Gallex or Galltrol) and *Bacillus subtilis* (Kodiak) received registration from the Environmental Protection Agency and most commonly used in USA. The alginate pellet formulation containing *Gliocladium virens*, developed in the Biological Control of Plant Diseases Laboratory (BPDL), United States Department of Agriculture (USDA), Beltsville, Maryland, USA for reduction of damping-off diseases of ornamentals caused by *Phytophthora* sp. and *Pythium ultimum* was the first product (GlioGard) to contain a biocontrol fungal agent to control soil-borne plant pathogens, registered for commercial use in USA. Two

TABLE 14.7 Commercial microbial biocides used in agriculture and forestry

Trade name of the biocide used for	*Biocontrol agent*	*Targeted against*	*Countries*
Pathogens	***Fungi***		
P.g. suspension	*Phlebiopsis (= Peniophora) gigantea*	Root and butt rot of freshly cut pine stumps (*Heterobasidion annosum*)	U.K., Europe
GlioGard (SoilGard)	*Gliocladium virens GL-21*	Damping-off of seedlings of ornamentals and vegetables (*Pythium* spp., *Rhizoctonia solani, Phytophthora* sp)	U.S.A
?	*Trichoderma viride*	Wound of fruit trees (*Chondrostereum purpureum*)	U.K., Europe
F-Stop	*Trichoderma harzianum*	Damping-off of seedlings, several rots and blights (*Rhizoctonia solani, Pythium* spp)	U.S.A
BINAB-T	*Trichoderma horzianum/ T. polysporum*	Wood decay (*Chondrostereum purpureum* and other basidiomycetes)	U.S.A
Aspire	*Candida oleophila*	Postharvest decay in citrus and apples (*Monilinia fructicola* and several other fungi)	U.S.A
	Bacteria		
Gallex or Galltrol	*Agrobacterium radiobacter* K-84	Crown gall of stone fruits (*Agrobacterium tumefaciens*)	U.S.A
Dagger-G	*Pseudomonas fluorescens*	Damping-off of seedlings of field crops and vegetables (*Pythium* spp., *Rhizoctonia solani*)	U.S.A
Kodiak	*Bacillus subtilis* GB03	Damping-off and soft rots of cotton and legumes	U.S.A
Bio-Save	*Pseudomonas syringae* (two strains)	Postharvest decay in citrus, apple, pear (*Monilinia fructicola* and several other fungi)	U.S.A
Insects pests	***Fungi***		
Metaquino, Metabiol	*Metarhizium anisopliae*	A range of insects, chiefly sugarcane pests	Brazil, Australia
Boverin	*Beauveria bassiana*	Several insects, mainly of forest trees	ex-USSR, China
	Bacteria		
Thuricide, Biotrol	*Bacillus thuringiensis* (several strains)	A range of insects including mosquito, mites	Several, mainly in U.S.A., Canada
MVP, M-Trak	Recombinant *Pseudomonas fluorescens* (*Bt* ICP genes)	A range of insects	U.S.A.
Weeds	***Fungi***		
Collego	*Colletotrichum gloeosporio-ides* f. sp. *aeschynomene*	Northern jointvetch (*Aeschynomene virginica*) in rice and soybean fields	U.S.A.
De Vine	*Phylophthora palmivora*	Milkweed vine (*Morrenia odorata*) in citrus orchards	U.S.A.
Casst	*Alternaria cassiae*	Sicklepod (*Cassia obtusifolia*) in peanut and soybean fields	U.S.A.
BioMal	*Colletotrichum gloeosporioi-des* f.sp. *malvae*	Roundleaf mallow (*Malva pusilla*) in small-grain fields	U.S.A.

other fungal biocides, containing *Trichoderma* spp (F-Stop and BINAB-T) were also developed in U.S.A. for commercial use against damping-off of seedlings, several rots and blights and wood decay. Two fungal products are registered for commercial use against diseases in U.K. and Europe.

Use of antagonists gave only a few economic successes and this is quite apparent in phyllosphere. This limited success in phyllosphere appears to be due to the simple reason that as compared to rhizosphere, phyllosphere is a highly stressed niche.

For insect pests, formulations of two fungi, *Metarhizium anisopliae* and *Beauveria bassiana* under the trade names Metaquino, (Metabiol) and Boverin are in commercial use for a range of insect pests of agricultural and forestry. Besides these, formulations of the bacteria, *Bacillus thuringiensis* and a recombinant *Pseudomonas fluorescens* are used at commercial scale to control insect pests (Table 14.7).

For biological control of weeds (Table 14.7) only fungal products could so far become successful on commercial scale. These include Collego and De Vine, obtained from *Colletotrichum gloeosporioides* f. sp. *aeschynomene* and *Phytophthora palmivora* respectively. These are used to control weeds of rice, soybean and peanut fields in USA.

Harmful Effects of Microbes on Agriculture

Of course the major source of detrimental microorganisms affecting plant growth is, plant pathogens. A range of microbes: viruses, bacteria, mycoplasma, fungi, protozoa and nematodes cause important groups of both, soil-borne and air-borne diseases. Study of the cause, mechanism and control of such diseases is the subject of independent discipline, plant pathology.

However, there are other ways also in which these microbes, particularly fungi and bacteria adversely affect the commercial value of plant products. They are responsible for **biodeterioration** of plant products, including agricultural produce. Some fungi contaminate feed/food of animals and humans due to production of potent **mycotoxins**. Microbes are largely responsible for biodeterioration of various agricultural produce (seed, vegetable, fruit etc.) during storage, transportation and marketing. Other materials that may be deteriorated by microbes include wood, paper, paint, leather, opticals, textiles etc. Such activities of microbes harmful to agriculture are of direct concern to environment. I would present these detrimental effects of such microbes on agricultural produce briefly here.

MICROBIAL BIODETERIORATION

Biodeterioration is the chemical or physical alteration of a product that decreases its usefulness for its intended purpose, caused by microorganisms or their enzymes.

It should not be confused with **biodegradation** which is a process of chemical breakdown of a substance to smaller products caused by microorganisms or their enzymes. Though both bacteria and fungi are involved in biodeterioration, the latter are relatively better known and more studied for mechanism of the process and factors which favour their growth in substrate. Both, cellulosic and non-cellulosic materials are attacked. Cellulosic materials include wood, used as timber or as wood products, such as paper and board and other plant fibres such as cotton, flax, jute, hemp and sisal used as rope, cordage, textiles and packing and filling materials. Non-cellulosic materials include plastic, glass, electric equipment, fuel, paints, photographic and paint films leather, glue etc. Fungi are known to biodeteriorate all these materials. Among fungi, mostly these are ascomycetes or deuteromycetes (Table 14.8). Wood is spoiled chiefly by basidiomycetes.

TABLE 14.8 Fungi involved in biodeterioration

Materials	*Characteristic genera involved in deterioration*
Wood	(a) Basidiomycetes: *Lentinus, Hydnum, Fomes, Ganoderma, Lenzites, Polyporus, Stereum, Polystictus, Poria, Serpula, Trametes,.* (b) Staining, non-basidiomycetous fungi such as *Ceratocystis, Alternaria, Aspergillus, Penicillium, Trichoderma, Aureobasidium, Cladosporium, Mucor, Rhizopus*
Paper	*Alternaria, Cladosporium, Trichoderma, Aspergillus, Penicillium, Mucor*
Textiles	*Aspergillus, Penicillium, Trichoderma, Humicola, Chaetomium, Cladosporium*
Leather	*Aspergillus, Penicillium*
Optical instrument	*Aspergillus, Penicillium, Scopulariopsis*
Plastic	*Aspergillus, Penicillium, Rhizopus, Mucor*
Paint and paint films	*Cladosporium, Aureobasidium, Phoma*
Foods (perishable as well as non-perishable).	Various fungi as the kinds of food are varied. Bacteria also become important in food poisoning (botulism) as fungi become important in toxins like aflatoxins of stored grains

As indicated earlier, I would restrict here to biodeterioration of agricultural produce-grains, vegetables and fruits as these are of direct concern to agricultural microbiology. Spoilage of other foods (highly perishable and semiperishable foods like bread, meat, egg, butter and other milk and dairy products by microbes is the subject of microbiology of foods.

Microbial Biodeterioration of Agricultural Produce

Microbes, chiefly bacteria and fungi contaminate our grains, fruits and vegetables during storage, transport and marketing spoiling their commercial value. These microbes bring about undesirable physical and chemical changes in these products making them unsuitable for consumption. Not only this, in some cases consumption of such products contaminated by these microbes is deadly poisonous. For instance, mycotoxins of cereals and peanuts produced by *Aspergillus* spp. are highly carcinogenic and deadly poisonous. Food value is altered due to changes in their carbohydrates, fats and protein contents. Such chemical changes are brought about by the **enzymes** of these microbes. Amylases, lipases as well as proteases are produced.

Biodeterioration of agricultural produce i.e. grains, fruits, vegetables (rhizomes, bulbs, tubers etc.) during transit and storage is of non-parasitic nature and entirely different from parasites of these produce, which bring about parasitic post-harvest diseases. Unlike weak parasites responsible for postharvest diseases of these agricultural products, the microbes responsible for biodeterioration are saprophytic and equipped with rather a relatively stronger enzymic equipment. They are biologically different from those responsible for diseases.

In India huge losses are incurred on account of different types of **postharvest diseases** and **postharvest microbial deterioration** and spoilage of agricultural products. These both together are studied under the field of **market pathology**. This field concerns with the problems encountered during picking, packing, transportation and storage of produce. About 20-30% average loss is caused in India due to postharvest diseases.

Environmental temperature, humidity, moisture content and nutrient levels of produce are important factors in their spoilage. Besides bacteria, fungal species largely responsible for spoilage belong to the genera *Aspergillus, Penicillium, Trichoderma, Fusarium, Aureobasidium, Rhizopus, Mucor, Chaetomium, Cladosporium, Stemphylium, Curvularia, Drechslera, Memnoniella, Stachybotrys, Nigrospora, Pithomyces* and *Epicoccum*.

Mycotoxins in Biodeterioration

Certain fungi synthesise poisonous secondary metabolites, known as **mycotoxins**. They induce death and also cause other toxic effects when contaminated feed/food is consumed by animals or human beings. Ingestion of mycotoxins causes intoxication, termed mycotoxicosis. Already in 1968 more than 100 toxin-producing mold species were known (Wright, 1968), mainly belonging to the fungal genera *Aspergillus, Penicillium* and *Fusarium,* and this list has since then considerably extended. Mycotoxin production can occur in most plant products, but cereal and oil seed crops are mostly contaminated. Peanuts, cotton seed, rice and corn are mainly contaminated.

According to a publication of World Health Organisation (1969), four groups of mycotoxins are considered to be causally associated with diseases in humans. These are: (i) aflatoxins, (ii) ochratoxins, (iii) zearalenone, and (iv) trichothecenes.

By far the most information on mycotoxin contamination is available for **aflatoxins**. Aflatoxin was first detected in the 1960's as a contaminant in groundnuts, and this is a commodity likely to be contaminated, with frequencies up to 49%. Also maize is often contaminated with frequencies up to 97%. In addition to these commodities, aflatoxin has been found in treenuts, copra and spices.

Common fungi producing aflatoxin are *Aspergillus flavus, A fumigatus, A. parasiticus* and *Penicillium islandicum*. However, aflatoxins are produced chiefly by *Aspergillus flavus* and *A. parasiticus,* when they grow on stored food products. Over 1,000 scientific papers on aflatoxins have appeared since their discovery in 1960.

Aflatoxins are amongst the most potent carcinogens known, as suspected cause of liver cancer in man. It is only when they pass to liver, they are converted into toxic molecules. Aflatoxins are derivatives of furanocoumarin, and although 17 compounds have been identified, aflatoxin B_1, B_2, G_1 and G_2 are the most commonly occurring in natural products.

Aflatoxins, like most mycotoxins, are low-molecular weight compounds (Table 14.9). They are heat stable so that they will survive most cooking conditions. Lactating animals ingesting aflatoxin contaminated feed excrete metabolites in the milk called **aflatoxin M_1** (Table 14.13).

TABLE 14.9 Chemical characteristics of aflatoxins

Aflatoxin	*Molecular formula*	*Molecular weight*	*Melting point*
B_1	$C_{17}H_{12}O_6$	312	268-269
B_2	$C_{17}H_{14}O_6$	314	286-289
G_1	$C_{17}H_{12}O_7$	328	244-246
G_2	$C_{17}H_{14}O_7$	330	232-240
M_1	$C_{17}H_{12}O_7$	328	299

The occurrence of aflatoxins (aflatoxin $B_1 + B_2 + G_1 + G_2$) in a number of plant products is shown in Tables 14.10, 14.11 and 14.12. It is evident that aflatoxin contamination occurs in crops from tropical and subtropical areas. This reflects the fact that aflatoxin-producing strains of *Aspergillus flavus* are found mainly in tropical-subtropical areas, where also the optimal temperature conditions for aflatoxin production are present. Contaminated crops can be shipped to other areas, and thus practically all countries of the world are faced with the problem of aflatoxin-contaminated food stuffs.

Ochratoxins are produced by a number of species within the genera *Aspergillus* and *Penicillium,* with *Penicillium viridicatum* being the dominant producer. Ochratoxins

TABLE 14.10 Occurrence of aflatoxin in maize (Stoloff, 1976)

Country	No. of samples	% contaminated	Range of concentration (μg/kg)
USA	1594	25	3-37
USA	2866	8.2	
USA (South East)	60	35.0	6-348
Thailand		35.0	average : 400
Uganda		40.0	average : 133
Philippines		97.0	average : 213
India	5		250-15, 600
France	461	0.2	52

TABLE 14.11 Occurrence of aflatoxin in other cereals (WHO, 1969; Stoloff, 1976)

Country	No. of samples	% contaminated	Range of concentration (μg/kg)
USA	66 (sorghum)	3.0	13-50
USA	157 (rice)	0.6	5
Uganda	69 (sorghum)	23.0	average: 152
Vietnam	139 (rice)	31.0	
ex-USSR	169 (wheat)	0.6	100
ex-USSR	138 (wheat)	17.4	10-333
ex-USSR (Kazakistan)	50 (wheat)	4.0	5-10

TABLE 14.12 Occurrence of aflatoxin in groundnuts (Stoloff, 1976)

Country	No. of samples	% contaminated	Range of concentration (μg/kg)
USA	361	15.0	5-50
Sudan	173	41.0	5-1,000
Thailand		49.0	average : 1,530
Uganda		17.0	average : 363

are a group of closely related derivatives of isocoumarin linked to an aminoacid, L-beta- phenylalanine. Of the nine toxins, only **ochratoxin A** is occurring in natural products. It contaminates mainly grain but has also been found in coffee, beans and peanuts. This is found mainly in temperatre arease of Europe and North America.

Zearalenone is a phenolic resorcyclic acid lactone, produced by a number of *Fusarium* spp., e.g., *Fusarium graminearum* and *F. moniliforme*. This is found predominantly in maize, though occurs in other cereals also.

TABLE 14.13 Occurrence of aflatoxin M_1 in milk (WHO, 1969)

Country	*No. of samples*	*% contaminated*	*Range of concentration (μg/kg)*
Belgium	68	61.8	0.02-0.2
DDR	36	11.1	1.7-6.5
FRG	260	45.4	0.05-0.54
FRG	419	18.9	0.05-0.54
India	21	14.3	up to 13.3
Holland	95	77.9	0.09-0.5
UK	278	30.6	0.03-0.5
South Africa	56	0	
USA	320	7.5	0.1-0.4
USA	302	63.6	up to 3.9

Trichothecenes possess the tetracyclic 12, 13- epoxytricho-hec-O-ene skeleton, and more than 30 trichothecenes have been identified, with T-2 toxin, nivalenol and deoxynivalenol being most commonly found. They are produced by a number of species belonging to the genera, *Fusarium, Cephalosporium, Myrothecium, Trichoderma* and *Stachybotrys.* Very few reports on these toxins have been published so far, and the understanding of these toxins as contaminants of foodstuffs is very indadequate. For details on mycotoxins the readers may go through the paper by Krough (1983).

15

ENVIRONMENTAL TRANSMISSION OF PATHOGENS

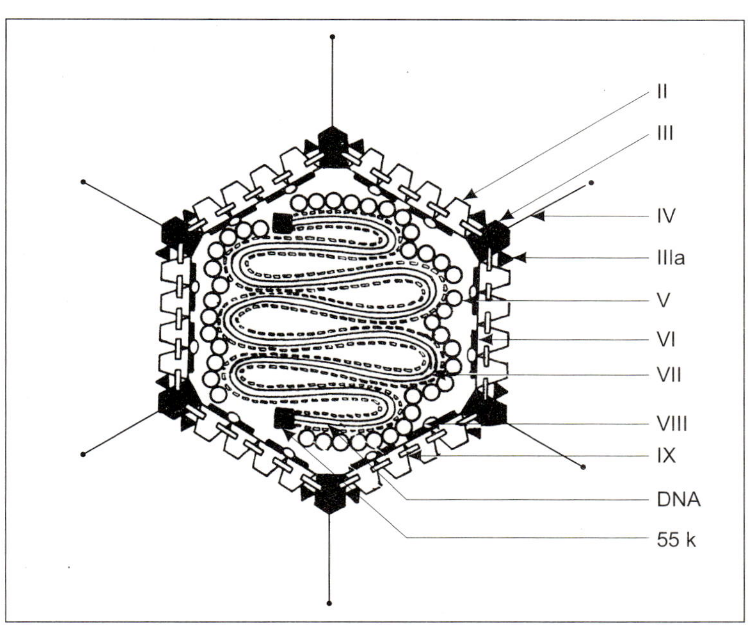

Chapter Outline

- *Transmission of Pathogens*
- *Bacterial Diseases of Man*
- *Viral Diseases of Man*
- *AIDS: A Death Warrant*
- *Ebola Haemorrhagic Fever*
- *Protozoan and Helminths Parasites*

Disease-causing microorganisms are called **pathogens**. The organism which is attacked and subsequently becomes diseased is called **host**. The disease is a complex process, an outcome of the host-pathogen interactions. **Infection** is the process in which microbe multiplies or grows in or on the host and in most cases results into the disease. The time between infection and the appearance of clinical symptoms is the **incubation time**.

The term **pathogenicity** refers to the ability of a parasite to gain entry into host's tissues and bring about a change (anatomical or physiological) resulting in a change of health and thus disease. The word is derived from **pathos** (Greek) meaning suffering. The term pathogen has same root and refers to an organism able to cause disease i.e. possessing pathogenicity. The symbiotic relationship betwen host and parasite is called **parasitism**. Parasites vary in their pathogenicity. Even in the same taxonomic species there may be non-pathogenic, less pathogenic or much pathogenic strains. Some as parasites of cholera, typhoid fever, plague are known to cause serious human diseases, whereas others like common cold viruses are less pathogenic. There are also **opportunistic pathogens** which exist as commensals in the body till normal defense mechanisms are suppresed, when they invade the tissue and act as pathogens. For instance, *Streptococcus pneumoniae* lives on surface of upper respiratory tract. *Pneumocystis carinii, Toxoplasma gondii* and *Cryptosporidium* spp., that are opportunistic otherwise, begin to invade the patients suffering from AIDS. Thus a shift in the body's delicate balance of controls may convert infection to disease. Several factors may alter the balance of microbes. Destruction of normal flora by indiscriminate use of antibiotics may lead to diseases.

A new concept has thus emerged in recent years. Traditionally, microbiologists believed that organisms were either pathogenic or nonpathogenic. This distinction has been blurred by the fact that normally benign organisms may be pathogenic when body defenses weaken or fail. The AIDS illustrates this concept very well. It is now recognised that pathogenicity is a function of the aggressive nature of the parasite as well as the level of resistance in the host.

Whereas pathogenicity is used in a qualitative sense, virulence used in quantitative sense giving a measure of the extent of pathogenicity of a microorganism. The term **virulence** is often used to express the degree of pathogenicity of a parasite. The term is derived from Latin **virulentus**, meaning full of poison. An organism such as the typhoid bacillus that invariably causes disease is said to be **highly virulent**, whereas *Candida albicans* which causes sometimes disease is called **moderately virulent**. Certain organisms, described as **avirulent** are not generally regarded as disease agents as for example the lactobacilli and streptococci found in curd etc. However, it should be noted that any microbe has the ability to change genetically and become virulent. *E. coli,* for example was long considered an avirulent commensal of humans, but certain toxin-producing strains, have been isolated that cause diarrhea and urinary tract diseases in humans.

The virulence depends in large part on two properties of a microbe: invasiveness and toxigenicity, important in human diseases. **Invasiveness** refers to the ability of a microbe to invade human cells and tissues and to multiply on or within them. **Toxigenicity** refers to the ability of a microbe to produce biochemicals, known as toxins that disrupt the normal functions of cells or are generally destructive to human cells and tissues.

TRANSMISSION OF PATHOGENS

Pathogenic microbes usually originate from an infected host (human or animal) or directly from the environment. A list of the important diseases of man caused by different microorganisms (Table 15.1) shows that pathogens may be transmitted by **direct** or close contact with the infected host, or **indirectly** through different factors of the environment. The main factors of transmission are **air/droplets, water** and **soil**. Besides these the pathogens are also transmitted via contaminated **food** or other inanimate objects, **fomites**. Examples of direct transmission are herpes virus, *Neisseria gonorrhoeae* (gonorrhea), and *Treponema pallidum* (syphilis).

Pathogens may come out of a host in respiratory secretions from nose and mouth or be shed on dead skin or in faeces, urine, saliva, or tears. These in turn may contaminate the air, water food or fomites. Airborne transmission can occur via release from the host in **droplets** (coughing-mucilage of nose and saliva) or through natural (surf at a beach) or human activities (cooling towers, showers). Virus transmission by the airborne route may be direct and indirect.

Microorganisms transmitted by the faecal-oral route are usually referred to as **enteric pathogens**, because they infect the gastrointestinal tract. They remain stable in water and food, and in case of enteric bacteria, are capable of growth outside the host under the right environmental conditions.

Water-related illnesses associated with pathogens are classified under different groups on the basis of precise way of the origin and subsequent transmission of the pathogens (Table 15.2).

BACTERIAL DISEASES OF MAN

Airborne Diseases

The airborne bacterial diseases are mainly diseases of the respiratory tract. They include important diseases as tuberculosis, diphtheria, whooping cough and primary

TABLE 15.1 Important microbial diseases of man

Disease	*Causal agent*	*Description of agent*	*Organs affected*	*Transmission/Vector*
BACTERIA				
Strep throat, scarlet fever	*Streptococcus pyogenes*	Gm (+) capsulated streptococcus	Upper resp. tract	Air
Diphtheria	*Corynebacterium diphtheriae*	Gm (+) rod	Blood, Skin, Upper resp. tract, Heart, Nerve fibres	Air
Pertussis (whooping cough)	*Bordetella pertussis*	Gm (–) rod	Upper resp. Tract	Air
Meningococcal meningitis	*Neisseria meningitidis*	Gm (–) diplococcus	Upper resp. tract, Blood, Meninges	Air
Haemophilus meningitis	*Haemophilus influenzae*	Gm (–) capsulated rod	Upper resp. tract, Meninges	Air
Flavobacterium meningitis	*Flavobacterium meningospecticum*	Gm (-) rod	Upper resp. tract, Meninges	Air
Tuberculosis	*Mycobacterium tuberculosis*	Acid-fast rod	Lungs, Bones, Other organs	Air
Pneumococcal pneumonia	*Streptococcus penumoniae*	Gm (+) capsulated diplococci in chains	Lungs	Air
Primary atypical penumonia	*Mycoplasma pneumoniae*	Mycoplasma (No cell wall), 0.2 µm diam.	Lungs	Air
Klebsiella pneumonia	*Klebsiella pneumoniae*	Gm (–) capsulated rod	Lungs	Air
Serratia pneumonia	*Serratia marcescens*	Gm (–) rod, Red pigment at 25°C	Lungs	Air
Q fever	*Coxiella burnetii*	Rickettsia, 0.45 µ diam.	Lungs	Air
Psittacosis	*Chlamydia psittaci*	Chlamydia 0.25 µm diam.	Lungs	Air
Botulism	*Clostridium botulinum*	Gm (+) spore-forming rod	Neuromuscular junction	Food, Water
Staphylococcal food poisoning	*Staphylococcus aureus*	Gm (+) staphylococcus	Intestine	Food, Water
Clostridial food poisoning	*Clostridium perfringens*	Gm (+) spore-forming rod	Intestine	Food, Water
Typhoid fever	*Salmonella typhi*	Gm (–) rod	Intestine, Blood, Gall bladder	Food, Water
Salmonellosis	*Salmonella* serotypes	Gm (–) rod	Intestine	Food, Water
Shigellosis	*Shigella* serotypes	Gm (–) rod	Intestine	Food, Water
Cholera	*Vibrio cholerae*	Gm (–) curved rod	Intestine	Food, Water
Brucellosis	*Brucella* spp.	Gm (–) rod	Spleen, Lymph glands	Food, Water
Anthrax	*Bacillus anthracis*	Gm (+) spore-forming rod	Blood, Lungs, Skin	Soil

(Table contd.)

(Table contd.)

Tetanus	*Clostridium tetani*	Gm (+) spore-forming anaerobic rod	Nerves at synapse	Soil
Gas gangrene	*Clostridium perfringens*	Gm (+) spore-forming anaerobic rod	Muscles, Nerves, Blood cells	Soil
Bubonic plague	*Yersinia pestis*	Gm (–) bipolar rod	Lymph nodes, Blood, Lungs	Rat flea
Relapsing fever	*Borrelia recurrentis*	Spirochete	Blood, Liver	Louse
Rocky Mountain spotted fever	*Rickettsia rickettsiae*	Rickettisia	Blood, Skin	Tick
Epidemic typhus (typhus fever)	*Rickettsia prowazekii*	Rickettisia	Blood, Skin	Louse
Endemic typhus (murine typhus)	*Rickettsia typhi*	Rickettisia	Blood, Skin	Flea
Scrub typhus	*Rickettsia tsutsugamushi*	Rickettisia	Blood, Skin	Mite
Rickettsialpox	*Rickettsia akari*	Rickettisia	Blood, Skin	Mite
Tickborne fevers	*Rickettsia conorii*	Rickettisia	Blood, Skin	Tick
Syphilis	*Treponema pallidum*	Spirochete	Skin, Cardiovascular organs	Sexual
Gonorrhea	*Neisseria gonorrhoeae*	Gm (-) diplococcus	Urethra, Cervix, Fallopian tubes, Epididymis, Eyes, Pharynx	Sexual
Chlamydial urethritis	*Chlamydia trachomatis*	Chlamydia	As above	Sexual
Ureaplasmal urethritis	*Ureaplasma urealyticum*	Mycoplasma	Urethra, Fallopian tubes, Epididymis	Sexual
Lymphogranuloma venereum	*Chlamydia trachomatis*	Chlamydia	Inguinal lymph nodes, Rectum	Sexual
Vaginitis	*Gardnerella vaginalis*	Gm (–) rod	Vagina	Sexual
Mycoplasmal urethritis	*Mycoplasma hominis*	Mycoplasma	Urethra, Fallopian tubes, Epididymis	Sexual
Leprosy (Hansen's disease)	*Mycobacterium leprae*	Acid-fast rod	Skin, Bones, Peripheral nerves	Contact
Staphylococcal skin diseases	*Staphylococcus aureus*	Gm (+) staphylococcus	Skin	Contact
Trachoma	*Chlamydia trachomatis*	Chlamydia	Eyes	Contact
Bacterial conjuctivitis (pink eye)	*Haemophilus* influenzae biotype III	Gm (–) rod	Eyes	Contact

(Table contd.)

(Table contd.)

VIRUSES			
Influenza	RNA	Resp.tract	Droplets
Adenovirus infections	DNA	Lungs, Eyes	Droplets, Contact
Respiratory syncytial disease	RNA	Resp. Tract	Droplets
Rhinovirus infections	RNA	Upper resp.tract	Droplets, Contact
Herpes simplex	DNA	Skin, Pharynx, Genital organs	Contact
Chicken pox (Varicella)	DNA	Skin, Nervous system	Droplets, Contact
Measles (Rubeola)	RNA	Resp. tract, Skin	Droplets, Contact
German measles (Rubella)	RNA	Skin	Droplets, Contact
Mumps (epidemic parotitis)	RNA	Salivary glands, Blood	Droplets
Small pox (Variola)	DNA	Skin, Blood	Contact, Droplets, Fomites
Warts	DNA	Skin	Contact
Yellow fever	RNA	Liver, Blood	Mosquito (*Aedes aegypti*)
Dengue fever	RNA	Blood, Muscles	Mosquito (*Aedes aegypti*)
Hepatitis A	RNA	Liver	Food, Water, Contact
Hepatitis B	DNA	Liver	Contact with body fluids
NANB hepatitis	RNA	Liver	Contact with body fluids
Viral gastroenteritis	Many RNA viruses	Intestine	Food, Water
Viral fevers	Many RNA viruses	Blood	Contact, Arthropods
Cytomegalovirus disease	DNA	Blood, Lungs	Contact, Congenital transfer
AIDS	Retrovirus (RNA)	T-lymphocytes	Contact with body fluids
Rabies	RNA	Brain, Spinal cord	Contact with body fluids
Polio	RNA	Intestine, Brain, Spinal cord	Food, Water, Contact
Slow virus disease	Prions	Brain	?
Arboviral encephalitis	Many RNA viruses	Brain	Arthropods
FUNGI			
Cryptococcosis	*Cryptococcus neoformans*	Lungs, Spinal cord	Air
Candidiasis, Vaginitis, Thrush, Onychia	*Candida albicans*	Intestine, Vagina, Skin, Mouth	Air, Sexual contact
Tinea pedis	*Trichophyton* spp.	Skin	Contact
Tinea capitis	*Microsporum* spp.	Skin	Contact

(Table contd.)

(Table contd.)

Tinea corporis, Tinea barhae	*Epidermophyton* spp.	Skin	Contact
Histoplasmosis	*Histoplasma capsulatum*	Lungs, Other organs	Air
Blastomycosis	*Blastomyces dermatitidis*	Lungs, Other organs	Air
Coccidiodomycosis	*Coccidioides immitis*	Lungs, Ears	Air
Aspergillosis, Otomycosis	*Aspergillus fumigatus*	Lungs, Ears	Air
PROTOZOA			
Amoebiasis	*Entamoeba histolytica*	Intestine, Liver	Water, Food
Primary amoebic meningoencephalitis	*Naegleria fowleri*	Brain, Lungs	Water
Giardiasis	*Giardia lamblia*	Intestine	Water, Contact
Trichomoniasis	*Trichomonas vaginalis*	Urogenital organs	Sexual contact
African sleeping sickness	*Trypanosoma brucei*	Blood, Brain	Tsetse fly (*Glossina*)
Leishmaniasis (Kala-azar)	*Leishmania donovani*	White blood cells, Skin, Intestine	Sand fly (*Phlemotomus*)
Toxoplasmosis	*Toxoplasma gondii*	Blood, Eyes	Domestic cats, Food
Malaria	*Plasmodium* spp.	Liver, Red blood cells	Mosquito (*Anopheles*)
Babesiosis	*Babesia microti*	Red blood cells	Tick (*Txode*)
Pneumocystosis (PCP)	*Pneumocystis carinii*	Lungs	Droplets
MULTICELLULAR PARASITES			
(A) **Flatworms**			
Chinese liver fluke (Animal host-snail, fish)	*Clonorchis sinensis*	Gall bladder, Liver	Fish consumption
Intestinal fluke (Animal host-snail)	*Fasciolopsis buski*	Intestine	Consumption of water plants
Lung fluke (Animal host-snail, crab)	*Paragonimus westermani*	Lung	Consumption of crabs
Liver fluke (Animal host-snail, cattle)	*Fasciola hepatica*	Liver	Consumption of water plants
Beef tapeworm (Animal host-cattle)	*Taenia saginata*	Intestine	Beef consumption
Pork tapeworm (Animal host-pig)	*Taenia solium*	Intestine	Pork consumption
Fish tapeworm (Animal host-copepod, fish)	*Diphyllobothrium latum*	Intestine	Pork consumption
Dwarf tapeworm (Animal host-None significant)	*Hymenolepis nana*	Intestine	Food, Contact
Dog tapeworm (Animal host-dog, other canines)	*Echinococcus granulosus*	Liver	Contact

(Table contd.)

(Table contd.)

(B) **Roundworms**			
Pinworm (Animal host-none significant)	*Enterobius vermicularis*	Intestine	Contact, Food, Clothing
Whipworm (Animal host-none significant)	*Trichuris trichiura*	Intestine	Food, Water
Round worm (Animal host-none significant)	*Ascaris lumbricoides*	Intestine, Lungs	Food, Water, Contact
Trichinosis (Animal host-pig)	*Trichinella spiralis*	Intestine, Muscles, Eyes	Pork consumption
Hookworm (Animal host-none significant)	*Ancylostoma duodenale, Necator americanus*	Intestine, Lungs, Lymph	Contact with moist vegetation
Filariasis (Animal host-mosquito)	*Wuchereria bancrofti*	Lymph vessels	Mosquito
Guinea worm (Animal host-copepod, fish)	*Dracunculus medinensis*	Skin	Food, Water
Eyeworm (Animal host-insects)	*Loa loa*	Eye, Connective tissues	Deerflies, Horseflies

TABLE 15.2 Different classes of water-related diseases of man

Class	*Origin and transmission*	*Examples*
Water-borne	Pathogens originate in faecal matter and transmitted by ingestion	Cholera, Typhoid fever
Water-washed	Pathogens originate in faeces and transmitted through contact due to inadequate sanitation or hygeine	Trachoma, Diarrhoea
Water-based	Pathogens originate in water or spend part of cycle in aquatic animals and come in direct contact with humans in water or by inhalation	Schistosomiasis, Legionella
Water-related	Pathogens with their life cycles associated with insects living or breeding in water	Yellow fever, Dengue, Filariasis, Malaria, Sleeping sickness

atypical pneumonia. Usually these diseases are spread where people are crowded and living conditions are not proper.

Tuberculosis

This is a disease of crowded populations, especially of lower socioeconomic groups. The causal organism is *Mycobacterium tuberculosis.* Often one member of a family is a source of infection to others.

The organism is a small rod that enters the respiratory tract and grows on the lung tissue. The symptoms are chronic cough, chest pains. high fever and a flow of thick expectorated matter called **sputum**, The latter may be rust-colored if the blood has entered the lung cavity. The incubation period is approximately 2-10 weeks for the primary infection, but six months may pass before the symptoms become fully recognised. In the infected lung. cells of macrophages, leukocytes and T -lymphocytes surround the parasite. and due to deposition of ca1cium salts and fibrous materials, a hard nodule-**tubercle** is formed. This may be visible in the patient's chest X-ray. The tubercle may break apart and spread bacteria to liver, bones and kidneys. The disease is then called **miliary tuberculosis**. The bacterium does not produce toxin, but due to rapid growth the tissues are consumed. Due to presence of fatty waxy materials in cell wall, the bacterium is much resistant to environmental changes. The organism is acid-fast, stained with acid-fast technique by Ziehl-Neelsen Carbolfuchsin stain. The cells stain red with this stain.

Bovine tuberculosis, caused by *Mycobacterium bovis* is equally dangerous in cows and humans. The organism is transmitted through milk. **Bacille Calmette Guerin** (BCG) is a preparation of an attenuated strain of *M. bovis* that is used in immunization programmes throughout the world. It is named for Albert Calmette and Camille Guerin who developed it in the 1920s.

Diphtheria

This is caused by *Corynebacterium diphtheriae,* first recognised in a material from a patients' throat in 1883 by Edwin Klebs. The bacterium is club-shaped, associated with leathery membranes, Gram-positive with numerous metachromatic granules in the cytoplasm. The disease is a major health problem in India. The organism is inhaled in respiratory droplets and, infects the upper respiratory tract near the tonsils. It grows rapidly and produces an **exotoxin**, the third dangerous chemical after botulism and tetanus toxins. The toxin interferes with the protein metabolism of cells as mucus, and leukocytes. The dead tissue accumulates forming a dirty grey pseudomembrane. The material is leathery and fibrous. In young children, there may be respiratory blockage.

In adults, the toxin spreads to blood stream causing damage to heart muscle, leading to cardiac weakness and heart failure. There may be paralysis. Mumpslike symptoms are common, and a severely swollen neck may be observed.

The disease is treated by antibiotics and antitoxins. Penicillins and horse serum toxins are used. Long-term protection is done by injecting diphtheria toxoid, that induces the formation of antitoxins. This is one of the DPT series; usually given to infants.

Meningococcal meningitis

This is caused by *Neisseria meningitidis,* a small Gram-negative diplococcus commonly called the **meningococcus**. The organism enters the body by droplets often from a carrier. It may pass rapidly through the mucuous membranes of upper respiratory tract into the blood stream. Here it multiplies quickly. Due to release of large amount of endotoxins, an endotoxin shock, called **meningococcemia** takes place. These events may occur within a short period of some hours of infection and substantial vascular damage may cause death. After the blood infection, the bacteria localise on the coverings of spinal cord and brain known as **meninges**. It causes severe headache and stiffening of the neck. There is also hemorrhagic rash on the skin. The spots (capillary thromobses) begin as bright red patches, turn to purplish and finally blue-black. The patient is often confused and delirius, and death may occur in 50 per cent of untreated cases.

Earlier treatment is desirable. Sulfonamide drugs, rifampin, and ampicillin are often used. Diagnosis is based on physical symptoms, and observation of Gram-negative diplococci in centrifuged samples of cerebrospinal fluid.

Some cases are complicated by hemorrhagic lesions that occur in the adrenal glands and cause hormonal imbalance and physiological disturbances. This condition, called the **Waterhouse-Friederishsen Syndrome**, may result from an allergic reaction involving immune complex formation and activation of complement.

Pneumococcal pneumonia

The word **pneumonia** in a broad sense includes a number of microbial diseases of the bronchial tubes and lung tissues. About 90 percent of bacterial cases are caused by Gram-positive diplococcus having a capsule and lancent-like pointed ends. This organism, commonly known as pneumococcus, was assigned in older literature to *Diplococcus pneumoniae.* However, in the latest edition of Bergey's Manual, it is described as *Streptococcus pneumoniae.*

It exists in the upper respiratory tract, usually acquired from another carrier by respiratory droplets or contact. The patient becomes usually susceptible when exposed to viral infections, allergic reaction, extensive surgery, malnutrition, excessive smoking or depression of immune system. The disease is characterised by high fever, sharp chest pains and bacterial infiltration of the lung tissue. Due to destruction of alveolar walls, blood seeps into the lung spaces and rust-colored sputum is expelled. If entire lobe of lung is involved, it is called **lobar pneumonia**. If both sides are infected, it is called **double pneumonia**. Scattered patches of infection in bronchial tree yield **bronchopneumonia**. Penicillin is the common drug.

The bacterium produces several virulence factors-hemolysin, leukocidin and hyaluronidase. Due to presence of capsule, it resists phagocytosis. On the basis of several serological tests, about a dozen of variants have been identified, which cause human infections.

The bacteria may be identified by isolating from the sputum and testing for alpha hemolysis on blood agar. The organism is also identified by bile-solubility test. A test known as the Neurfeld Quellung reaction (Fig. 15.1) helps to establish the capsular type.

Whooping cough (Pertussis)

This is a serious disease of young children. The bacteria grow in the lining of upper respiratory tract, causing disintegration of cells and their accumulation in the airways. Due to blockage of bronchi with mucus and debris, difficulty in breathing occurs leading to oxygen starvation or **hypoxia**. The high-pitched "whoop" on rapid inspiration results from the narrowing of tubes. A child may suffer ten to fifteen attacks of coughing during the day and generally falls asleep exhausted.

The causal organism, *Bordetella pertussis,* was first isolated by Jules Bordet and Octave Gengou in 1906 and became known as Bordetella-Gengou bacillus. It is a small, fragile, Gram-negative rod. The bacterium passes from one person to another by respiratory droplets. Early treatment with penicillin derivatives and erythromycin is recommended as prophylactic measures. The vaccine is produced by killing *Bordetella pertussis* with merthiolate in a manner first described by Pearl Kendrick and

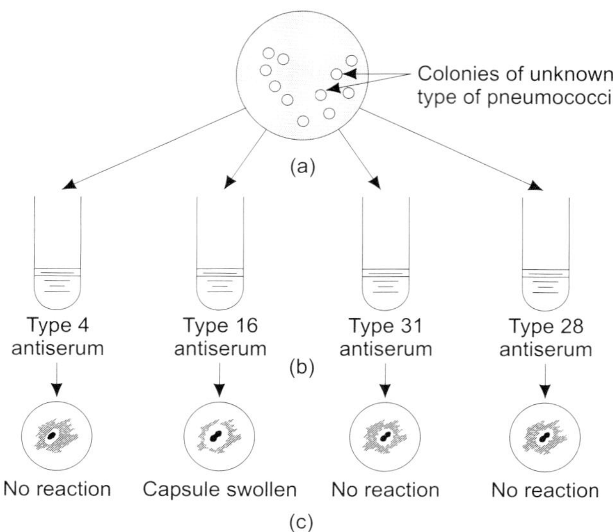

FIG. 15.1

The Neurfeld quellung (German, swelling) reaction. Pneumococci are cultivated on blood agar (a), and samples of colonies are transferred to tubes that contain various types of pneumococcal antiserum (b). When samples of each mixture are placed on slides with alkaline methylene blue (c), the capsule is seen to swell around those pneumococci whose antigens have reacted with the antiserum. In the illustration, the pneumococci are identified as Type 16 *Streptococcus pneumoniae*

Grace Eldering in 1939. The intact bacteria were then mixed with diphtheria and tetanus toxoids in the DPT preparation.

Primary atypical pneumonia (PAP)

The organism was located in the early 1940s by Monroe Eaton at Harvard University, as a tiny viruslike agent on agar media supplemented with blood. The organism was named as **Eaton agent**, and the disease described as **primary** (as was not due to any secondary complication of previous infection) **atypical** (as the amount of fluid in the lung and involvement of the alveoli was minimal as compared to typical bacterial pneumonia) **pneumonia**. By 1957 it was found that PPLO (pleuropneumonialike organism) of cows was, identical with the Eaton agent of humans, and the name *Mycoplasma pneumoniae* was given to the causal organism.

Mycoplasma pneumoniae is one of the smallest of the free-living pathogenic organisms, measuring about 0.2 μm in size, and lacks a cell wall, thus not susceptible to penicillin. They are pleomorphic, passed by respiratory droplets in crowded conditions. This disease resembles viral pneumonia. There is fever, fatigue, dry cough due to lack of sputum. The disease, however, is not fatal and also called **walking pneumonia**, Antibodies produced during infection aggultinate human Type O red blood cells at low temperatures. This gave rise to a **cold agglutinin screening test** (CAST) (Fig. 15.2) used to test a patient's serum, if the disease is suspected.

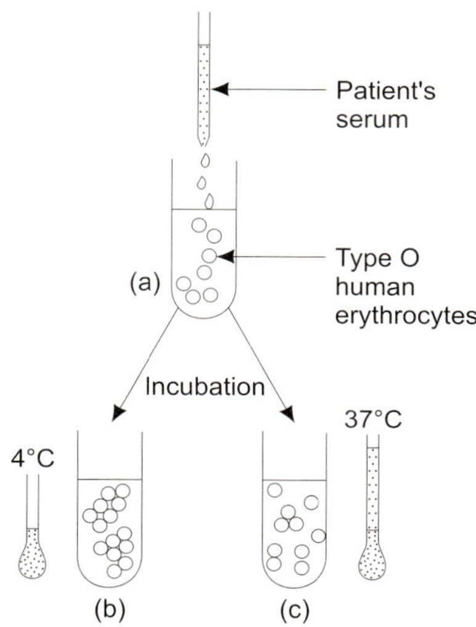

FIG. 15.2

The cold agglutinin screening test (CAST). (a) serum from a patient is mixed with Type O human erythrocytes. (b) At 4°C, the erythrocytes will clump strongly, (c) at 37°C there will be little or no agglutination

Foodborne and Waterborne Diseases

Some life-threatening diseases will be explored.

Botulism

This is most dangerous of the food-borne diseases. The exotoxin of the bacteria is very powerful. For instance, an ounce of the purified material may eliminate the entire population of many countries. The bacterium, *Clostridium botulinum* exists as spores in the intestines of many fish, birds and barnyard animals as cows and horses, and also in human intestines. The spores reach the soil in manure, sewage and organic fertilisers and may be attached to harvested foods. When placed in anaerobic conditions as improperly processed cans or jars they germinate and multiply as Gram-positive rods which produce exotoxin.

Bacteria are of little consequence as they do not grow further in the body. However, this is the exotoxin which causes harm. It is a high molecular weight protein, which is activated by the enzyme trypsin in the intestine and is absorbed in the blood stream. Within hours the patient begins to feel paralysing symptoms. There is blurred vision, impaired speech, difficulty in swallowing and chewing and respiratory distress. The limbs have little tone and become flably, a condition called **flaccid paralysis**. At the synaptic junction of the nerves and muscles, the toxin inhibits the release of the neurotransmitter, **acetylcholine** (Fig. 15.3). Nerve stimulation to the muscle therefore ceases and the diaphragm fails to contract. Death by respiratory paralysis may occur within a day or so.

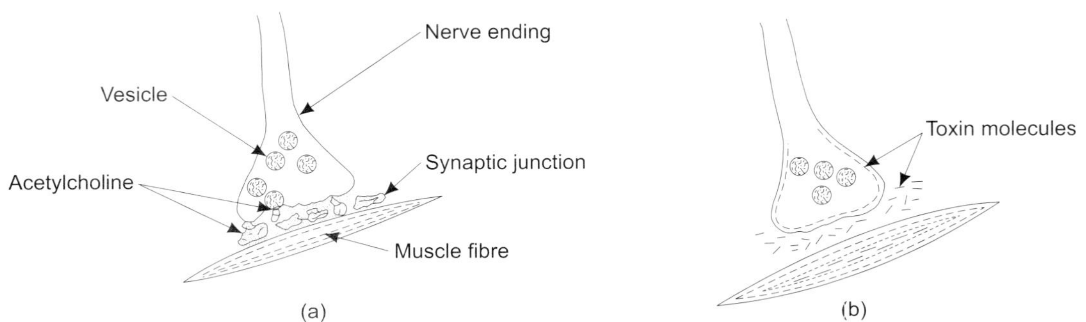

FIG. 15.3

The mechanism of action of the botulism toxin. (a) in the normal condition, acetylcholine is released from vesicles in the nerve ending to assist the transfer of impulses to the muscle, (b) in the pathological condition, the botulism toxin enters the nerve fiber and binds to the inner surface of the nerve membrane, thereby preventing the release of acetylcholine from the vesicles

Antibiotics are of no help in botulism, as the patient is intoxicated rather than infected. Large doses of botulism **antitoxin** are given. There are at least eight types of *Clostridium botulinum* distributed in soils. In different animals they cause different types of botulism. These are **fodder disease** in cattle and **limberneck** in fawl (due to ingestion of toxin from silage and feeds), **wound botulism** in humans (due to active growth in dead tissue of wounds) and **sudden infant death syndrome** (SIDS) (due to germination of spores in infants' intestines).

Clostridial food poisoning

This is the second most common food-borne disease, caused by *Clostridium perfringens*. The symptoms are due to a toxin that increases the water secretion in the latter part of small intestine, leading to diarrhea. There are abdominal pains but no fever or blood involvement. The toxin may be produced under anaerobic conditions in canned foods as well as the intestine. Death is rare, and severe cases treated with antitoxin and penicillins.

Typhoid fever

It differs from botulism or other food poisonings in that patient ingests bacteria rather than toxins and soon becomes infected rather than poisoned. The causal organism, *Salmonella typhi* is a Gram-negative rod that is very resistant to environmental conditions. Thus it may remain alive for long periods in fresh water and foods.

The chief source of disease is the human carrier. The recovered patient may continue to harbor bacteria which are shed in the feces for long periods. Typhoid Mary (1906) demonstrated this fact. Contaminated sewage is another source. When it mixes with drinking water due to broken sewer lines or natural disasters, epidemics may follow. Flies also transport bacteria from sewage to food or water. Since bacterium is acid-resistant it passes through stomach and enters small intestine, where it multiplies quickly. There it produces deep ulcers, constipation, and bloody stools are the indication of bacterial penetration to blood vessels of intestinal walls. The fever rises gradually, and skin covered with bright red rose spots.

The disease was previously treated with chloramphenicol. However, resistant strains of bacteria have developed. At present amoxicillin and sulfam-ethoxazole-trimethoprim have been substituted. The bacterium has some preference for gallbladder, and antibiotics are not effective in this organ. In such cases gall bladder is to be removed. It should be noted that this bacterium is notable exception to the more caustic body fluid (the bile), which inhibits many bacteria otherwise in the body.

Chemically killed bacteria are used in the vaccine for typhoid fever with about 100,000 dead organisms in each dose. The danger of the disease may be ascertained by testing the water samples for some indicator organisms as *E. coli,* a normal intestinal inhabitant. If these indicators are located in water (due to mixing of contaminated sewage with drinking water), it indicates that *Salmonella* may also be present and an epidemic may follow. The organism may also be identified by isolating it from urine, blood, stool or bone marrow cultures. In 1896, Fernand Widal devised the **Widal test**, an aggultination method in which typhoid fever antibodies are detected by mixing the patients' serum with *Salmonella* cells and observed for clumping. These days extracts of flagella or cellular antigens of bacterium rather than cells are used for this test.

Cholera

There is severe dysentery, vomiting and violent cramps with much loss of fluid. Eyes become shrouded in grey and sink into orbits. Skin becomes dry, wrinkled and cold and muscular cramps occur in the arms and legs. There is continuous thirst but patient is unable to hold fluid. The blood is thickened and urine production stops. Without urine, meabolic wastes accumulate and tissues suffer toxic damage. The sluggish blood flow to the brain stops oxygen to this vital organ and patient enters into coma and soon dies.

The causal organism, *Vibrio cholerae,* a Gram-negative rod was first cultivated by Robert Koch in 1883. The bacterium has two recognised **biotypes:** the classical - *V. cholerae* and the **El. Tor** biotype. The latter produces a hemolysin that disrupts sheep red blood cells, but the classical biotype does not. The El Tor also exists in carriers, not found in the classical form. The bacteria must enter the gastrointestinal system in large numbers since the stomach acid kills most of the bacteria. The surviving organisms enter the intestine and move about through the mucous coating. They adhere strongly to the tissue and produce powerful toxins. The toxin consists of three chains of protein that stimulate the production of the enzyme, **adenyl cyclase**. This enzyme increases the level of adenosine monophosphate (AMP) which leads to abundant secretion of fluid by the intestine. The presence of infection also retards water absorption through the wall.

The critical treatment is to restore the water balance in the body. Intravenos injections of tetracycline are useful in control of bacteria. Immunization with preparation of dead bacteria provides protection for about six months. Prevention methods include sanitation, personal hygiene and care in food preparation.

Soilborne Diseases

Tetanus

This is among the most dangerous of human diseases. The bacterium, *Clostridium tetani* exists in air, water, animal and human intestine and especially the soil. The bacteria enter a wound in very small numbers and produce second most powerful toxin known to science. The toxins provoke sustained and uncontrolled contractions of the muscles, and spasms occur throughout the body; the patient experiences a violent death.

The bacterium exists as spores in the intestines of many animals and man. However, intestinal infection does not occur. The organism is excreted in the feces to the soil, from where the spores are introduced to the anaerobic environment of dead necrotic tissue in a deep puncture. Wound contamination may also occur. Rusty nails are a threat because spores may cling to the rough edges of the nail.

The symptoms develop very rapidly. Within hours of entry, the spores revert to vegetative cells which produce several toxins including **tetanospansmin**, a high molecular weight protein. The toxin spreads through the blood stream and along the axons of the nerve cells to the nerve endings where it inhibits the removal of acetylcholine in the synapse. There is muscle contraction. There is muscle stiffness, especially in the jaw and swallowing muscles. Soon the back arches, the jaws clench and a fixed smile comes over the face. These symptoms give the disease its traditional name of **lockjaw.**

It is surprising that the bacterium is saprobe rather than parasite as long they live only in dead anaerobic matter. They are motile Gram-positive rods. that may exist for many years in soil as spores. The disease incidence is high in India. where ear piercing is common and where umblicial stump of newborns is sometimes dressed with soil.

Immunisation to tetanus may be achieved by injecting **toxoid** consisting of formaldehyde-treated toxin that is precipitated with alum. This is used in the DPT programmes, and booster injections are recommended every ten years. For suffering persons antitoxins, to neutralise the remaining toxins, and penicillins are given.

VIRAL DISEASES OF MAN

Viruses pathogenic on humans are atificially classified into four groups (Table 15.3) mainly on the basis of organs on or in which typical symptoms are produced. These groups are as follows:

Pneumotropic diseases, involving respiratory tract, such as influenza and common colds.

Dermotropic diseases, involving mainly skins, as measles, chicken pox, small pox etc.

Viscerotropic diseases, involving mainly internal organs, as yellow fever, dengue fever, hepatitis etc., and

Neurotropic diseases, involving mainly central nervous system, such as rabies, polio, encephalitis etc.

TABLE 15.3 Classification of human viral diseases based on the tissue affected

Group	*Tissues affected*	*Important diseases*
Pneumotropic	Respiratory system	Influenza, respiratory syncytial disease, adenovirus infections, rhinovirus infection
Dermotropic	Skin and subcutaneous tissues	Chicken pox, herpes simplex, measles, mumps, small pox, molluscum contagiosum, German measles
Viscerotropic	Blood and visceral organs	Yellow fever, dengue fever, infectious mononucleosis, cytomegalovirus disease, infectious hepatitis, serum hepatitis, sandfly fever, Colorado tick fever
Neurotropic	Central nervous system	Rabies, polio, encephalitis, slow virus disease

In each of the above four groups, other organs may also be involved. We shall explore some of the common diseases in each group.

Influenza

This is an acute disease of the upper respiratory tract. The virus is a helical virion containing single-stranded RNA and is enveloped. The envelope contains a series of projections, or spikes as shown in Fig. 15.4. These spikes contain the enzymes **hemagglutinin** which allows the virion to bind the red blood cells, and **neuraminidase** which dissolves the cell membrane during replication. Both enzymes are said to be antigens because they induce the body to form antibodies that neutralise the virion. The hemagglutinin spike is composed of three sets of HA_1 and HA_2 polypeptides. The neuraminidase spike consists of four NA polypeptides. The nucleocapsid is segmented and gives the appearance of a double helix due to association of internal proteins (NP, P_1, P_2, and P_3).

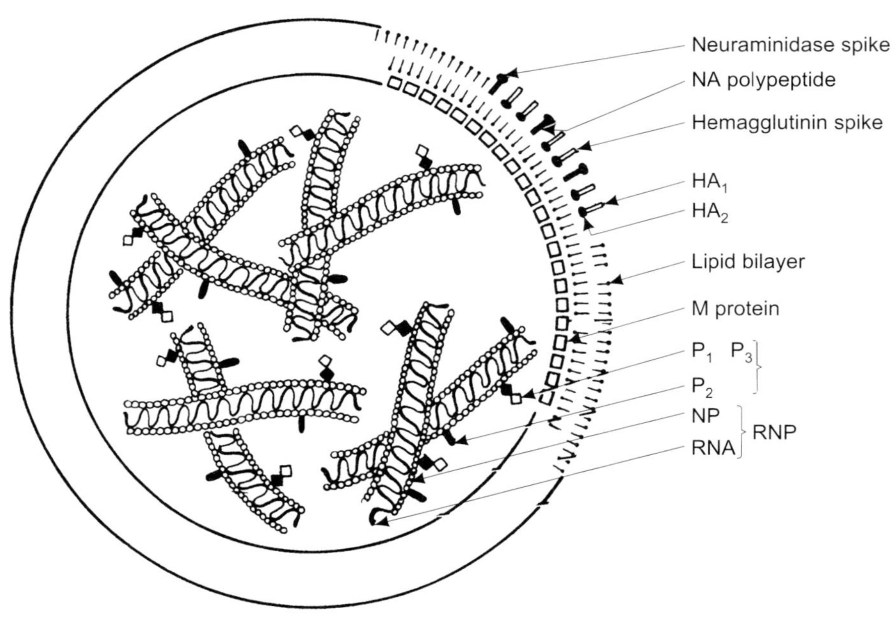

FIG. 15.4

Schematic model for influenza virus particle. The nucleocapsid is segmented and gives the appearance of double helix owing to association of internal proteins (**NP, P_1, P_2,** and **P_3**) with the single-stranded RNAs. The hemagglutinin spike is composed of three sets of **HA_1** and **HA_2** polypeptides. The neuraminidase spike consists of four **NA** polypeptides

There are three distinct types of influenza virus: Types A and B cause worldwide epidemics called pandemics, and Type C, which occurs sporadically. Each type is known for its **antigenic variation**, a process in which changes occur in the protein structure of the capsid as well as in the hemagglutinin and neuraminidase. This results into a myriad of variants that are unrecognised by the antibodies produced in previous infection. One therefore suffers innumerable cases of influenza. Also, this phenomenon prevents the development of an effective vaccine. The nomenclature for influenza viruses is derived from the changing antigenic pattern. Virologists name the viruses according to their type (A, B or C), the animal of origin if other than humans, the location of first isolation, the strain number, and the year of isolation. It is thus common to see such names as A/Victoria/3/75 or A/Swine/Wisconsin/15/30.

Infection causes sudden chills, fatigue, headache, general aches and pains mainly in the back and legs. The fever may rise to 103°F - 104°F over a 24 hr period, with severe cough. Nose obstructs, throat becomes dry, chest tight. Transmission occurs usually by bits of respiratory mucus and salvia called **droplets**. There may be secondary bacterial infections - *Streptococcus* and *Staphylococcus* invading and damaging the epithelium. The disease is diagnosed by testing patient's serum for neutralising antibodies. If the titer or concentration of antibodies increases during the course of infection, the disease is confirmed. Hemagglutination tests are also made.

Influenza A is sometimes treated with amantadine, a synthetic drug which may block the viral penetration to cells. In epidemic state, influenza may result into **Guillain-Barre syndrome** (GBS) causing nerve damage and poliolike paralysis, and Reye's syndrome-involving liver and brain especially of young children.

Adenovirus infections

Adenoviruses are a collection of at least 35 types of icosahedral virions having double-stranded DNA. The virion is 60-90 nm in diameter, having a dense central core and an outer coat, the capsid. The capsid is composed of 252 capsomeres - 240 hexons (making the faces and edges of equilateral trinagles) and 12 pentons (making the vertices). In capsid there are four additional minor proteins (IIIa, VI, VIII, IX) associated with the hexons or pentons in stoichiometric amounts (Fig. 15.5). The core proteins include proteins V and VII associated with the viral DNA.

The viruses owe their name from the adenoid cells from which they were first isolated in 1953. They multiply in the nucleus of infected cells causing visible nuclear granules or inclusions, which are made up of numerous virions arranged in a crystalline pattern. Adenoviruses cause diseases of respiratory tract, eyes and meninges. Respiratory infection are transmitted by droplets. Because of their stability to drying contaminated inanimate surfaces may play a significant role in transmission of adenoviruses. Like influenza they also cause mild horseness, cough, and loss of apetite called **anorexia**. The terms **common cold** and **croup** are often used for infections. Type 8 adenovirus is the chief cause of kerato-conjunctivitis-eye inflammation, tearing, sensitivity to light and swelling. Type 7 causes meninges in the spinal cord coverings.

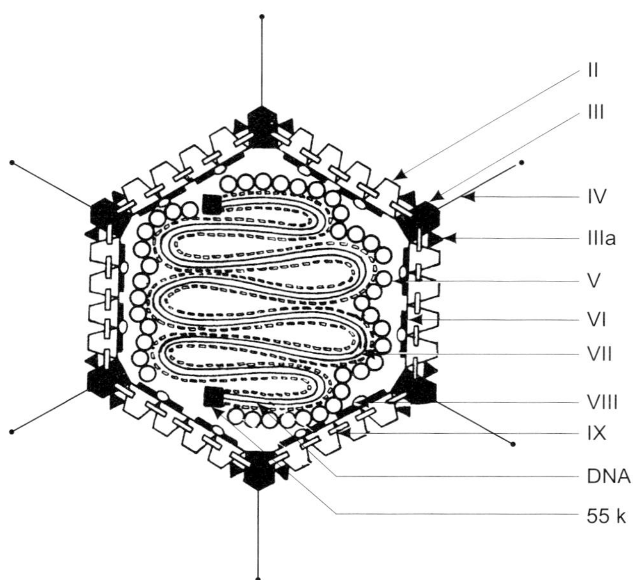

FIG. 15.5

Model of an adenovirus particle, showing the apparent architectural interrelationships of the structural proteins (**roman numerals**) and the nucleoprotein core in the virion. The hexon (**II**), penton base (**III**), and fiber (**IV**) and the hexon-associated proteins (**IIIa, VI, VIII,** and **IX**) make up the capsid. Proteins **V** and **VII** are core proteins associated with the viral DNA, the 55k protein is covalently linked to the 5' end of the DNA

Rhinovirus infections

They are a large group of more than 100 small RNA viruses with icosahedral symmetry. **Rhinos** means nose. The virion consists of a single molecule of RNA (2.3×10^6 – 2.8×10^6 daltons) and a capsid containing 60 copies of four different polypeptide chains. They are involved in respiratory infections. They multiply in upper respiratory tract causing injury to mucosal cells. There is headache, sneezing, stuffy nose and sore throat, a condition called **acute coryza**, but more commonly as **head cold**. There is burning of eye and nasal membranes with little fever.

Of the many types, only 15-20 cause most infections in humans. Vitamin C and chicken soup are useful. Interferons are being developed. For instance zinc ions are found to mask viral proteins and inhibit multiplication.

They are generally transmitted through droplets ie. by inhalation of contaminated aerosols. However, it may also be due to contact and through fomites.

Chicken pox (Varicella)

This is among the most communicable of diseases caused by varicella-zoster virus. The virus belongs to herpesvirus group. It is a double-stranded virion with icosahedral symmetry and 162 capsomeres. In adults the same virus causes **herpes zoster (shingles)**.

Virus is transmitted chiefly by droplets and skin contact. There is in the beginning fever, anorexia and headache, when virus multiplies in the respiratory tract. When it passes to the blood-stream and localises in the peripheral nerves and skin, it multiplies rapidly in cutaneous tissues. This results in the development of fluid-filled, tear-drop shaped skin lesions called **vesicles**. The vesicle contains large amount of virus-laden, highly infectious fluid. Later, after 3-4 days, they break open, forming crusts which become dry. The lesions are larger than those of small pox. The disease becomes serious, if lungs (in which pneumonia may develop) or brain (encephalitis) are involved. Secondary infections of skin by bacteria may also occur. Adenine arabinoside-A (Ara-A), a nucleic acid substitute has shown good results in the treatment. In Japan, a vaccine is also developed.

Herpes zoster (shingles) in adults is a painful disease in which the virus multiplies in the ganglia along the spinal cord and then passes down the nerves of trunk to the skin. There is superficial tingling and burning sensations in the skin with blotchy red patches.

Herpes simplex

Herpes simplex is a collection of diseases. It may occur as cold sores in the mouth, eczema of skin, encephalitis in new-borns, or urinogenital infections in adults. The virus is also linked with human tumors. Figure 15.6 summarizes these infections.

There are two major types of herpes simplex virus: Type 1, designated as **HSV-1**, infecting usually areas above the waist. whereas Type-2, **HSV-2**, often found below waist. Both are large DNA viruses with icosahedral symmetry and envelopes with spikes. The particles have a diameter from 180 nm to 200 nm. The DNA-protein core consists of double-standed DNA wrapped around an associated protein, as one the spindle of a spool (Fig. 15.7). The envelope contains a number of glycoproteins, the capsid is composed of 162 elongated hexagonal prisms, the capsomeres and at least four unique proteins; and the so-called integument consists of about eight distinct polypeptides.

The viruses mutliply rapidly in the nucleus of infected cells forming inclusions called **Lipshutz bodies**, an aid in viral diagnosis. The virus remains for a long period of time in the body causing recurrent infections inspite of the presence of antibodies. The Greek word **herpes** means to creep, indicating the spreading nature of disease.

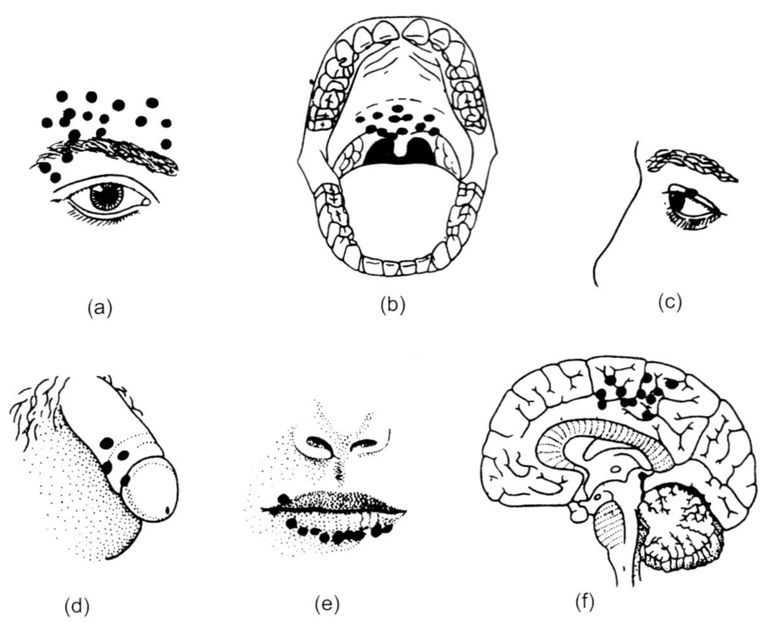

FIG. 15.6

The variety of herpes simplex infections. Note the infections of skin (a), gingivostamatitis of the pharynx (b), lesions of eye (c), genital sores of penis (d), canker sores about mouth (e), and lesions of the brain (f)

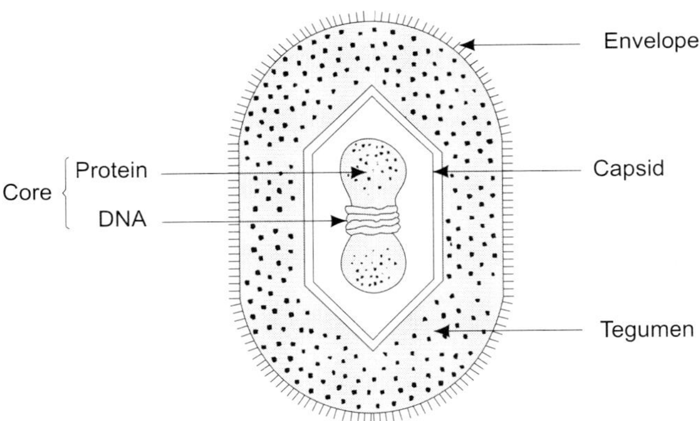

FIG. 15.7

Model of virion of herpes simplex virus. The DNA-protein core consists of DNA wrapped around an associated protein as on the spindle of a spool. The envelope contains a number of glycoproteins. The capsid is composed of at least four unique proteins; and the so-called tegument consists of around eight distinct polypeptides

In adults, herpes simplex infections appear as closely grouped, thin-walled vesicles, or cold sores which occur repeatedly in the same general area of skin or mouth (Fig. 15.6). In children, the disease manifests itself as cold sores or as a more serious condition, **gingivostomatitis**. Type 1 herpes simplex virus also induces a type of nonepidemic encephalitis in children. Viral multiplication in the brain tissue leads to lesion formation. Transmission occurs by several means including contact with saliva or contaminated utensils.

Adults commonly suffer a genital form of herpes simplex caused by HSV-2, and lesions develop on external genital organs of the male (Fig. 15.6) and internally along the vaginal wall and cervix in females. In this case transmission is by sexual contact. The child borne by infected female may develop herpes encephalitis as the virus is transmitted from the vaginal fluids.

In recent years, the acronym TORCH has been coined for diseases with congenital significance - T for toxoplasmosis; R for rubella; C for cytomegalovirus infection and H for herpes simplex. The O represents other diseases. HSV-2 is found to be associated with cervical tumors.

A nucleic acid analog, iododeoxyuridine (IDU), replacing thymine in DNA of the virus produces a nonsense viral genome. This is applied topically to reduce side effects and used for herpes simplex infections of skin as eye lesions and cold sores. Adenine arabinoside (Area-A) also known as Vidavabine (Viva-A) is also applied as topical ointment. Acyclovir (Zovirax) was licenced in 1982 for use against all herpes viruses, especially for genital simplex. The herpes virus is one of the most common viruses in the environment.

Measles (Rubeola)

This is a highly communicable respiratory disease of children, whose symptoms develop on skin. Rubeola is derived from the Latin **rube**, meaning red-indicating the red rash of the disease. In the beginning there is fever, coughing and sneezing, the most communicable phase of the disease. As the virus localises in subcutaneous tissues the skin breaks out in a blotchy rash. The rash begins behind the ears at the hairline then covers the face and trunk. The first evidence of disease appears along the gums and on the wall of pharynx. There develop red patches with central white lesions. These occur concurrently with the fever and are known as **Koplik spots**, after the American physician, Henry Koplik who first described them. This is an important symptom.

The virus is a helical RNA virion, closely related to the mumps and RS viruses. Average size is 125-250 nm (range 100-800 nm), nucleocapsid diam. 18 nm, single-stranded RNA of mol. wt. of $5\text{-}6 \times 10^6$ daltons. It contains envelope spikes with hemagglutinin, a factor used in identification by the **hemagglutination-inhibition**

(HI) **test**, There are no neuraminidase spikes on envelope. Measles may lead to many complications. Due to secondary infection of damaged epithelium of respiratory tract, bacterial pneumonia may also develop. There is also evidence that measle virus may be related to multiple sclerosis, diabetes and encephalitis.

The earliest vaccine for measles was developed in 1954 from chemically inactivated viruses by John Enders and Thomas Peebles. It was, however, replaced in 1960s by another vaccine containing attenuated viruses. At present the children are vaccinated after 15 months of age.

Mumps

This is a disease of children's salivary glands, especially the parotid glands - thus technical name, **epidemic parotitis**. Disease is generally transmitted by droplets, contact and fomites. The virus is present in urine, blood and cerebrospinal fluid. The disease may begin on one side of the mouth but both glands are affected in most cases. This virus occurs only in humans. It is single-stranded RNA - containing helical virion, and closely related to measles and RS viruses. In all respects the virions are similar to measles virus except that there are also neuraminidase spikes on envelope. There are hemagglutinin-containing spikes on the envelope. It is usually a benign disease without any complications in children. There develop antibodies. In adults, mumps represents a threat to reproductive organs, and some males may develop **orchitis**, i.e. swelling of testicles to three or four times of normal size. Sperm count may be reduced. Though rare, in females, **oophoritis** (lower back pain and enlargement of ovaries) may develop.

The mumps vaccine was developed in 1967 from live viruses of Jeryl Lynn strain grown in duck embryo tissues. The virus may multiply in other tissues, in beta cells of pancreas, causing **pancreatitis** or in brain tissues causing **mumps encephalitis**.

Small pox (Variola)

It has a complex architecture without obvious symmetry. The virus is a brick-shaped DNA particle, largest in size - 270 nm. The nucleocapsid is surrounded by a swirling series of fibers and has no envelope. The particles have round corners and a central dense region with crescentic dense areas on each side. The surface layer is made of threadlike, double-ridged, beaded structures. In section, there appears a central nucleoid with a dumbbell-shaped dense core composed of regularly arranged, DNase-sensitive, electron dense threadlike structures. The nucleoid is surrounded by lipoprotein membranes. Between the nucleoid and the outer viral coat is an ellipsoidal body which causes a prominent central bulging of the virion. The particle is composed of about 3.2% DNA associated with spermidine, 91.6% protein, 5% lipid and 0.2% non-DNA carbohydrate (present in the membrane glycoproteins).

The early symptoms are chills, fever and general prostration. With the fall in temperature, pink-red spots called **macules** appear, first on the scalp and forehead, then on neck and extremities and finally on trunk. The spots now turn to pink pimples or **papules**, and later fluid-filled **vesicles**. The technical name, **variola** is derived from the Latin **varus** meaning pimple. In the final stage, the vesicles advance to **pustules**, covering the body surface. When they break open, pus comes out. They are deep in the skin rather than on surface as in chicken pox. The pustules may form a soft crust, which falls of leaving a pitted scar, or **pock**. The name small pox was used to distinguish the pock from the larger lesions of syphilis or varicella (chicken pox).

Transmission is by contact with the skin or any of the body fluids including blood, urine or droplets. In 1798, an English physician, Edward Jenner observed that milkmaids and others working with dairy cattle contracted a closely related but mild form of pox called cowpox, or **vaccinia** from the Latin **vacca** means cow. Apparently those who had recovered from cowpox did not contract small pox. Jenner, therefore, substituted material from a cow pox lesion for the smallpox material and established the process of vaccination. His method became so famous that Napoleon ordered his army to be vaccinated in 1806.

Introduction of the cowpox virus to the blood stream induces the body to manufacture antibodies which neutralise both the cowpox and smallpox viruses as they are partially similar (Fig. 15.8). The World Health Organisation (WHO) received in 1967 funds for an attempt at global eradication of smallpox. By a process of **Surveillance containment**, pox patients were isolated, and every known case was

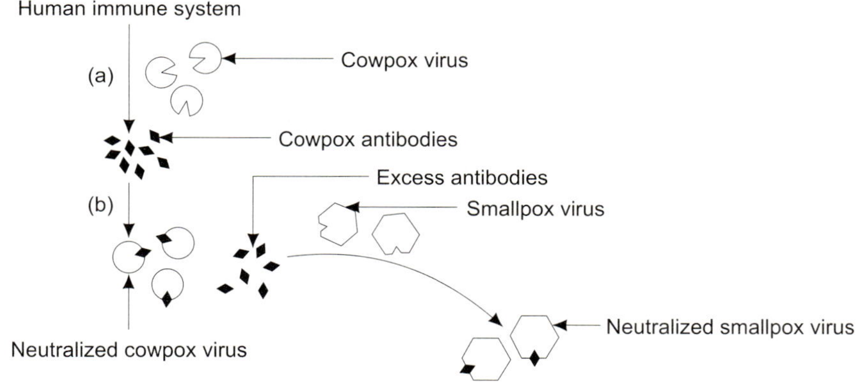

FIG. 15.8

How cowpox antibodies immunize to smallpox. (a) when cowpox viruses enter the body, the human immune system produces cowpox antibodies, (b) these antibodies neutralize the cowpox viruses, and the excess continue to circulate in the blood system, (c) when smallpox viruses enter the body, these same antibodies also neutralize the smallpox viruses and prevent the disease from becoming established

vaccinated. The last epidemic occurred in 1974 in Bangladesh, and in 1976 in Ethiopia. By October 1977, the WHO claimed that the last case had been isolated. On October 26, 1979 the announcement was made that smallpox had been eliminated from the earth, the first such claim made for any disease. The virus is said to exist only in four research laboratories of U.S.A., Russia, U.K. and Japan, permitted by the WHO.

Rabies

Rabies has the highest mortality rate of any human disease once the symptoms have fully materialised. It occurs in most animals in nature from dogs and cats, to horses and rats, to skunks and bats. It is equally fatal to animals and humans.

The virion is RNA-containing enveloped particle, round on one end and flattened on the other to give the appearance of a bullet. The virus enters the tissue during a bite, skin wound, or abrasion from an infected animal. Though, disease associated with saliva, the urine, lymph, blood or milk may be equally infectious. In 1978, a woman died of rabies in Boise, Idaho, due to cornea transplant from an infected person.

The incubation period for rabies varies according to the number of virions introduced to the wound and the wound's proximity to the central nervous system. It may vary from six days to one year, and generally shorter in children than in adults. The virus multiplies in the muscle tissue and then spreads rapidly to the neural pathways. Early signs of rabies are tingling, burning or coldness at the bite site alongwith fever and headache. Later, due to increased muscle tone, patient becomes alert, aggressive and show unusual behaviour. Due to paralytic effect in pharyngeal muscles, difficulty in swallowing is felt. Salivation becomes profuse and saliva drips from the mouth as it cannot be swallowed. This symptom together with brain degeneration increases the person's reactions to the sight, sound or thought of water. Traditionally, the disease has been called **hydrophobia,** meaning "fear of water". The paralysis spreads and death occurs within a few days due to respiratory inhibition.

Dogs are usually infectious for ten days before they show rabies symptoms. Due to this a dog may be held for observation after a bite. If ten days pass without any event it may be assumed that the dog was not contagious at the time of bite. If the dog dies immediately after the bite, its brain tissue sample could be inoculated to laboratory mice that should show symptoms within three weeks if the dog was rabid. A person suffered from animal bite should be precautionary treated as if the animal were rabid. The wound is flushed thoroughly with soap and any antispetic be applied. Serum containing anti-rabies antibodies should be injected to the base of wound. Thus an interferon is helpful. A tetanus shot may also be given if not given during last ten years or so.

A combination of vaccine and antibodies are then injected to neutralise the virus before it reaches the central nervous system. Early vaccines contained viruses from chick embroys or rabbit nerve tissue. But these were found to be allergic. Later, a vaccine was produced from viruses cultivated in duck embryo tissue inactivated with propiolactone. Duck embryo vaccine (DEV) was injected into the abdominal fat at a 45° angle on each 14 successive days or more, if needed. This virus is absorbed slowly from among the fat cells and exposes the body to virus for a longer period. The injections were painful due to the sensitivity of abdominal muscles.

Recently, a new rabies vaccine, **Merieux human diploid cell vaccine**, is shown to produce high amount of antibodies after only three injections in the arm. The viruses are cultivated in human embryonic lung cells, and since human tissue is used, allergic responses are minimum. This vaccine, licensed in 1980 in U.S.A. is now used in other countries including India. It is an expensive vaccine. The old and new methods of vaccination are shown in Fig. 15.9.

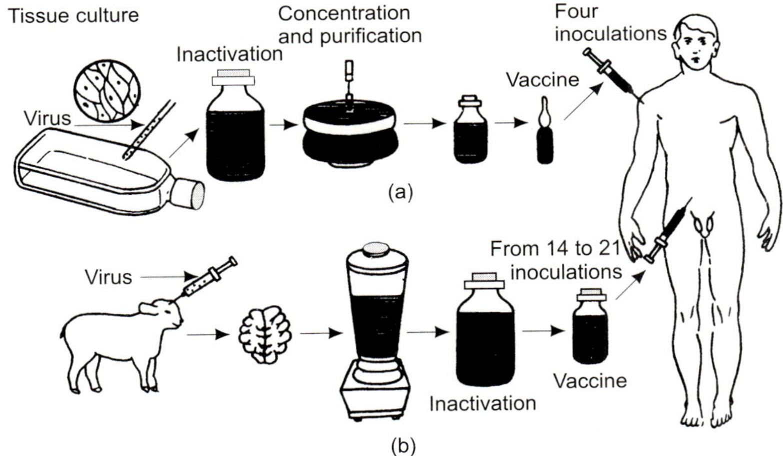

FIG. 15.9

A comparison of new and old vaccination methods in rabies. (a) the new method involves cultivation of rabies viruses in human tissue culture followed by inactivation of the viruses, concentration, and purification. The vaccine is then injected into the muscle. (b) in the old method, viruses were cultivated in brain tissues as shown, or in fertilized egg cultures. The tissue was then ground in a blender and the viruses were inactivated. This method required many successive inoculations into the abdominal fat

Rabies in animals occurs in two forms:

Furious form, characterised by violent symptoms as the animal becomes wide-eyed, drools, and attacks anything in sight. It becomes paralysed, lapses into coma, and dies.

Dumb form, recognised by deep lethargy and animal dies suddenly of paralysis.

Polio

This is one of the smallest known viruses. It has a diameter of about 27-30 nm. It has single-stranded RNA (mol. wt. 2.5×10^6 daltons i.e. 7.7 Kb) with an icosahedral capsid. It multiplies rapidly in nerve cells, producing over 10,000 new particles during a single cycle of about six hours.

The virus enters the body through contaminated food or water. It first multiplies in the tonsils and the lymphoid tissues of gastrointestinal tract. There may be influenza, nausea, vomitting, and intestinal cramps. Disease may end at this point. However, in some cases, it progresses, and the virus passes to the bloodstream. It then localises on the gray matter of the spinal cord. The word polio is taken from the Greek **polios** meaning gray. Meningitis may occur and paralysis experienced the muscles of arms, legs, trunk and other areas controlled by spinal cord. This condition is often called aseptic meaningitis. The most serious form of polio, is the one in which the virus infects the stem of the brain, known as the **medulla**. This conditions is called **bulbar polio**, due to bulblike appearance of the medulla. The nerves of the upper body torso are affected. Swallowing becomes difficult and paralysis occurs in the tongue, facial muscles and neck.

Three types of polio virus are known to exist. Type 1, the **Brunhidle** strain, causing a major number of epidemics and moderate cases of paralysis. Type II, the **Lansing** strain, occurring sporadically but inducing greatest percentage of paralytic cases. Type III, the **Leon** strain, also occurring sporadically but virus confined to intestine, rarely causing paralysis.

Jonas Salk and his group in 1940s prepared huge quantities of viruses and inactivated them with formaldehyde. The vrius was then injected into the body to producle antibodies. Albert Sabin and his group developed a vaccine containing attenuated viruses prepared by multiple passages through tissue cultures. This vaccine may be taken orally and is in current use. It produces more antibodies, providing a longer exposure to the virus due to low multiplication rate of viruses in intestine. Modern vaccines incorporate all three types of polio viruses, said to be trivalent.

AIDS: A DEATH WARRANT

The first of the few cases of AIDS were reported in 1981 by the Centre for Disease Control, Atlanta, USA. Isolated cases of AIDS were identified as early as 1976. Serological data suggest that AIDS virus was not present in the human population before 1979. There is suspicion that this virus may be a monkey virus which was first transmitted to man in Africa and then to USA and Europe. Millions of Americans are now infected with this virus. The disease is prevalent in Australia, South Africa, Central Africa, different parts of Europe, the Caribbean and Japan. Reports of the incidence of AIDS are increasing from various parts of India also. The disease that was once restricted only to the west, is gradually appearing here also. Medical researchers have been examining blood samples of the suspected individuals.

Today, there are about 5 to 10 million infected persons in, world who could have the disease themselves or could infect others. More than 80% of the suspected patients of the world are in USA alone. Till 1986 the total number of patients who have died from AIDS in the UK is about 200 and in the USA about 20,000. AIDS kills ruthlessly, deceptively, remorselessly and surely. Mortality approahces 100% in four years. On an average, most of the infected persons are men, making 92.5%, followed by women, 6.5% and children 1%. The vast majority of the patients are homosexuals, making about 75% that is today the major cause of death of single man in USA in the age group of 25 to 44.

The following figures show an average incidence of AIDS in different categories of persons:

Men	—	92.5%
Women	—	6.5%
Children	—	1%
Homosexuals	—	75%
Intravenous drug users	—	15%
Blood transfusion recipients	—	1.5%
Haemophiliacs	—	0.6%
Heterosexuals	—	0.9%
Others	—	7%

From the above figures, it is evident that vast majority of the patients are homosexuals and males.

In 1981 while working in hospital near Los Angeles, Dr. Gottlieb was dismayed to treat in a short period of three months as many as four cases of an extremely rare infection of the lungs-pneumocystis pneumonia, an infection that he would have never come across during his life period. It was an "opportunistic infection"—an infection that invades the body when its defense system collapse. And to his utter surprise they all turned out to be avowed homosexuals. After the publication of these cases, reports of similarly afflicted young homosexuals started pouring in at the Centre for Disease Control in Atlanta, USA. Most of the patients were sexually active. It was, however, later found that all the subjects were not homosexuals and a few were intravenous drug abusers who often shared needles, or those who had received blood or its products (like factor VIII); where the donor was an AIDS patient. It was shown to occur in haemophiliacs. The female partner of an AIDS patients or infant born of an AIDS mother may also develop the disease.

It could therefore be assumed that this disease was not infectious by casual contact. The infection is not spread by touch or by contamination of food and drinks as in the case of typhoid, cholera and jaundice nor is it airborne as in lung tuberculosis and respiratory infections. None of the health service staff in the world had contracted AIDS from a patient.

It is now established that AIDS is prevalent amongst male homosexuals. The association of AIDS with the life style is conspicuous. The risk of contracting the disease depends on living in high risk areas, anal receptive intercourse, multiple sexual partners exposure to men in high risk area (high risk patients are those who remain asymptomatic throughout their lives, while carrying the virus; a patient may be suspected to have AIDS if he shows unexplained fever muscle pains, night sweats and persistent lymphnode enlargement), multiple episodes of sexually transmitted diseases like syphilis, gonorrhea, genital herpes, hepatitis B, giardiasis etc., intravenous drug abuse and use of recreational drugs like cocaine, amphetamine, methaqualone amylnitrite etc. Anal receptive intercourse helps probably through anal deposition of semen and trauma to the rectal mucosa inflicted in the process, allowing systemic access of retrovirus and the semen. Multiple doses of semen are said to be immunosuppressive. Multiple episodes of sexually-transmissible diseases are also immunosuppressive. The nitrites facilitate entry of infective agents into the system through rectal mucosa by their vasodilating effect.

AIDS thus spreads through blood in (i) homosexuals, (ii) intravenous drug abusers, (iii) female sexual partners of any of 1 or 2, (iv) through open wound, (v) transfusion of whole blood, plasma or platelet transfusion, (vi) use of improperly sterilised or unsterilised syringes and needles, (vii) infants get from their infected mothers.

AIDS (Acquired Immune Deficiency Syndrome) is presumed to be caused by a virus, which has been isolated from the infected patient's blood, lymph glands, brain tissue, cerebrospinal fluid, tears, bone marrow cells, cell free plasma, saliva and semen. The virus responsible for AIDS has been given different names, such as Aids associated Retrovirus (ARV), Lymphoadenopathy associated virus (LAV) and Human T-cell Lymphotropic Virus Type-III (HTLV-III). HTLV-III has also been recently given the name Human Immunodeficiency Virus (HIV). HIV is a retrovirus i.e. belongs to virus family Retroviridae. HIV is said to exist in the form of several strains in human blood. The other two viruses infecting the lymph system of man are, Human T-cell lymphotropic virus-I (HTLV-I) and HTLV-II, that are oncogenic, causing adult T-cell leukaemia and lymphoma, and hairy T-cell leukaemia respectively. HIV is shown to act as a cofactor in the development of a tumor disease in man, the Kaposi's sarcoma caused by Cytomegalo virus (a member of Herpesviridae).

Proteins, GP120 and GP41 together make up the outer skin of HIV virus (Fig. 15.10). The genetic material is RNA which is enclosed by another group of protein, P24 that makes up the inner core of the virus. The genetic information contained in RNA of the virus is reversed by the enzyme reverse, transcriptase. The enzyme is used to translate its genetic information into DNA, which is then inserted into the genes of human cells infected by the virus.

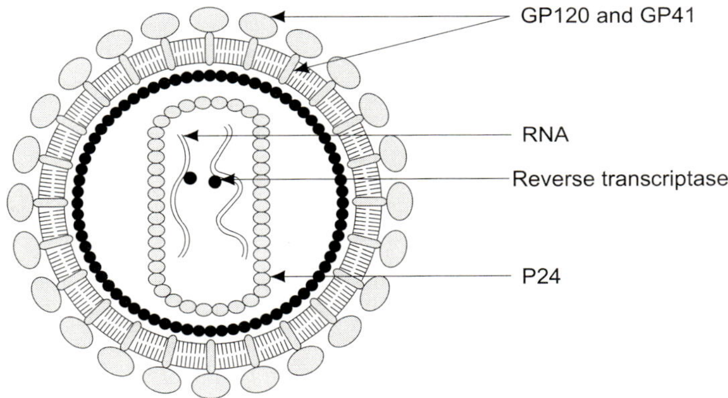

FIG. 15.10
Model showing structure of HIV

Recent studies on blood samples, taken from Madras have shown the presence of antibodies, and researchers at AIIMS, Delhi presume that there may be an Asian strain of HIV. It is likely that HIV exists as several strains in different parts of the world. HIV-1 is supposed to be the American strain i.e. it has typical characteristics found largely in blood samples of patients from the U.S.A., Europe and Central and Eastern Africa. HIV-2 perhaps occurs largely in blood samples of patients from West Africa. Perhaps the Indian isolate may be an Asian strain of the virus.

Human beings have a disease defence mechanism, an immune system. The invading microbes are generally killed by this system. The important component of this system are a type of blood cells called lymphocytes. There are T and B types of lymphocytes, of which the former initiate and prime the immune response. The AIDS virus has a liking for T-lymphocytes-T helper (T_4) lymphocytes causing a decline in the number of helpter T cells and a change in the T_4 (helper): T_8 (suppressor) lymphocyte ratio. It multiplies inside them, and finally disintegrates them. The liberated virus particles enter fresh lymphocytes to repeat the cycle, till all lymphocytes are killed. Thus the immune system of the patient collapses and passes into defenseless state. This state is called AIDS.

The AIDS syndrome could be clearly divided into two stages. First, there is an infection which causes the immunosuppression. The excess of suppressive activity is thought to be responsible for the resulting immunodeficient state which characterizes AIDS. Second, such immunodeficient people are susceptible to microbial attack and it is the inability to counter infection which leads to death. It should be emphasised that only 5-20% of those infected by HIV become permanently and grossly immunodeficient, thus evidence of HIV infection is not synonymous with AIDS. AIDS patients suffer from a variety of diseases where pneumonia caused by the protozoan *Pneumocystis carinii* is common. One form of cancer, Kaposi's sarcoma is seen more frequently in AIDS than in non-AIDS people, and this cancer is linked to Cytomegalovirus. Whatever the final cause, clinically recognisable AIDS has a 100% fatality rate.

The period from which the patient is exposed to infection to the development of the full-fledged clinical picture is generally between 9 to 300 months. In blood transfusion related cases, this period is likely between 4 to 14 months.

Persons having AIDS infection can be classified into three categories as follows:

(1) **Healthy "carriers".** This is the **first** and the commonest stage. They have no any symptoms. They may remain so throughout their lives. They are capable of infecting others. They themselves are living at top of a volcano. No one knows when it will blow up and how many of them will develop the deadly disease. There are about a million carriers in USA alone; 5 to 10% of these will develop the full blown disease within 5-7 years.

(2) **Prodromal phase.** This is also called AIDS related complex (ARC) or Lymph node syndrome (LNS) or Lymph adenopathy syndrome (LAS). This group has a **milder** form of the disease with fever, diarrhoea, weight loss and enlarged lymph glands but they show unequivocal evidence of depression of immune system. Sometimes the prodrome may be associated with some fungal infections of skin or inside (mucosa) the mouth, warts and herpes. This phase may last for 1-30 months (even upto 5 years). It is seen that all cases of prodromal phase may not progress to a full-fledged disease, but some 20-30% do in the next three years or so.

(3) **The end stage.** This is the full blown picture of AIDS. The patient dies of fulminant and uncontrollable infections. The immune system completely collapses, never recovers. The patient may survive a year or two but not longer than four years. Presence of HIV may encourage the development of skin tumor-Kaposi's sarcoma caused by Cytomegalovirus.

The picture comprises symptoms of (i) **respiratory system** where lung infection is the commonest presentation of AIDS. This is seen as dry cough, breathlessness on exertion, vague chest pain and unexplained fever. Fungi and bacteria may also infect lungs, (ii) **digestive system** in which there are low volume frequent stool, loss of weight, abdominal pain and discomfort, (iii) **neurological system** where the symptoms are caused by infection and patient may suffer from headache, personality changes and congnitive changes, loss of memory, changes in consciousness level and convulsions, and (iv) **malignancy** where the commonest one is Kaposi's sarcoma- a skin malignancy. This is caused by a virus Cytomegalo virus. Advanced stage of AIDS infection acts as a cofactor for the development of this oncogenic virus. This has a tendency to invade the body allover. Other types of malignancy include that of lymph glands (Non-Hodgkins B cell Lymphoma) and carcinoma of the mouth and the rectum.

The laboratory tests help in diagnosis of AIDS. Blood samples from the patient are put to virological, serological and immunological tests. HIV could be isolated from blood. Detection of antibodies against HIV in serum of AIDS - affected and high risk patient shows indirect evidence. Immunosuppression state can be shown by depletion of T-helper lymphocytes and alteration of T-helper to T-suppressor lymphocytes in circulating blood.

There are no curative methods to control AIDS. The only ways are (i) a prophylactic vaccine against the AIDS virus and (ii) to check the spread by development of tests. Work going on HIV may result in near future in the development of a vaccine. Attempts are being made to use bone marrow graft and transfusion of immune cells from healthy individuals. One drug, suramin is being tried. Tests are also being made with Rivavirin (broad spectrum antiviral). The drug Azidothymidine is also being tried and also a mineral compound HPA-23.

The spread of the disease can be greatly checked by development of tests which can identify persons harbouring AIDS virus. Blood donors should be routinely checked and rejected if found positive. Factor VIII-a product of stored blood used in treatment of hemophiliacs-is now made safe by heating that kills the virus. Mothers with AIDS are advised against pregnancy. An educational campaign to bring about consciousness about danger of promiscuity, intravenous drug abuse and homosexuality, should be given priority. Press and other media also may play important role.

Alarming Increase in AIDS Cases

According to WHO reports there is possibility of anything between five to thirty lakh new AIDS cases across the world over the next five years. In the year 1987 there was swift and dramatic rise in the number of AIDS cases. There were 58,235 cases till August 1987, that rose to 73,747 by the end of December 1987, a 27% rise. Till Feb.

1988, this figure soared to 81,433 from 133 countries. According to WHO, there should be roughly 150,000 cases and between five and ten million persons may be currently infected with HIV. The US Public Health Service estimates that by 1991, some 270,000 cases of AIDS will have occurred in the USA. In Europe, where 27 countries have reported 10,177 AIDS cases, most country are said to be facing its rising form. WHO estimates between half to one million persons in Europe infected with HIV. Highest rates per million population are found in Switzerland, France and Denmark. By the end of 1988, some 25,000 new cases would have occurred in Europe.

In Africa, the number of countries reporting AIDS has arisen over the past year. By February 1988, 41 African countries had reported 9,788 cases. In Asia, 19 countries have reported 233 cases of AIDS. Oceania including Australia and NewZealand has reported 826 cases.

Cat's Claw, Possible Cure for AIDS

Recently a climbing plant, *Uncaria tomentosa* (Rubiaceae) has been claimed to be the elusive cure for this deadly scourage. The plant is found in the jungles of Amazon, South America and better known as cat's claw by indigenous people as the plant is covered with hairy claw-like spines. This plant is credited with magical powers. Its antiinflammatory properties are already known and is now being hailed as the new miracle drug that could cure cancer and even AIDS in 21st century.

Cat's claw has been used successfully to treat inflammations associated with gastritis and arthritis. It is now known to strengthen the body's immunological system. A Peruvian, Manual Moreyra claimed to have cured a malignant tumour in his head by the plant preparation. Dr. Roberto Inchastegui, President of the Peruvian Committee for Sexually Transmitted Diseases and AIDS in the Amazon city of Iquitos, claims to have cured seven AIDS patients with a preparation made from cat's claw and other jungle plants. Several biologists and consultants at hospitals in Peru accept that this plant has antiinflammatory effects and is a powerful immunological stimulant. For instance, Dr. Rumberto Muro at the Lima Naval Hospital confirmed that several patients who were seriously ill with full-blown AIDS are now merely carriers of the HIV virus, after undergoing treatment with the plant. Dr. Armando Luza of the naval hospital in Peru bas successfully treated a patient. According to a neurosurgeon, Fernando Cabieses, an expert on medicinal plants, tests conducted in Europe have shown that cat's claw contain substances that inhibit growth of cancerous cells. Researchers at the Innsbruck University Pharmacological Institute in Austria discovered that the plant contains six alkaloids with powerful therapeutic effects. In some parts of Central Europe the plant is being used for patients suffering from allergies, cancer and AIDS. The plant has already become a highly profitable item marketed and exported under 30 different, brand names. Oscar Schuler's family has been traditionally associated with this wonder drug plant. It is preferred to sell it powdered in capsules. Cat's claw comes in many shapes and forms (powder packed in small bags for infusions, in drops, tablets and capsules). It has become a lucrative business in the United States and Europe, and Peru may not even be asked to pay for a license to sell its own *Uncaria.* In 1986, the plant's oxindalic alkaloids were patented in the U.S.A. by a group of European scientists. According to some, there may develop Intellectual Property Rights problems and some experts insist that the American patent is only valid in the U .S.A. because it was never registered in Peru or in any other country. It seems that the humble cat's claw could, in future, become a source of discord between nations, as well as a beacon of hope for millions of patients around the world.

EBOLA HAEMORRHAGIC FEVER

This deadly disease has recently shaken again the Central Aftrica, killing hundreds of lives in Zaire. The mortality rate is 50 to 90% and no prophylactic (vaccine) or therapeutic method of control could be developed so far.

The disease manifests itself as high fever with headache. Severely infected person develops muscle pain, weakness, sore throat followed by vomitting, diarrhoea, rashes, kidney and liver afflications and internal as well as external bleeding. Incubation period ranges between 2-21 days. After entering the body, the virus attacks almost every organ jamming it with congealed blood and disintegrate. There is uncontrolled bleeding from nose and gum. Delirium and death follow even during treatment.

The disease spreads through direct contact with infected person (contagious) or through body fluids and secretions. The killer disease is caused by a virus, which is a worm-shaped filovirus native to rain forests of Central Africa, also home of HIV. Ebola-related filoviruses were first isolated from cynomolgus monkeys, *Macacca fascicularis,* imported into the U.S.A. from Philipines in 1989. It is not known how monkeys became infected, from bitting insect or from animal or plant.

The Ebola virus has been classified as Level 4 Pattagen which can be studied only in controlled, protected environment. Even AIDS virus is classified at Level 3.

Ebola epidemic in Zaire first broke out in 1976, when all the 90% deaths were linked to contaminated syringes and needles, Out of 300 persons infected by the virus, about 275 died in single village in Zaire. Like AIDS, Ebola virus may also spread to other parts of the world.

Primary cause of the epidemic is said to be ecological and environmental degradation by man. There are three types of Ebola virus, (i) Ebola Zaire – the most deadly that broke out in 1976, (ii) Ebola Sudan, isolated around the same time, and (iii) Ebola Reston, appeared in quarantines of USA. So for two epidemics originated in hospitals with poor hygiene. The use of contaminated needles is believed to be the most important factor in the outbreak of this disease. In 10 April 1995 epidemic a surgical patient became infected by medical personnel.

PROTOZOAN AND HELMINTHS PARASITES

Besides bacteria and viruses, protozoa (unicellular microorganisms of Kingdom Protista) and Helminths of Kingdom Animalia (roundworms, flatworms, tapeworms, and lukes) also are causal agents of several important diseases of man. The study of such human parasites is a separate discipline called **Parasitology**. Among protozoa, the two very common human parasites are *Giardia lamblia* and *Entamoeba histolytica*, whereas among Helminths, *Ascaris lumbricoides* and *Taenia saginata* are frequently encountered. This is beyond the scope of the text of this book to describe the etiology, transmission, characteristics of the parasites and control of various diseases caused by protozoa and helminths. However, the characteristics of some human parasites are presented in Table 15.4 Similar information on human parasites of this group is also given in Table 15.1.

TABLE 15.4 Characteristics of some important human parasites

Organism	*Infective form*	*Mode of transmission*
Giardia lamblia	Cyst	Person-person, Waterborne, Foodborne
Estamoeba histolytica	Cyst	As above
Cyclospora cayetanesis	Sporulated oocyst	Waterborne, Foodborne
Ascaris lumbricoides	Embryonated egg	Oral from soil contact
Tricuris trichiura	Embryonated egg	Waterborne, Foodborne
Necator americanus	Filariform larva	Skin penetration
Ancylostoma duodenale	Filariform larva	Skin penetration, Oral from soil contact
Taenia saginata	Cysticercus	Ingestion of undercooked beef, waterborne? (intermediate host, cattle)

Notes on some common diseases caused by such human parasites are given here.

Amoebiasis is characterised by painful ulcers of human intestine. The causal organism is *Entamoeba histolytica* (Sarcodina). It enters the body as cyst and passes through stomach acid, as shown in Fig. 15.11. In small and large intestines the amoebae emerge and begin to feed on tissue. This feeding form is known as **trophozite**. Lesions coalesce to form deep ulcer, causing appendicitislike pain. If not

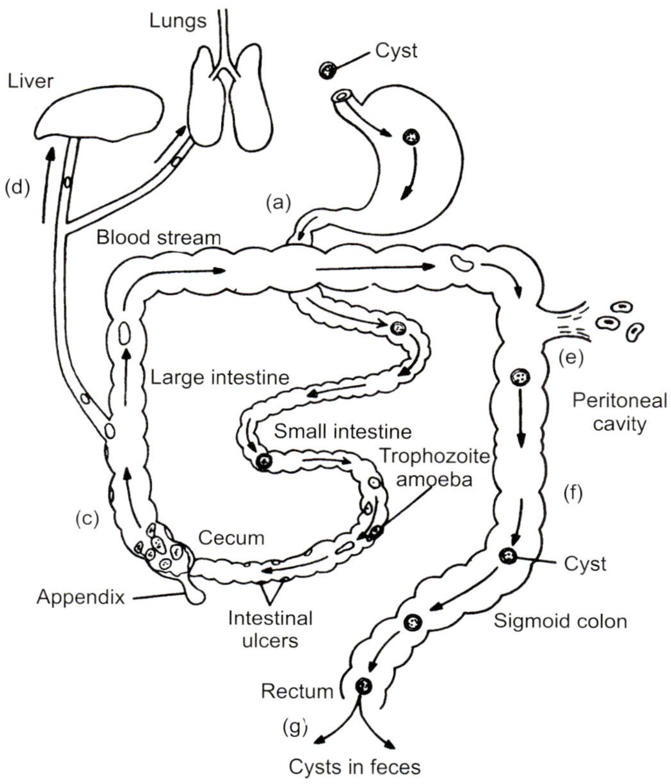

FIG. 15.11
The course of amoebiasis, caused by *Entamoeba histolytica*

controlled, the amoebae eventually reach the blood vessels of intestinal wall. Now the stool becomes bloody, causing the problem of **parasitemia**. If amoebae localise in the lung or liver, these may be fatal. Amoebiasis may be treated with paromomycin and metronidazole and tetracycline.

Roundworm disease is commonly caused by *Ascaris lumbricoides,* the largest of intestinal nematodes. Tropical and subtropical areas are the common foci of this disease. The nematode is like an earthworm. The female is over a foot long and the male somewhat smaller with a curled tail. Heavy infection may cause intestinal obstruction due to presence of tightly compacted masses of the parasite. The female is a prolific egg-producer, sometimes producing 200,000 eggs per day. After fertilisation the eggs are excreted to soil where they hatch to larvae. These attach to plants from which are again ingested (Fig. 15.12). In many parts of the world human feces - night soil is used as a fertiliser to vegetables. This helps in the spread of the parasite. Water pollution from the soil and contact with contaminated fingers are

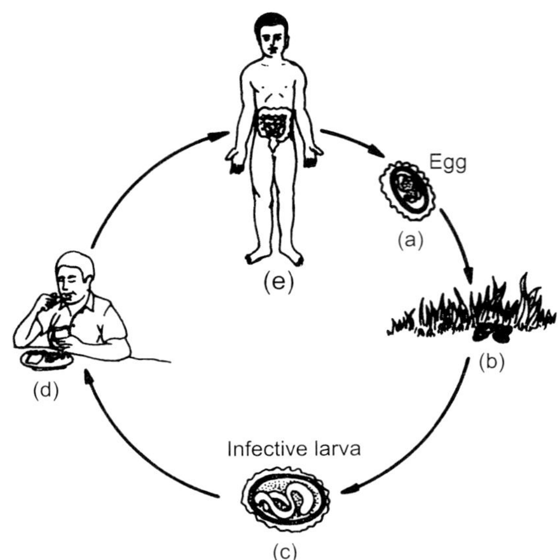

FIG. 15.12
Life cycle of *Ascaris lumbricoides*

other ways of transmission. Roundworms larvae may also pass from intestine to other parts of the body. They may be passed to lungs causing a pneumonialike disorder.

Hookworm disease is caused by two nematodes: the old world hookworm. *Ancyclostoma duodenale,* and the New World hookworm, *Necator americanus.* Of these the former is common in Europe, Asia and the U.S.A. Both are about 10 mm long with a set of hooks or sucker devices, which allow them to attach firmly to intestine. They may induce ulcers and consume blood, causing amemia. Hookworm eggs pass into soil where the larvae emerge as long **rhabditiform** larvae. These convert to hairlike **flariform** larvae. The flariform larvae attach to skin during contact with moist vegetation in the soil and penetrate the skin layers to the bloodstream. They localise in the lungs and are carried up the bronchi, from where they are swallowed. Infection then occurs in the intestine (Fig. 15.13).

In **beef and pork tapeworm diseases** human is the definitive host for both the beef tapeworm, *Taenia saginata* and the pork tapeworm *T. solium.* The former is about 25 feet long, whereas the latter about 8 feet long with up to 2,000 proglottids. The main damage is caused to intestine, causing obstruction. The life cycle involves the release of gravid proglottids to soil, from where they are consumed by cattle or pigs (Fig. 15.14). Embryos are released from the eggs and these are passed to the muscle of animal where they encyst. These may then be consumed in poorly cooked beef or pork.

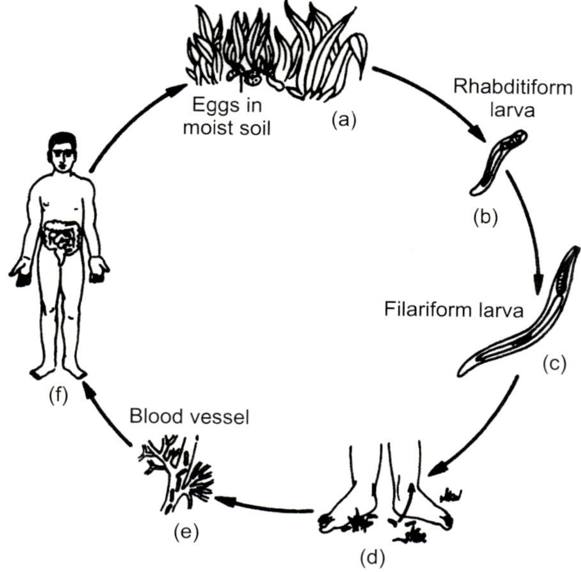

FIG. 15.13

Life cycle of hookworms, *Ancylostoma duodenale* or *Necator americanus*

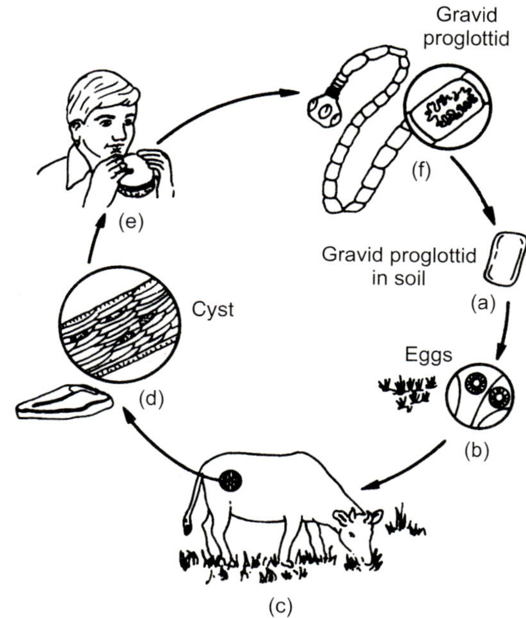

FIG. 15.14

Life cycle of the beef tapeworm, *Taenia saginata*

16

INDICATOR MICROORGANISMS IN POLLUTED WATERS

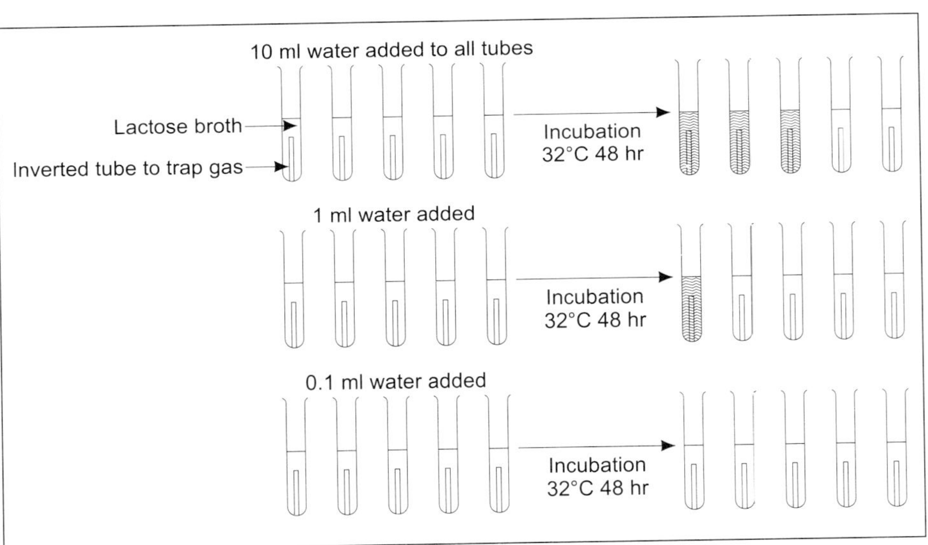

Chapter Outline

- *Microbial Components of Water*
- *What is an Indicator?*
- *Groups of Indicator Microbes*
- *Water Treatment*

There are two major types of water: (i) **ground water** originates from deep wells and subterranean springs. This is virtually free of bacteria due to filtering action of soil, deep sand and rock. However, it may become contaminated when it flows along the channels. (ii) **surface water** is found in streams, lakes, and shallow wells. The air through which the rain passes may contaminate the water. Other sources of contamination are the various types of establishments and agricultural farms etc. by the sides the water flows. Possible sources of microbial contamination of a body of water are soil and agricultural run off, farm animals, rain water, industrial waste, discharges from sewage treatment plants and storm water run off from urban areas.

In water microbiology the water is **polluted** when it contains a chemical or biological poison or an infectious agent. **Potability** refers to the drinkability of water. When potable, it is fit for drinking. When unpotable it is unfit due to some contaminant or pollutant.

MICROBIAL COMPONENTS OF WATER

The types of microorganisms differ in unpolluted, polluted and marine waters. We shall briefly examine such differences. Typical microorganisms in these water environments are shown in Table 16.1.

In **unpolluted water** of mountain lake or stream there are usually low organic nutrients. The number of bacteria is very much limited, a few thousand per ml. Actinomycetes are typical. Other microbes are also shown in Table 16.1

TABLE 16.1 Microflora of different types of waters

Unpolluted	*Polluted*	*Marine*
Actinomycetes	Coliform bacteria	Halophilic microbes
Yeasts	*Escherichia coli*	Psychrophilic microbes
Bacillus spores	*Desulfovibrio* sp.	Diatoms
Clostridium spores	*Clostridium* sp	Dinoflagellates
Cellulose digesters	Fecal streptococci	Mold spores
Autotrophic bacteria	Protozoan cysts	*Pseudomonas* sp.
Euglena	Blue-green algae	Foraminiferans
Paramecium	Enteric viruses	Luminous microbes

Autotrophic bacteria are common alongwith free-living protozoans as *Euglena*, *Paramecium* and various amoebae.

In **polluted waters**, there are large amounts of organic matter from sewage, feces and industrial complex. The microbes are usually heterotrophic. The digestion of organic matter by these organisms is incomplete, due to which there accumulate acids, bases, alcohols and various gases.

The major type of bacteria are **coliform bacteria**, the Gram-negative nonspore-forming bacilli usually found in the intestine. This group includes *E. coli* and species of *Enterabacter*. They ferment lactose to acid and gas. Noncoliform bacteria-*Streptococcus, Proteus* and *Pseudomonas* are also present.

Under some conditions, the polluting microbes multiply rapidly and consume most of the available oxygen. For instance, nutrients enter the river from sources like sewage treatment plants or urban/suburban runoff. Thus river suddenly develops a high nutrient content. Under these conditions algae may bloom rapidly. This leads to depletion of oxygen in water. There is very little oxygen available to the protozoa, small animals, fish and plants. Due to this non-availability of oxygen, a layer of dead organisms, mud and silt accumulate at the bottom, and anaerobeic species of *Clostridium, Desulfovibrio* etc. will flourish.

They produce gases. One gas, H_2S combines with lead or iron to give a precipitate which makes the mud black and the water poisonous. Due to complete depletion of oxygen, the suspended bacteria die in their own waste products. There is hardly any life in water at this stage. The gas bubbles from the anaerobes in the mud break the surface. Such processes lead to death of a river.

The **marine water** is high in salt content and only halophilic microbes survive. Since temperature is very low, most of them are psychrophilic. Diatoms and protozoa, such as dinoflagellates are important components in food chain. In the off-shore **oceanic zone**, photosynthetic organisms are present. These are diatoms and dinoflagellates. Most marine microbes are found along the shoreline or **littoral zone** where nutrients are in abundance. Some unusual types are also present on the ocean floor in the **benthic zone** and even at the bottom of several miles-deep trenches, the **abyssal zone**.

WHAT IS AN INDICATOR?

Water is infact a vehicle for the transfer of a wide range of diseases of microbial origin. The important bacterial diseases include typhoid fever, cholera and bacterial dysentry that are generally transmitted when human feces from carriers or patients contaminate the water. Viral diseases transmitted by water include virus A hepatitis and polio. Many protozoa form cysts which survive for long periods in water. The causal agents of amoebiasis, giardiasis etc. are important concerns in water pollution.

It is a tedious and time-consuming task to perform regular examination of environmental samples, including water for the presence of intestinal or other pathogens. It has rather become a practice to look first for certain microorganisms whose presence indicates (provides a clue) that pathogenic microbes may also be present in the sample. Such microorganisms are known as **indicators** or **indicator microorganisms**. The presence of indicators in the sample prohibits its use for drinking, because its consumption would lead to enteric diseases. The indicator concept depends on the fact that certain nonpathogenic bacteria occur in faeces of all warmblooded animals. These bacteria can easily be isolated and quantified by simple bacteriological methods. Detection of these bacteria in water means that faecal contamination has occurred and suggests that enteric pathogens may also be present.

For instance, **coliform bacteria**, the normal inhabitants of intestines of warmblooded animals are excreted in great numbers in faeces. In polluted water, coliforms are found in densities roughly proportional to the degree of faecal pollution. Their absence from water indicates that water is bacteriologically safe for consumption, whereas their presence indicates the presence of other kinds of disease-causing microbes.

GROUPS OF INDICATOR MICROBES

Indicator microorganisms have also been used to assess the efficacy of food processing and water and wastewater treatments. An ideal indicator organism of faecal contamination should meet the criteria listed in Table 16.2. There is not a single group that can meet such requirements, and, therefore, various groups of microbes have been used as indicators. Concentrations of indicator bacteria found in wastewater and faeces are shown in Table 16.3.

TABLE 16.2 Criteria for an ideal indicator organism of faecal contamination

1.	It should be useful for all types of water.
2.	It should be present wheresoever enteric pathogens are present.
3.	It should have a reasonably longer survival time than the hardiest enteric pathogen.
4.	It should not grow in water.
5.	Testing method should be easy to perform.
6.	Its density should have some direct relationship to the degree of faecal pollution.
7.	It should be a member of the intestinal microflora of warmblooded animals.

Total Coliforms

The coliform group (*Escherichia, Citrobacter, Enterobacter* and *Klebsiella* species) is relatively easy to detect. This group includes all aerobic and facultatively anaerobic, Gram-negative, non-spore forming, rod-shaped bacteria that produce gas upon lactose fermentation in prescribed culture media within 48 hr. at 35°C. The group has been used as standard for assessing faecal contamination of recreational and drinking waters.

TABLE 16.3 Estimated levels of indicator bacteria in raw sewage

Organism	*CFU/100 ml*
Coliforms	10^7–10^9
Faecal coliforms	10^6–10^7
Faecal streptococci	10^5–10^6
Enterococci	10^4–10^5
Clostridium perfringens	10^4
Staphylococcus (cogulase +ve)	10^3
Pseudomonas aeruginosa	10^5
Acid-fast bacteria	10^2
Bacteroids	10^7–10^{10}

Three standard methods are commonly used to identify coliforms in water. These are the membrane filter (MF), the most probable number (MPN), and the specific tests.

The membrane filter (MF) test may be used in the field also. A special, collecting bottle is held against the water current to collect the water. A measured amount of water (usually 100 ml) is passed through a membrane filter (pore size 0.45 μm) that traps bacteria on its surface. The membrane is then transferred to a thin absorbent pad saturated with a specific medium to allow growth of the microbe. For coliforms, a modified Endo medium is used. The culture is incubated at 35°C for 18-24 hours.

For **most probable number (MPN)** water in 10 ml, 1 ml and 0.1 ml amounts is inoculated into lactose broth tubes (Fig. 16.1). The tubes are incubated and coliform organisms may be identified by their production of gas from the lactose. By referring to a MPN table (Table 16.4), a statistical range of the number of coliform may be determined by observing how many broth tubes showed gas. This method does not detect total number of bacteria in the water nor it locates noncoliforms like *Salmonella*. However, it indicates the presence and quantity of coliforms.

Among **specific tests (coliform counts)** the most frequently used indicator organism is the normally nonpathogenic coliform bacterium *Escherichia coli*. Positive tests for *E. coli* do not prove the presence of enteropathogens, such as *Salmonella* and *Shigella*, but do establish the possibility. *E. coli* is more numerous and easier to be grown. For fecal contamination of water, using *E. coli*, as an indicator organism, a three-stage test procedure is followed (Fig. 16.2).

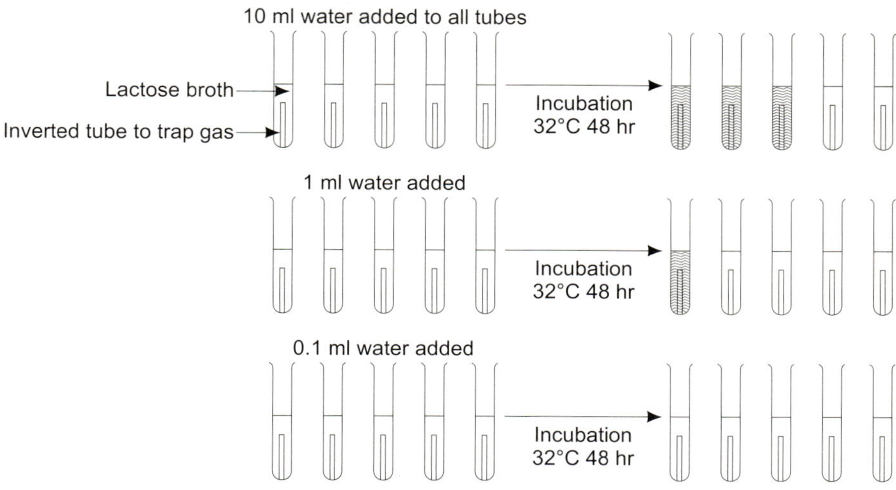

FIG. 16.1

The most probable number (MPN) test. Water in 10 ml, 1.0 ml, and 0.1 ml amounts is added to five tubes of lactose broth, each containing a small inverted tube to trap any gas that may evolve. The fifteen tubes are incubated at 32°C for 48 hours and examined for the presence of gas in the inverted tubes. In the example shown, three tubes containing 10 ml of water and one tube containing 1 ml of water show gas. By reference to an MPN table (Table 16.4), it is determined that the most probable number of coliform bacteria in the water was 12 per 100 ml of water

TABLE 16.4 MPN of coliform organisms in 100 ml of a water sample

No. of positive lactose broth tubes			*No. of coliform organisms*	*No. of positive lactose broth tubes*			*No. of coliform organisms*
10 ml	*1 ml*	*0.1 ml*		*10 ml*	*1 ml*	*0.1 ml*	
0	0	0	0	3	0	0	8
0	1	0	2	3	0	1	11
0	1	1	4	3	1	0	12
1	0	0	2.2	4	0	0	15
1	0	1	4	4	0	1	20
1	1	0	4.4	4	1	0	21
2	0	0	5	5	0	0	38
2	0	1	7	5	0	1	96
2	1	0	7.6	5	1	0	240

(i) **presumptive test.** lactose broth tubes are inoculated with undiluted or appropriately diluted water samples. The tubes showing gas formation are marked as

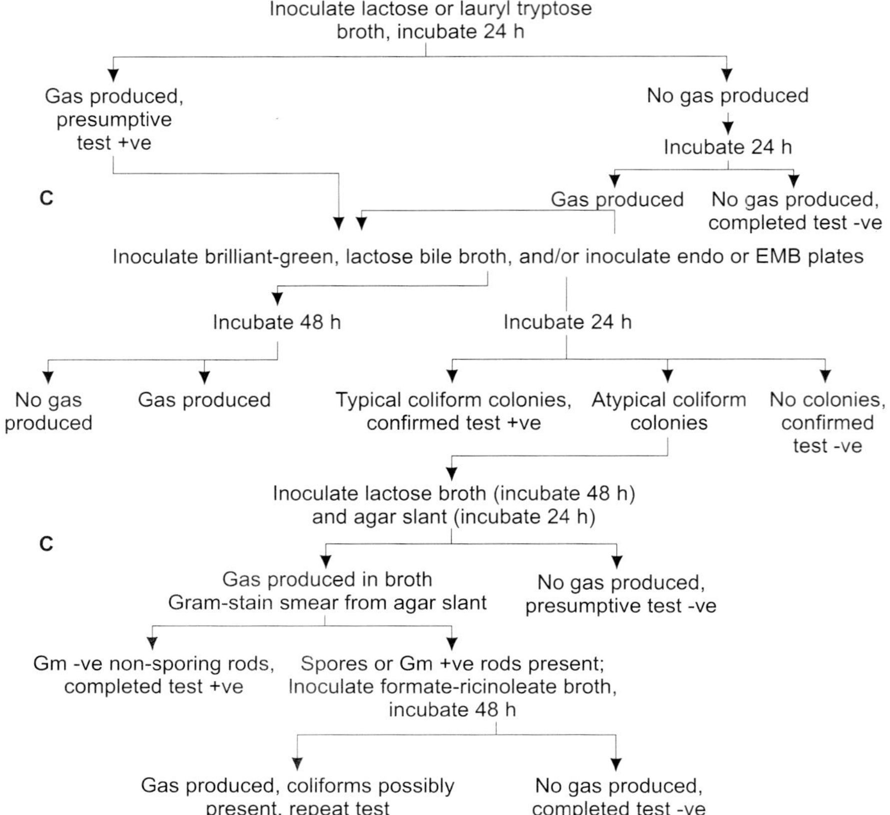

FIG. 16.2

An outline of the water quality testing procedure for determining the number of coliforms

positive and used to calculate the most probable number (MPN) in the sample. Gas formation in inverted test tubes (Durham tubes) indicates fecal contamination by coliforms. (ii) **confirmed test.** gas formation in lactose broth at 37°C is characteristic not only of *Salmonella, Shigella* and *E. coli* strains but also to the nonfecal coliform *Enterobacter aerogenes* and some *Klebsiella* spp. In second stage, the presence of enteric bacteria is confirmed by streaking samples from the positive lactose broth culture onto a medium, such as eosin methylene blue (EMB) agar. Fecal coliform colonies on this medium acquire a greenish metallic sheen, *Enterobacter* spp. form reddish, and nonlactose fermenters form colourless colonies respectively. (iii) **completed test.** the confirmed test can be accomplished by using brilliant green lactose-bile broth (BGLB). If BGLB is used, it is then sub-cultured onto EMB. Subculturing colonies showing a green metallic sheen on EMB into lactose broth incubated at 35°C should produce gas, completing a positive test for fecal coliforms.

Faecal Coliforms

Although the total coliform group has served as the main indicator of water pollution for many years, many of the organisms in this group are not limited to faecal source. Some other methods, have, therefore, been developed to restrict the enumeration to coliforms that are more clearly of faecal origin—that is, the **faecal coliforms**. The two genera (faecal coliforms), *Escherichia* and *Klebsiella* are differentiated in the laboratory by their ability to ferment lactose with the production of acid and gas at 44.5°C within 24 hours. However, in tropical waters, both coliform and faecal coliform, occur frequently. Thus, new indicators are needed for tropical waters. Some have suggested the use of *E. coli* as an indicator because it can be easily distinguished from other members of the faecal coliforms (e.g. absence of urease and presence of β-glucuronidase). Faecal coliforms are detected by methods similar to those for coliforms.

Faecal Streptococci

They are a group of Gram-positive Lancefield group D streptococci and belong to the genera *Enterococcus* and *Streptococcus*. Several species of *Enterococcus* and two *Streptococcus* are used as indicators in water industry. A faecal coliform/faecal streptococci (FC/FS) ratio of 4 or more indicates a contamination of human origin, whereas a ratio below 0.7 indicates animal pollution. Both, membrane filtration and MPN methods are used for the isolation of faecal streptococci.

Clostridium perfringens

This is a sulphite-reducing, anaerobic, spore-forming, Gram-positive, rod, exclusively of faecal origin. The spores are very heat resistant, which limits its usefulness as an indicator. However, it could serve as an indicator of past pollution. Other anaerobic bacteria, such as *Bifidobacterium* and *Bacteroides* can also be used as potential indicators because they are primarily associated with humans.

Heterotrophic Plate Count (HPC)

HPC provides an assessment of the numbers of aerobic and facultatively anaerobic bacteria in water. This group includes Gram-negative bacteria belonging to the genera,

Pseudomonas, Aeromonas, Klebsiella, Flavobacterium, Enterobacter, Citrobacter, Serratia, Acinetobacter, Proteus, Alcaligenes, Enterobacter and *Maraxella*. However, these counts themselves have no or little health significance, because they have been frequently isolated from surface waters and groundwaters and also widespread in soil. Although, the HPC is not a direct indicator of faecal contamination, it does indicate variation in water quality and potential for pathogen survival and regrowth.

Bacteriophages

Due to their constant presence in sewage and polluted waters (approx 10^2–10^3 CFU/100 ml of raw searage), the use of bacteriophages as appropriate indicators of faecal pollution has been suggested. They have also been proposed as indicators of viral pollution. Their presence in water samples denotes the presence of bacterial hosts. Two groups of bacteriophages have been used: (i) **somatic coliphage**, which infects *E. coli* strains through cell wall receptors, and (ii) **F-specific RNA coliphage** which infects strains of *E. coli* and related bacteria through the F^+ or sex pili. Both, plating and MPN methods can be used to detect coliphages.

Standards and Criteria for Indicators

Bacterial indicators such as coliforms have been used for development of **water quality standards**. The criteria used for such standards are different for different countries of the world. Concerned agencies of the particular country enforce the use of such standards. For instance in U.S.A., there are both, state and federal standards for water quality. They make recommendations for acceptable levels of indicator organisms and these are to be legally enforced and also serve as guide to indicate any water quality problem. There are drinking water criteria for tap and bottled waters enforced by European Union also. In our country also, standards have been formulated by the concerned agencies.

WATER TREATMENT

We have seen that water and sewage harbour a variety of organisms responsible for diseases in man. It is thus necessary to remove or rather interrupt the disease cycle in nature through treatment of water and sewage. Water purification and sanitary sewage disposal are two major methods for interrupting disease cycle. In **water purification** organisms are prevented from reaching the body, whereas in **sewage disposal**, they are removed from body waste products.

The various steps in the purification of drinking water in municipal water supplies are shown in Fig. 16.3. It may be seen that there are **three** basic steps in the purification of drinking water: sedimentation, filtration and chlorination.

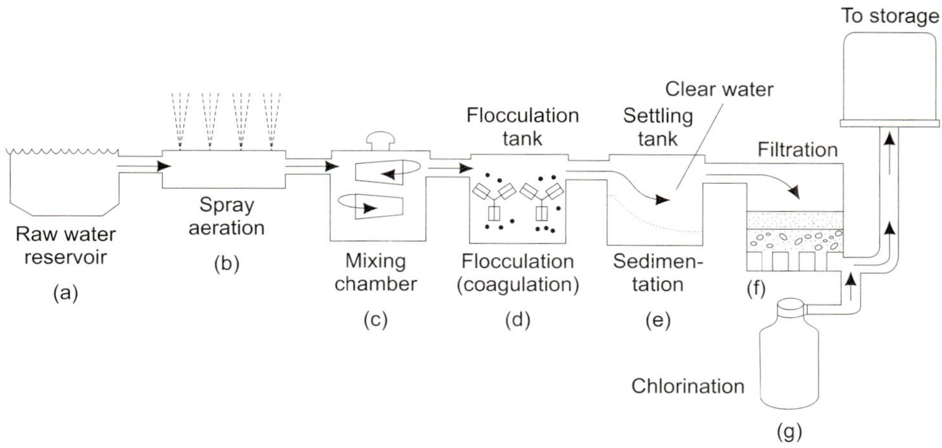

FIG. 16.3

Steps in the purification of municipal water supplies. (a) in the reservoir, large objects are removed. (b) the water is then sprayed in the air to increase its oxygen content. (c) next the water is piped to a mixing chamber where flocculating agents are added. (d) the flocculating agents are churned in the water and large jelly like masses or flocs form (e) the flocs settle to the bottom of the sedimentation tank. (f) the water is then filtered, and (g) chlorinated before being piped off to storage tanks

(1) **Sedimentation.** The objective is to remove bulky objects such as leaves, sand and gravel particles and the materials that have run off from the soil. Sedimentation is done in large reservoirs or in a restricted area of a settling tank. Aluminium sulphate (alum) or iron sulphate may be added to hasten sedimentation process. This is done in a mixing chamber (c) in Fig. 16.3. After mixing, the chemicals form jelly like masses of coagulated material called **flocs**. They fall through the water and cling to organic particles and microorganisms, dragging a major portion to the bottom sediment in a process called **flocculation** (Fig. 16.3d). Coagulated material is allowed to settle in a settling tank.

(2) **Filtration.** This process removes the remaining microorganisms from the water. Although, different types of filter material are available, mostly a layer of sand and gravel is utilised to trap the microorganisms. A **slow sand filter** (Fig. 16.4) uses several feet deep layer of fine sand particles. Within the sand there forms a layer of microorganisms which acts as an additional filter. This layer is called a **Schmutzdecke** or dirty layer. To cleanse the filter, the top layer is removed and replaced with fresh sand. A **rapid sand filter** (Fig. 16.5) contains coarser particles of gravel. The schmutzdecke does not develop; the

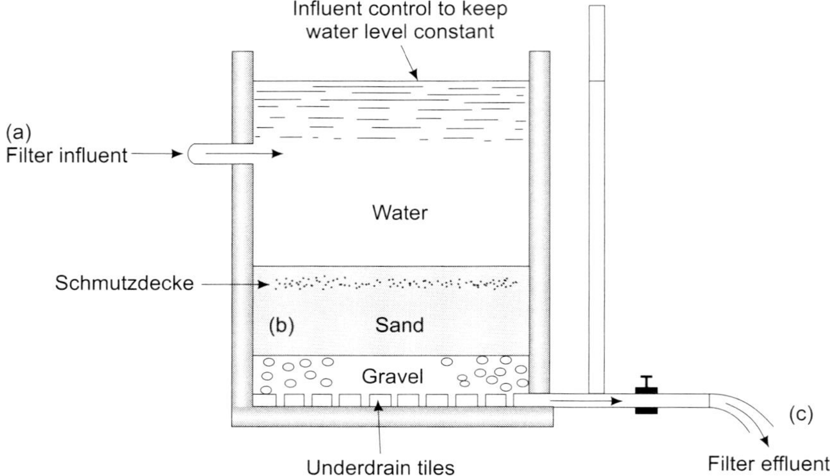

FIG. 16.4

Schematic diagram of the slow sand filter. (a) water enters through the filter influent, (b) passes through the fine sand and gravel, and (c) exits through the filter effluent

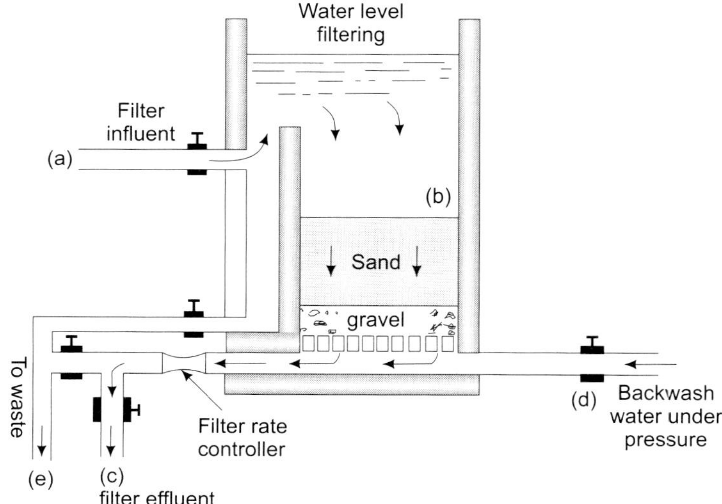

FIG. 16.5

Schematic diagram of the rapid sand filter, (a) water enters through the filter influent and (b) passes over the coarse sand and gravel where it is filtered, (c) the water then exits through the filter effluent drain, (d) to wash the filter, water is introduced, under pressure through the back wash valve, the contaminated water passes up through the filter and out the waste water valve (e)

filtration rate is higher (over 200 million gallons per acre per day). This type of filter is generally used in municipal supplies.

(3) **Chlorination.** In this final step, chlorine gas is added to the water. This gas is an active oxidant which reacts with organic matter in the water. Gas is added until any residue is present. A residue of 0.2 ppm-1.0 ppm of water is often the standard used. Under these conditions, most remaining microbes die within 30 minutes.

17

DOMESTIC WASTES AND WASTE TREATMENT

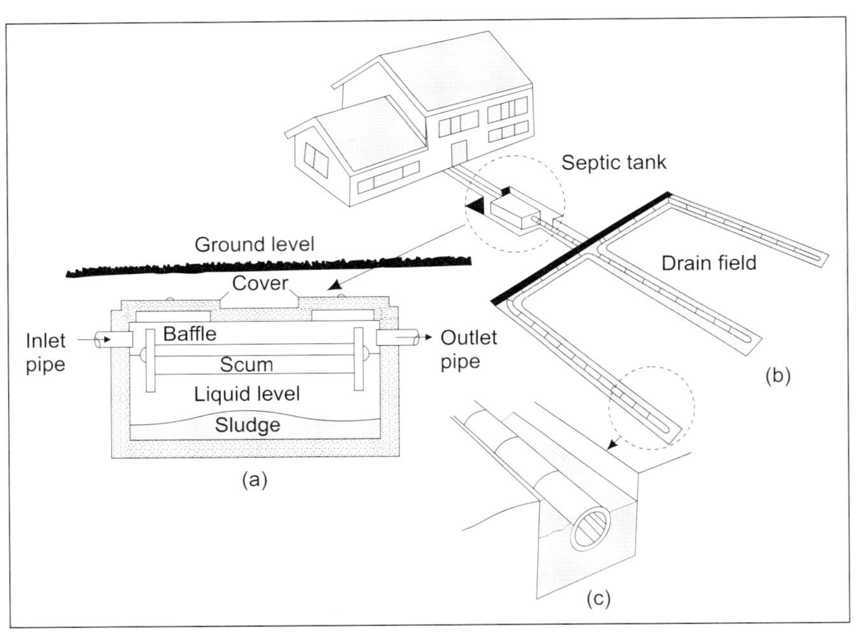

Chapter Outline

During ancient times human wastes were simply discharged into the nearest body of water, such as lake, stream, river, or ocean. In rural areas of many developing countries, this practice still continues. As the human population size grew, these water bodies became degraded, affecting aquatic wildlife due to depleting levels of oxygen. Moreover waterborne diseases began to threaten human lives also. Diseases like cholera became very common. This prompted the proper collection and disposal of domestic wastes of populated areas. Processes were developed to treat this waste before its disposal in natural water bodies. Sewage treatment is a relatively modern practice which virtually began at the turn of 20th century.

COMPONENTS OF DOMESTIC WASTEWATER

Domestic wastewater or **sewage** is primarily a combination of human faecal matter, urine and graywater. Graywater results from washing, bathing, and meal preparation. Water from various industries and business establishments may also enter the system. Major organic and inorganic constituents of untreated domestic sewage are shown in Table 17.1.

TABLE 17.1 Typical constituents of untreated domestic sewage

Components	*Concentration (mg/l)*		
	Low	*Moderate*	*High*
Solids (total)	350	720	1200
Dissolved (total)	250	500	850
Volatile	105	200	325
Suspended solids	100	220	350
Volatile	80	165	275
Setteable solids	5	10	20
BOD	110	220	400
Organic carbon (total)	80	160	290
COD	250	500	1000
Nitrogen (total as N)	20	40	85
Organic	8	15	35
Free NH_3	12	25	50
Nitrites	0	0	0
Nitrates	0	0	0
Phosphorus (total as P)	4	8	15
Organic	1	3	5
Inorganic	3	5	10

Modified from Metcaff and Eddy (1991)

The amount of organic matter in domestic wastes determines the degree of biological treatment required. Three tests are used to assess the amount of organic matter: total organic carbon (TOC), biochemical oxygen demand (BOD), and chemical oxygen demand (COD). The major objective of domestic waste treatment is the reduction of BOD, which may be in soluble or solid state.

Pathogenic microorganisms are invariably present in domestic wastewater (Table 17.2). Their chief source appears to be the excreted matter from infected persons.

TABLE 17.2 Microorganisms typically reported from untreated domestic wastewater

Microorganisms	*Concentration (per ml)*
Total coliform	10^5–10^6
Faecal coliform	10^4–10^5
Faecal streptococci	10^3–10^4
Enterococci	10^2–10^3
Shigella	Present
Salmonella	10^0–10^2
Clostridium perfringens	10^1–10^3
Giardia cysts	10^{-1}–10^2
Cryptosporidium cysts	10^{-1}–10^1
Helminth ova	10^{-2}–10^1
Enteric virus	10^1–10^2

(Modified from Metcaff and Eddy, 1991)

Sewage is the used water supply containing domestic waste together with human excrement and wash water and industrial waste, including acids, greases, oils, animal matter, vegetable matter and storm waters. The basic principle in sewage treatment is that water is separated from the waste while the solid organic matter is biodegraded by microorganisms to simple compounds like nitrates, sulphates, carbonates, carbon dioxide, methane etc. Knowledge of microbiology of sewage is central to maintenance of quality of environment. Sewage treatment is done both, at small as well as large scale.

SMALL SCALE SEWAGE TREATMENT

This is managed on small scale as in individual homes, rural areas, and quite a few residents of town and small cities. They depend on cesspools, septic tanks (pit toilets or outhouses), or oxidation ponds for waste disposal.

Cesspools

In many homes, human waste is dumped into cesspools. A cesspool is an underground construction, consisting of concrete cylindrical rings with pores in the walls of ring (Fig. 17.1). Water passes out the bottom. and through the pores into the surrounding sand, whereas the solid waste accumulates at the bottom. Anaerobic bacteria in the compacted sludge layer at the bottom of cesspool digest the organic matter. The breakdown products diffuse into the ground. After few years it may become necessary to pump out the sludge layer (if it becomes thicker) and to clean with strong acid. Chemicals that can digest the grease should also be added to avoid blockage of the pores. Dried bacterial spores as those of *Bacillus subtilis* are available in stores. These may be added to accelerate the sludge digestion. Addtion of yeasts at intervals is also useful.

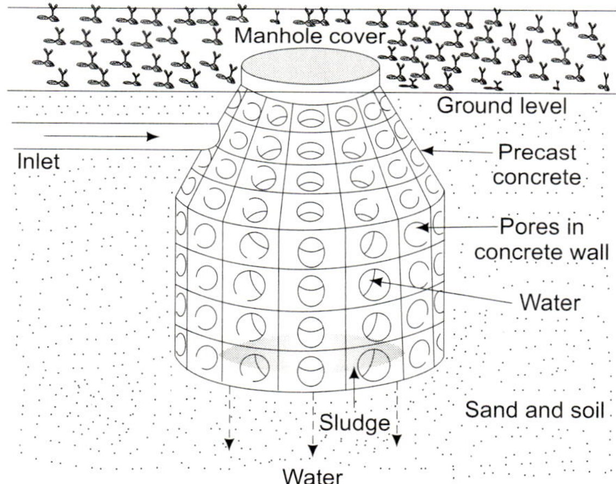

FIG. 17.1

Operation of the cesspool. Waste enters through the inlet pipe, and solid material falls to the bottom of the cesspool to form sludge. Water passes through the open bottom and through pores in the wall of the cesspool into the surrounding soil

Septic Tanks

These are also used in some homes. A septic tank (Fig. 17.2) is an enclosed concrete box into which waste flows from the house. The organic matter accumulates at the bottom of tank whereas the water rises to the outlet pipe to flow to a distribution box. The box is then separated into pipes which empty into the surrounding area. The tank is to be pumped out regularly as there is no absorption of the digested organic matter into the earth. Small towns collect the sewage into large ponds called **oxidation**

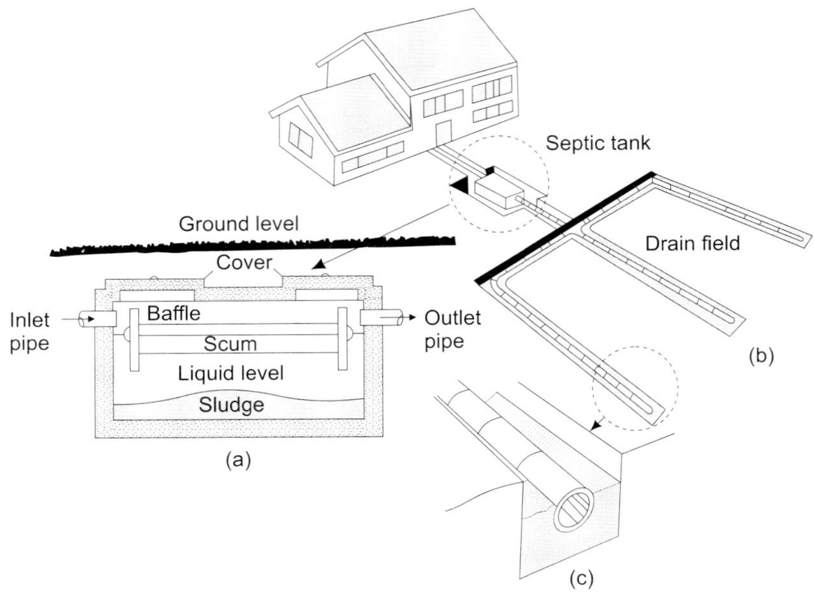

FIG. 17.2

Operation of the septic tank. Waste enters the enclosed septic tank (a) where solids fall to the bottom and accumulate as sludge, which in time decomposes. Relatively clear water passes through the outlet pipe and is distributed by a distribution box (b) into a buried field of clay pipes that open to the soil (c)

lagoons. The sewage is left in lagoon for a couple of months during which aerobic bacteria digest the organic matter in the water and anaerobic organisms breakdown the sedimented material.

Oxidation Ponds

Sewage lagoons are often referred to as **oxidation** or **stabilisation ponds** and are the oldest of the sewage treatment systems. Since they require a minimum of technology and are relatively of low cost, they are most common in developing countries. However, biodegradable organic matter and turbidity are not as effectively reduced as in modern activated sludge treatment systems. Oxidation ponds, used for simple, secondary treatment in rural areas are usually no more than a hectare in area and just a few meters deep. They are infact natural "stewpots" where wastewater is detained while organic matter is degraded. A period of 1-4 weeks or sometimes longer is required for complete decomposition of organic matter.

There are four categories of oxidation ponds, which are often used in series: aerobic ponds, aerated ponds, anaerobic ponds, and facultative ponds: **Aerobic**

ponds are naturally mixed, shallow to allow effective light penetration to stimulate algal growth which promotes subsequent O_2 generation. The wastewater is detained generally for 3-5 days. **Anaerobic ponds** may be 1 to 10 m deep and require a relatively long (20-50 days) detention time and do not require much aeration. Often generating small amounts of sludge, they serve as a pretreatment step for high-BOD organic wastes. **Facultative ponds** are most common for sewage treatment. Sewage is treated by both aerobic and anaerobic processes. The ponds are 1 to 2.5 m deep and subdivided in three zones: an upper aerated zone, a middle facultative zone, and a lower anaerobic zone (Fig. 17.3). The detention time varies between 5 and 30 days. **Aerated ponds** are mechanically aerated, may be 1-2 m. deep and have detention time of less than 10 days. Treatment depends on the aeration time, temperature as well as the composition of sewage.

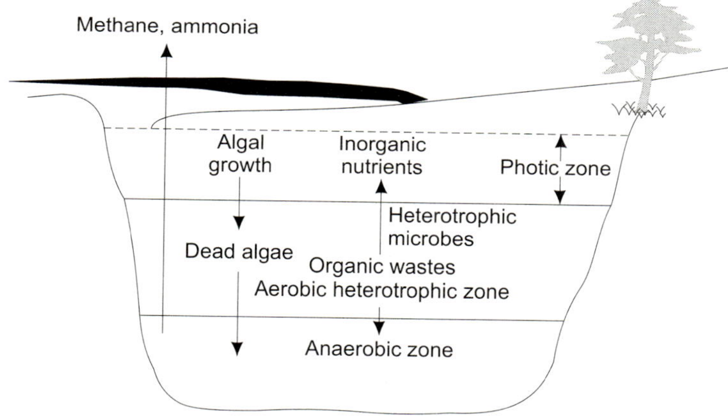

FIG. 17.3

A schematic of an oxidation pond showing the biological processes involved in this form of sewage treatment. The photic zone is highly productive. Dead algae fall into an aerobic heterotrophic zone where they are partially degraded, and then fall with dead bacteria into the anaerobic bottom layer of the pond

MODERN LARGE SCALE SEWAGE TREATMENT

The primary objective of sewage treatment is the removal and degradation of organic matter under controlled conditions. Municipal plants are equipped for a mechanized sewage treatment that handles massive amounts of daily generated waste and garbage. A simplified view of such a waste treatment facility is shown in Fig. 17.4. The overall process can be divided into **three** steps: the primary, secondary and tertiary treatments (Fig. 17.5). These different treatments to sewage depend on the quality of the effluent deemed necessary to be achieved to permit the maintenance of acceptable water quality. **Primary treatments** rely on physical separation of large debris, followed by sedimentation; **secondary treatments** rely on microbial biodegradation to further reduce the concentration of organic compounds in the effluent; and the **tertiary treatment** is usually a physicochemical process that removes turbidity caused by the presence of nutrients (e.g. nitrogen), dissolved organic matter, metals or pathogens. Municipal treatment plants are not capable of dealing with industrial wastes containing toxicants, such as heavy metals.

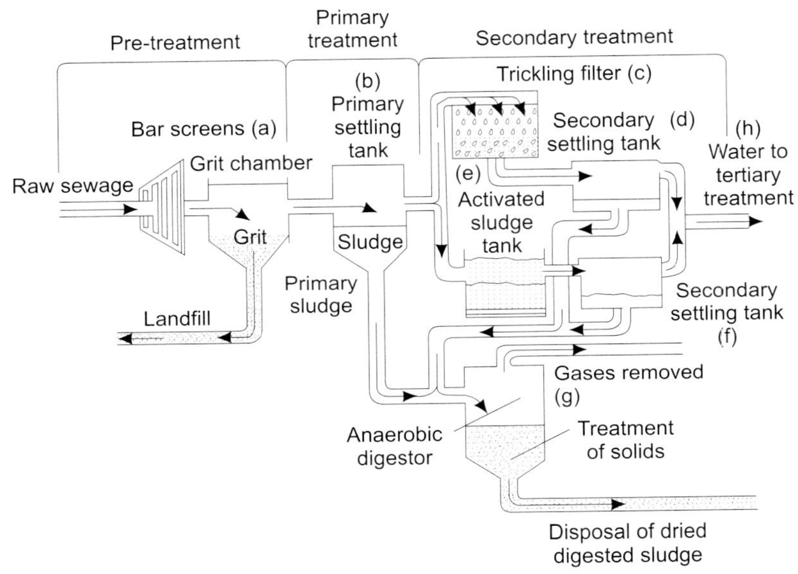

FIG. 17.4

A schematic view of a waste treatment facility. (a) sewage is initially pretreated with a bar screen to remove grit. (b) the sewage is then piped to a settling tank where organic waste passes out to a sludge tank. This is primary treatment. (c) in the liquid phase of secondary treatment, microorganisms digest the soluble organic matter as the water percolates through a filter. (d) the water is separated from the microorganisms and passes out, while the sedimented material flows to the sludge tank. (e) in the solid phase of secondary treatment, sludge is treated in an activated sludge tank with thorough aeration. (f) this is followed by separation and flow of sedimented material to the anaerobic sludge tank. (g) in the anaerobic sludge tank, sludge from three processes (b, d and f) is held for several weeks, during which anaerobic bacteria break down the sludge to usable end-products. (h) the water from the settling tanks may be further processed in tertiary treatment

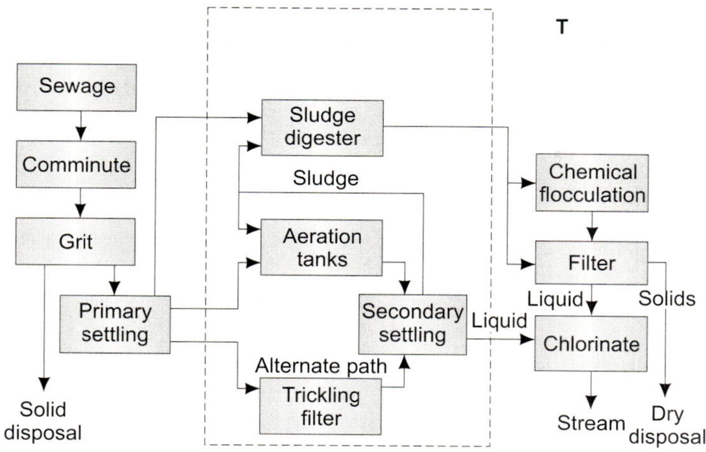

FIG. 17.5

Flow chart of the stages of sewage treatment. Primary treatment is mainly physical, secondary mainly biological and the tertiary one mainly chemical

Primary Treatment

This removes suspended solids in settling tanks or basins. Solids are drawn off. Raw sewage is piped into huge open tanks where it is screened to remove solid fraction. This solid material is then subjected to anaerobic digestion in landfill or composting. The liquid portion is then passed into sludge tanks for further treatment. Flocculating materials as aluminium or iron sulphate are also added to raw sewage to trap microbes and debris to the bottom as in the sedimentation process of water purification. These materials are added to sludge tanks also. In the settling tank suspended organic solids settle to the bottom as **sludge** or **biosolids**. The resulting sludge is referred to as **primary sludge.**

Secondary Treatment (Microbial Biodegradation)

To achieve an acceptable reduction in the BOD, secondary treatment by a variety of means is necessary (Table 17.3).

TABLE 17.3 Efficiency of various types of sewage treatment

Treatment	*BOD (% reduced)*	*Suspended solids(% removed)*	*Bacteria(% reduced)*
Sedimentation	30-75	40-95	40-75
Septic tank	25-65	40-75	40-75
Trickling filter	60-90	0-80	70-85
Activated sludge	70-96	70-97	95-99

In this treatment a small portion of the dissolved organic matter is mineralised, and the large portion is converted to removable solids. By now the original sewage BOD is reduced to 80-90 per cent. Secondary treatment relies on microbial activity, may be aerobic or anaerobic, and is conducted in large variety of devices. Since this step is a microbial process, accidental introduction of a toxic chemical is very dangerous. Various devices used in this treatment are as follows:

Oxidation ponds

Oxidation ponds or stabilisation ponds and lagoons are used for simple secondary treatment in rural areas or industrial units. Heterotrophic bacteria degrade sewage organic matter within ponds, producing cellular material and mineral products that support the growth of algae. Oxygen produced by algae compensate for the poor O_2 conditions created by heterotrophic bacteria. The pond should be shallow, 10 feet deep, to maximise the euphotic zone for algal growth. Bacterial and algal cells settle at the bottom of pond. Effluents containing oxidised products are periodically removed.

Trickling filter

This system is a simple and expensive film-flow type of aerobic sewage treatment device (Fig. 17.6) Sewage is distributed by a revolving sprinkler suspended over a bed of porous material. The sewage slowly percolates through this porous bed and the effluent collected at the bottom. The porous material of the filter bed becomes coated with a dense, slimy bacterial growth, mainly composed of *Zooglea ramigera* and similar

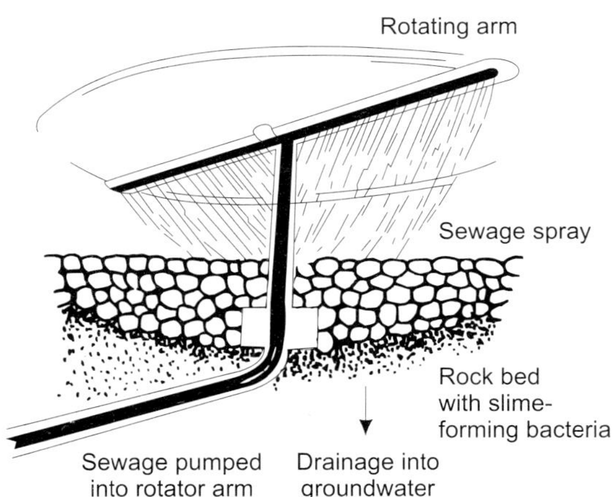

FIG. 17.6

A trickling filter, where the sewage is sprayed over rocks. The bacterial film on the rocks aerobically decomposes the dissolved organic matter. The drainage may go through the soil column to the groundwater or drain into a nearby river or ocean

slime-forming bacteria. This slimy matrix thus generated harbors a heterogeneous microbial community, including bacteria, fungi, protozoa, nematodes and rotifers. The most frequent bacteria are *Beggiatoa alba, Sphaerotilus natans, Achromobacter* spp, *Flavobacterium* spp, *Pseudomonas* spp and *Zooglea* spp. This community absorbs and mineralises dissolved organic nutrients in the sewage reducing the BOD of the effluent. Porous bed allows aeration. Sewage may be passed through two or more trickling fieters or recirculated through the same filter.

Biodisc system

The **biodisc system** or **rotating biological contactor** is a more advanced type of aerobic film-flow treatment system. Here, closely spaced discs, usually made of plastic are rotated in a trough containing the sewage effluent. The discs are partially submerged and become coated with a microbial slime similar to that developing in trickling filters. Continuous rotation of the discs keeps the slime well aerated and in contact with the sewage. Microbial growth on the disc surfaces is sloughed off gradually and removed by subsequent settling. When the film becomes so thick that O_2 and nutrients fail to reach the inner portion of film, the innermost microbes die, causing detachment of the film. The system is used in communities for treatment of domestic and industrial sewage effluents.

Conventional activated sludge

This process is also known as **aeration-tank digestion**. This process is very widely used aerobic suspension type of liquid waste treatment system. After primary settling, the sewage is introduced into an aeration tank and mixed with a bacteria-rich slurry known as **activated sludge**. Air injection and/or mechanical stirring provides aeration. The rapid development of microbes is also stimulated by reintroduction of most of the settled sludge from a previous run to a secondary settling tank, where water is siphoned off the top of tank and sludge is removed from the bottom. Some of the sludge is used as inoculum for incoming primary effluent (activated sludge). The remainder of the sludge, **secondary sludge** is removed (Fig. 17.7), and the process derives its name from this inoculation with activated sludge. Vigorous development of heterotrophic microbes has taken place during the holding period. The populations include Gram-negative rods, predominantly *Escherichia, Enterobacter, Pseudomonas, Achromobacter, Flavobacterium* and *Zooglea* spp; other bacteria including *Micrococcus, Arthrobacter,* various coryneform and mycobacteria; *Sphaerotilus* and other large filamentous bacteria; and low numbers of filamentous fungi, yeasts and protozoa. Bacteria occur in free suspension and as flocs. The flocs are microbial biomass held together by slime. Due to extensive microbial metabolism of the organic compounds in sewage, a large proportion of dissolved organic substrates is mineralised and another portion converted to microbial biomass. In advanced stage,

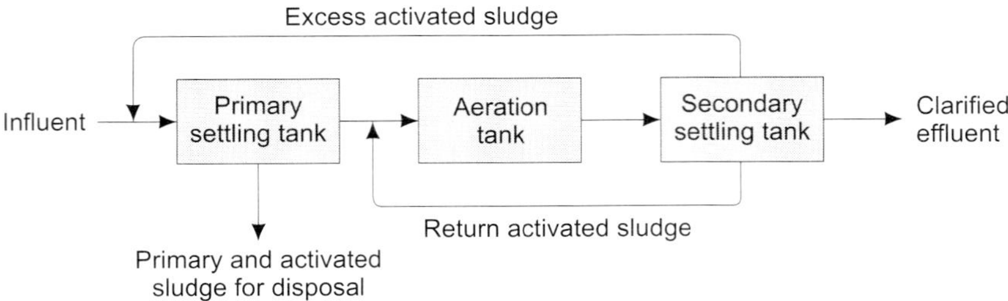

FIG. 17.7
Flow of materials through an activated sludge secondary sewage treatment system

biomass associated with flocs is removed by settling. Poor settling may cause **bulking** of sewage sludge caused by proliferation of filamentous bacteria like *Sphaerotilus, Beggiatoa, Thiothrix,* and *Bacillus* and filamentous fungi like *Geotrichum, Cephalosporium, Cladosporium* and *Penicillium.* A portion of the settled sewage sludge is recycled for use as the inoculum for incomming raw sewage, but the rest goes for additional treatment by composting or anaerobic digestion. Activated sludge process reduce the BOD of the effuent to 10-15% of that of the raw sewage. Intestinal pathogens are also reduced. Non-pathogenic bacteria proliferate in number.

An important characteristic of the activated sludge process is the recycling of a large proportion of the biomass. This results into a large number of microbes that oxidise organic matter in a relatively short time.

Non-conventional activated sludge process

Conventional activated sludge processes can be modified for removal of **nitrogen** and **phosphorus.** In conventional process these elements are removed during tertiary treatment of sewage. In this way, modified activated sludge process removes nitrogen and phosphorus employing biological rather than chemical processes. **Nitrogen**, by activated sludge process is removed by encouraging nitrification followed by dentrification. The conventional activated sludge system can be modified to encourage denitrification as follows:

Single sludge system. This comprises a series of aerobic and anaerobic tanks in lieu of a single aeration tank. Methanol or settled sewage serves as the carbon source for denitrifiers.

Multisludge system. Carbonaceous oxidation, nitrification, and denitrification are carried out in three separate systems. Methanol or settled sewage serves as carbon source for denitrifiers.

Bardenpho process. This consists of two aerobic and two anoxic tanks followed by a sludge settling tank. Tank 1 (anoxic) is used for denitrification, with wastewater used as carbon source. Tank 2 (aerobic) is utilised for both carbonaceous oxidation and nitrification. Mixed liquor from this tank, which contains NO_3, is returned to tank 1. The anoxic tank 3 removes the NO_3 remaining in the effluent by denitrification. Finally tank 4 (aerobic) is used to strip the N_2 gas that results from denitrification, thus improving mixed liquor settling.

Phosphorus is removed by two systems:

A/O (aerobic/oxidation) process. The A/O process consists of an anaerobic zone (detention time 0.5-1 hr) upstream of the conventional aeration tank (detention time 1-3 hrs). Under anaerobic conditions, microbes release stored phosphorus to generate energy, which is used for uptake of BOD from sewage. When aerobic conditions are restored microbes exhibit phosphorus uptake levels above those normally required to support cell maintenance, synthesis etc. Excess phosphorus is stored as polyphosphates within the cell. The sludge containing the excess phosphorus is then wasted.

Bardenpho process. This also removes nitrogen as well as phosphorus by nitrification–denitrification process.

Tertiary Treatment

This conventional treatment is designed to remove pollutants and mineral nutrients, especially N_2 and P salts by chemical process. Activated carbon filters are normally used in their removal from secondary-treated effluents. To prevent eutrophication, phosphate is removed from sewage by precipitation as calcium, aluminium or iron phosphate. **Breakpoint chlorination** is a process to remove NH_3. In big cities, a highly advanced tertiary water treatment system integrates several of the tertiary treatment processes.

Modern large scale sewage treatment, however involves additional processing of sewage after tertiary treatment. This includes removal of pathogens, sludge processing, and landfarming.

Removal of Pathogens

The subject of removal of pathogenic microbes by activated sludge and other wastewater treatment processes has been reviewed from time to time. Enteric bacterial pathogens can be effectively removed by these processes. These include viruses, *Salmonella, Giardia* and *Cryptosporidium*. With progressive treatment of raw sewage, their numbers reduced significantly during successive steps, being almost completely eliminated in final treatments. When compared with other biological treatments as trickling filters, activated sludge process is relatively efficient in reducing the numbers of pathogens in raw sewage.

Sludge Processing

The sludge or **biosolids** resulting from the various stages of sewage processes (both the primary sludge and the activated sludge generated by secondary processes) is also treated further for stabilisation of their organic matter and reduction of water content. Primary sludge contains 3.8%, whereas secondary sludge 0.5 to 2% solids. Organic matter stabilisation prevents the odour formation and decreases the number of pathogens. Reduction of the water content reduces the weight of sludge making its transport easier.

Sludge treatment typically involves several steps (Fig. 17.8). **Thickening** reduces the volume of sludge. This is achieved by allowing the solids to settle in a tank or by centrifugation. **Digestion**, a microbial process results into stabilisation of the organic matter and some elimination of pathogens due to higher temperatures. Sludge can be

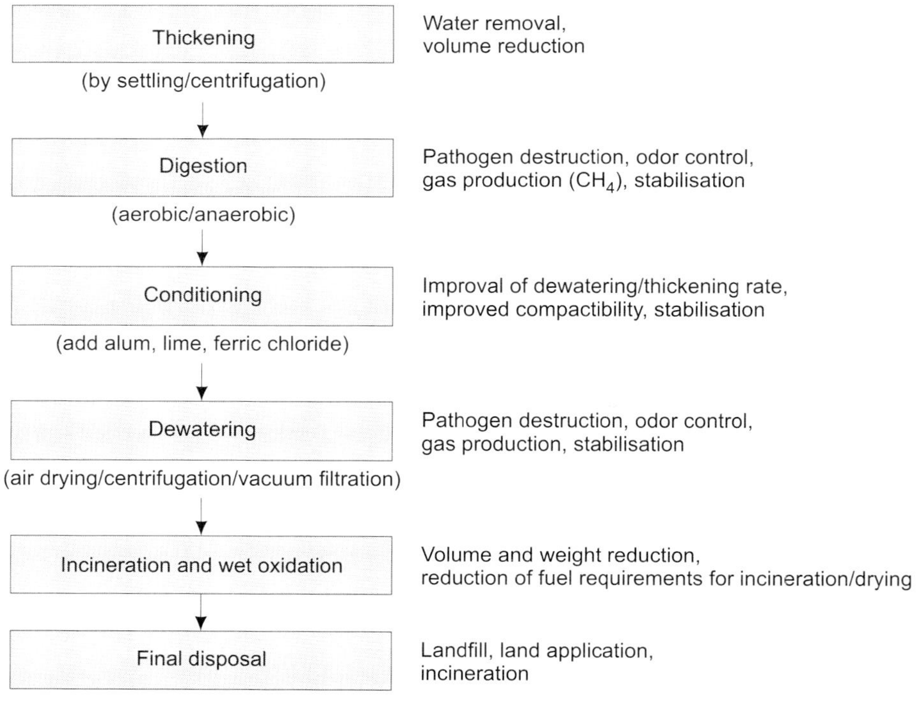

FIG. 17.8

Sludge treatment processes flow chart

digested both anaerobically and aerobically. **Anaerobic digestion** is the most common treatment, extending over a period of 2-3 weeks in large covered tanks of sewage treatment facility. Methane is produced during this process, which can be used as energy source. Four groups of bacteria are involved during this complex process. For **aerobic digestion**, air is supplied in open tanks. Aerobic microbes degrade the sludge resulting in reduction of solids. Solids are reduced further during **conditioning** in which chemicals (alum, lime, ferric chloride) are added to aggregate suspended particles. This is followed by **dewatering**, achieved by several methods such as air drying, centrifugation or vacuum filtration.

Landfarming

This is the practice of disposing biosolids produced by wastewater plants on agricultural land. The biosolids may be added to soil as solids or liquids. This practice adds nutrients and water to the soil. Liquid sludge can be injected into the soil from tanker trucks. Before adding to the soil, sludge may be treated by **processes which significantly reduce pathogens (PSRP)**. For application on land where food crops are grown, the sludge may be treated further by **processes to further reduce pathogens (PFRP)**.

LAND APPLICATION OF WASTEWATER

Although treated domestic wastewater is usually discharged into waterbodies, it may also be disposed of via land application-sometimes for crop irrigation and sometimes as a means of additional treatment or disposal. The three basic methods used in the application of treated wastewater include, low rate irrigation, overland flow, and high-rate infiltration. With **low-rate irrigation**, sewage effluents are applied by sprinkling or by surface application. This is useful for small communities requiring large (5-6 hectares/1000 people) areas. In **overland flow** method, wastewater effluents are allowed to flow for a distance of 50-100 m along a 2-8% vegetated slope and are collected in a ditch. This is used for clay soils with low permeability and infiltration. In **high-rate infiltration**, also referred to as **soil aquifer treatment (SAT)** or **rapid infiltration extraction (RIX)** wastewater treatment is done at loading rate of more than 50 cm/week (in overland flow method it is 5-14 cm/week, whereas in low-rate irrigation, 1.5 to 10 cm/week). The treated water, most of which has percolated through coarse-textured soil, is used for groundwater recharge and may be recovered for irrigation. This system requires less land than low-rate irrigation or overland flow methods.

SOLID WASTE DISPOSAL

Major cities generate two major types of solid wastes: (i) **municipal solid waste (MSW)**, which includes the inert part such as glass, metal and plastic, as well as degradable organic waste such as food, animal and plant residues, kitchen scraps, paper and other household and industrial garbage, and (ii) **sewage sludge** or **biosolids** derived from treatment of liquid wastes, animal waste from cattle feed lots, and poultry and swine farms. In rural areas these may be recycled into land as fertiliser. However, in urban areas they pose an environmental problem.

Following microbiological methods have been developed to treat these wastes by way of using microbial biodegradation.

Sanitary Landfills

The material is placed in a landfill to allow it to decompose. Both, organic and inorganic solid wastes are deposited together in a low-lying land. To avoid, foul odour and attraction of insects and rodents, each day's waste deposit is covered over with a layer of soil, creating a **sanitary landfill**. A developed landfill can be used for construction and recreation purposes. For 30-50 years, the organic matter undergoes slow, anaerobic microbial decomposition and the resulting products, CO_2, H_2O, CH_4, low mol. wt. alcohols, and acids diffuse in the surrounding air and water. Thus landfill settles down.

Various methods are used for making sanitary landfills. These are, (i) **area method** where a bulldozer spreads and compacts solid waste; cover material is hauled in and spread at the end of operation, (ii) **trench method**, in which the waste collection truck deposits its load in a trench where it is spread and compacted, and (iii) **ramp variation**, where solid wastes are spread and compacted on a slope.

Old-style sanitary landfills (Fig. 17.9) were constructed at locations chosen more for convenience or budgetary concerns than for any environmental considerations. It may be abandoned pit located near a river, mine etc. In such landfills conditions were favourable to pollution, particularly water pollution caused by the leachates produced by the infiltration of water through the water material. Besides contaminated water, such landfills release pollutants also into the air. Anaerobic microbial activities generate greenhouse gases, such as N_2O, CH_4 and CO_2.

A **modern-style** landfill (Fig. 17.10) is designed to meet exact standards for of all materials, including leachates containment and gases. It should have minimum impact on the environment, with special reference to groundwater protection. Geology and soil type, groundwater considerations as water table depth etc. are taken into account. Fresh garbage is covered daily with a layer of soil. The high

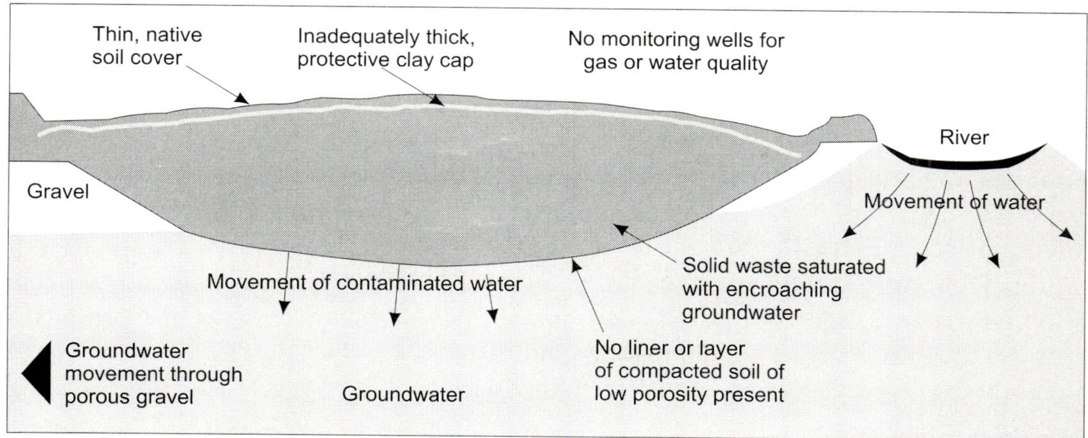

FIG. 17.9

A schematic of old-style sanitary landfill, where location is chosen more for convenience or budgetary concerns than for any environmental considerations. Here, an abandoned gravel pit located near a river exemplifies an all too common siting arrangement

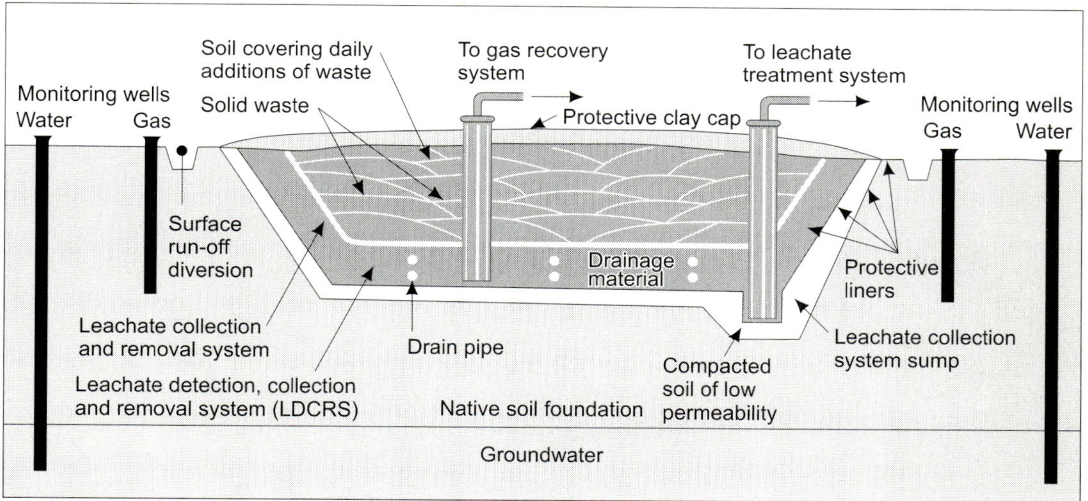

FIG. 17.10

Cross-section of a modem sanitary landfill showing pollutant monitoring wells, a leachate management system, and leachate barriers. In contrast to the old-style site (Fig. 17.9), the modern sanitary landfill emphasizes long-term environmental protection. In addition, whereas landfills were formerly abandoned when full, a modern landfill is monitored long after closure

temperatures (50°C) generated during the decomposition of buried or composted solid waste inactivates most of the enteric pathogens present in the waste.

Composting

Here the organic part of waste is biodegraded by composting. In this process, the waste is degraded by aerobic, mesophilic and thermophilic microbes. Inorganic fraction is separated either at source or by using magnetic separators. Organic fraction is ground up, mixed with sewage sludge and/or bulking agents (shredded newspaper, wood chips) and then composted. This balances C : N balance. Microbial composting converts the waste into a stable, sanitary, humus-like product. The various composting methods are used. These include the following:

Window method

Solid waste is arranged in long rows and covered to allow decomposition. Material is turned over repeatedly.

Static pile or aerated pile method

Here composting rates can be enhanced by forced aeration. Waste is arranged in piles and forced aeration is used to supply extra-O_2. Perforated pipes are buried inside the compost pile and the air is then pumped (Fig. 17.11).

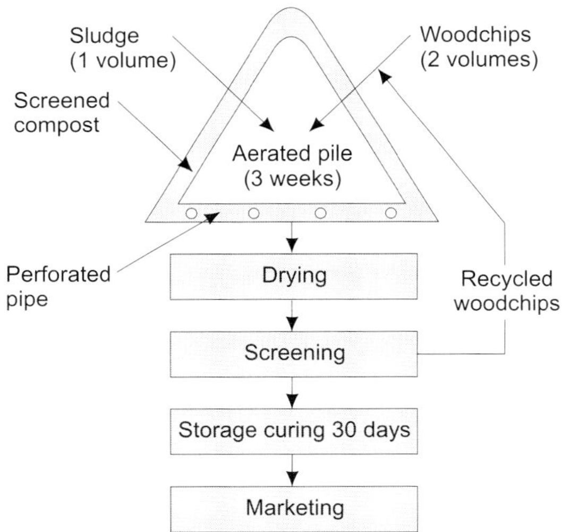

FIG. 17.11

An outline of static aerated pile composting method

In-vessel or mechanical or enclosed-reactor composting method

It is a continuous feed approach. It may incorporate the feature of window and/or static pile methods of composting. It uses a reactor that permits control of the environmental parameters. Reactor is like an industrial fermenter. Here composting is complete within 2-4 days. But this method is expensive.

At present, most of the operational facilities use static pile composting or window method. These two methods yield stable products, and are low in cost. In compost of domestic garbage and sludge, several microbial species that come from soil, water and human fecal matter are present. High moisture content of the compost favours the growth of bacteria than fungi. Mesophiles are the first to appear and as the temperature rises thermophiles begin to grow. Thermophiles that are dominant in compost are bacteria, like *Bacillus stearothermophilus, Thermomonospora* spp., *Thermoactinomyces* spp. and *Clostridium thermocellum,* and the fungi, such as *Geotrichum candidum, Aspergillus fumigatus, Mucor pusillus, Chaetomium thermophile, Thermoascus auranticus* and *Torula thermophila*. The reactor is maintained at thermophilic temperatures. Maximum composting occurs at 50-60°C, moisture should be 50-60% water content, the C : N ratio must not exceed 40 : 1, the lower nitrogen will not allow the growth of enough microbial biomass.

18

CONTROL OF MICROORGANISMS

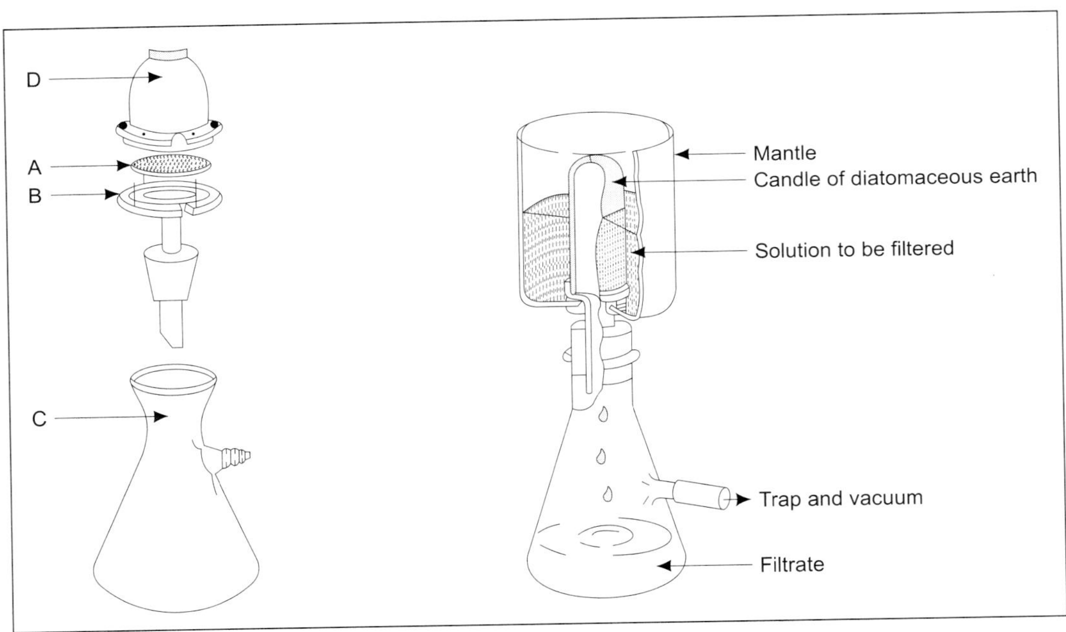

Chapter Outline

- *Physical Control*
- *Chemical Control*
- *Chemotherapeutic Agents*
- *Sites of Action of Antibacterial Agents*
- *Antibiotic Sensitivity Assays*

The control of microorganisms is necessary for good health. In this chapter I would present a brief account of some of the established methods of their control. There are three principal kinds of agents used in microbial control: (i) **physical agents**, that are used exclusively on objects outside the body, (ii) **chemical agents**, which are used on inanimate objects as well as on the body surface and (iii) **chemotherapeutic agents**, which are most often used inside the living body of diseased person to cure the disease.

PHYSICAL CONTROL

The chief agent used in this type of control is **heat**, applied in different forms. In addition, there are some **other methods** also in which different types of physical agents are used.

Heat

The aim is often sterilisation. This term implies the removal of all life forms and this is an absolute term. Following are the common ways in which heat is used in control.

Direct flame

This method is used in the process of **incineration**, employed in the laboratory to sterilise the bacteriological loop and needle before removing a sample from a culture tube and after preparing a smear. The tip of tube is also flamed to destroy microbes which may contact it.

Hot-air oven

This instrument (Fig. 18.1) utilises **dry heat** for sterilisation. The material is put at 160°C for a period of two hours to kill the bacterial spores and other microbial structures. This method is oftenly used to sterilise dry powders, oily substances and glasswares.

Boiling water

Here the objects are immersed in boiling water. The moist heat sterilises the objects. This heat penetrates better and rapidly than dry heat. Since temperature is 100°C, bacterial spores may require two hours exposure for destruction. Moist heat kills microorganisms by coagulating and denaturing their proteins. The object must be completely immersed in boiling water.

Autoclave

The principle used here is to increase the pressure of steam (gas) in a closed system that increases its temperature. The water molecules become more aggregated that

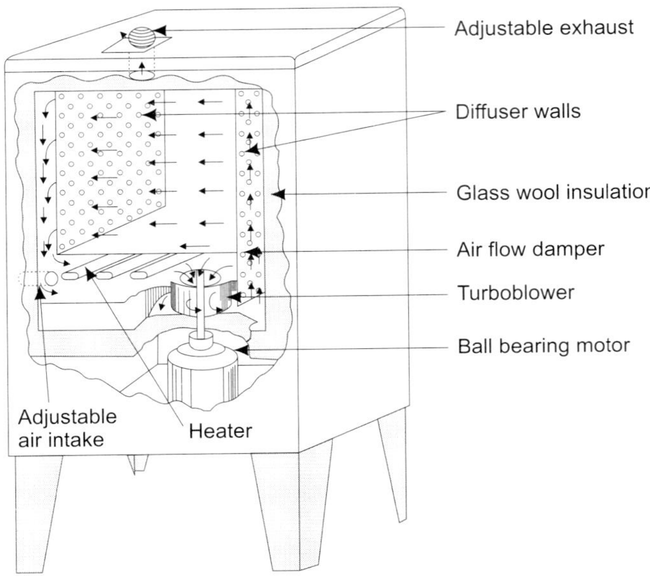

FIG. 18.1

Schematic diagram of the hot-air oven. The motor at the bottom forces dry hot air at 160°C into the chamber from the right. The air moves over and through materials in the chamber and exits at the left whereupon it passes back to the motor for reheating and recirculating

increases their penetration considerably. This principle is used to reduce sterilising time in the laboratory instrument called the **autoclave**. It contains a sterilising chamber into which articles are placed and a steam jacket where steam is maintained (Fig. 18.2). A special valve increases the pressure to 15 Ib/sq. inch above normal atmosphere pressure. The temperature rises to 121.5°C and superheated water molecules rapidly conduct heat into the microorganisms within about 15 min. Culture media, glassware, metalware etc. are sterilised by this method.

Fractional sterilisation

It is also called **tyndallisation** (after its developer, John Tyndall), and **intermittent sterilization** as it is a stop-and-start operation. Sterilisation is achieved by a series of events. During first day objects are exposed to free-flowing steam at 100°C for 30 min. which kills all organisms except bacterial spores. When left overnight the spores germinate into vegetative cells. These are killed during the second day's exposure to steam for 30 min. Again, the material is cooled, and the few remaining spores germinate, which are killed on the third day exposure. This method is particularly important in modern microbiology, for sterilising those objects which could not be sterilised by autoclave. The instrument used is **Arnold steriliser**.

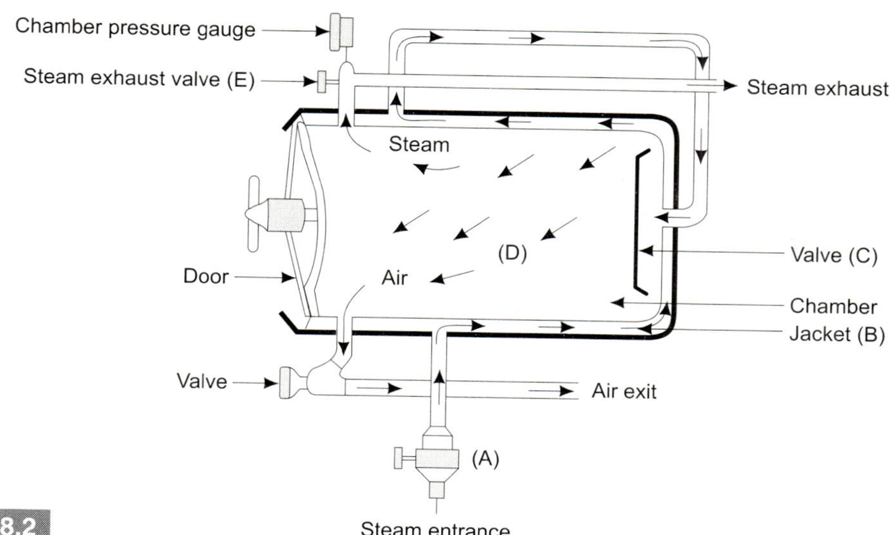

FIG. 18.2

Operation of the autoclave. Steam enters through the port (A) and passes into the jacket (B). After the air has been exhausted through the vent, a valve (C) opens to admit pressurized steam (D) that circulates among and through the materials, thus sterilizing them. At the conclusion of the cycle, steam is exhausted through the steam pipe (E)

Pasteurization

This is not supposed to provide sterilisation. Here the bacterial population of a liquid as milk is merely reduced, and the organisms which may cause human diseases are destroyed. Spores are not affected. One method for milk pasteurization is **holding method**. It involves heating at 62.9°C for 30 min. It kills both tubercle bacterium, *Mycobacterium tuberculosis* and the Q fever agent. Other methods are the **flash pasteurization** at 71.6°C for 15 seconds, and the **ultrapasteurization** at 82°C for 3 seconds.

Hot oil

Some physicians and dentists use hot oil at 160°C for one hour for sterilisation of the instruments.

Other Methods

Physical methods other than heat, are as follows:

Filtration

It is a mechanical device for removing microorganisms from a solution. The organisms are trapped in the pores of the filter, and the filtrate is decontaminated or possibly sterilised. There are several types of filters used in microbiology laboratory. Inorganic filters are typified by the **Seitz filter**, which consists of a pad of asbestos mounted in a filter flask (Fig. 18.3). Porcelain and ground glass may also be used. There are organic filters also. The organic molecules of the filter attract the organic components of the microorganisms. An example, the **Berkefeld filter** (Fig. 18.4), utilises the substance-diatomaceous earth, composed of the skeletal remains of marine algae, diatoms. Third type of filter is **membrane filter** which received general acceptance. It consists of a pad of organic compound such as cellulose acetate or polycarbonate mounted in a holding device. Membrane filters are available in various pore sizes according to the microorganisms to be trapped. This method can be used for quantitative estimation of microbes in a given sample. The filter pad is put on nutrient culture medium where cells grow.

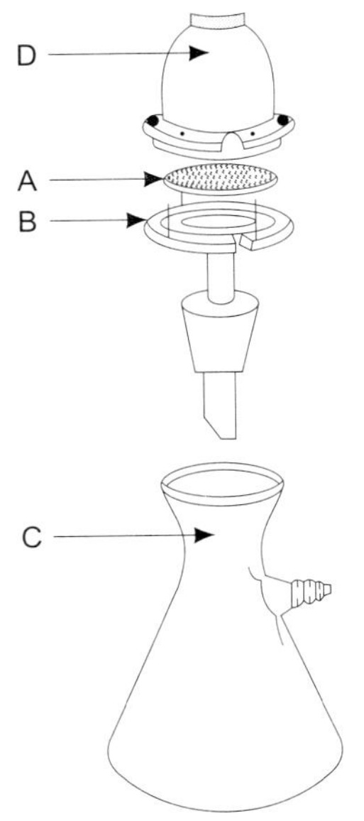

FIG. 18.3

The Seitz filter which consists of an asbestos pad (A) mounted on a wire screen (B). The device is assembled and mounted by the stem into a suction flask (C). Fluid is poured into the cylindrical cup (D) and microorganisms are trapped in the asbestos disc

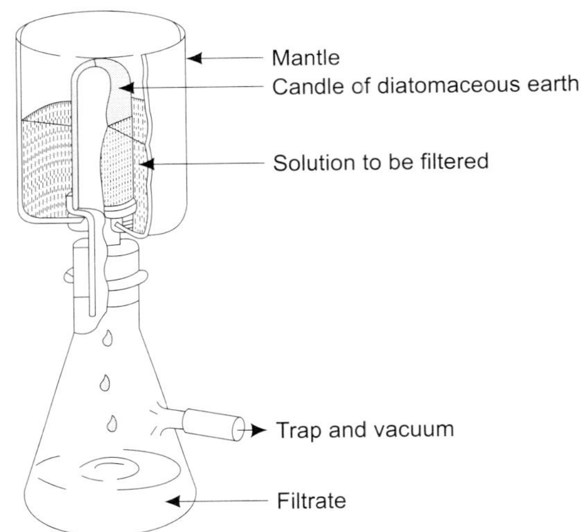

FIG. 18.4

The Berkefeld filter system where diatomaceous earth is shaped into a hollow candle. The solution is poured so that it surrounds the candle. As the solution passes through into the hollow centre, the organisms are trapped in the diatomaceous earth

Ultraviolet light

UV light has a wavelength between 100 and 400 nm, and the energy at about 265 nm is most destructive to bacteria. Fig. 18.5 shows the spectrum of visible and invisible

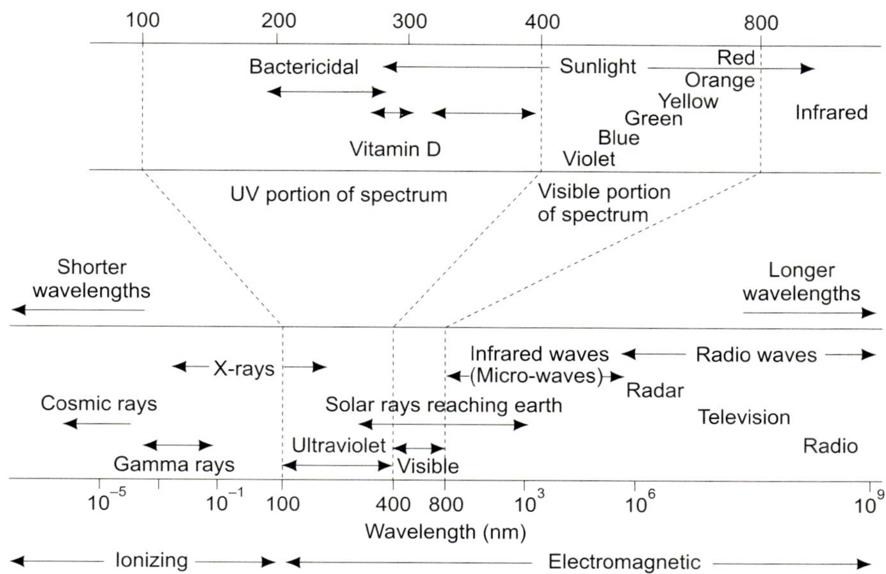

FIG. 18.5

The ionizing and electromagnetic spectrum of energies. A complete spectrum is presented at the bottom of the graph and the ultraviolet and visible sections are expanded at the top. Note how the bactericidal energies overlap with the UV portion of sunlight. This may account for destruction of microorganisms in the air and in upper layers of soil

energies. Exposure to UV damages the DNA. UV radiations are used to reduce air contamination.

Other radiations

The spectrum of energies with wavelengths less than that of UV includes—**X-rays** and **gamma rays** which can also be used to destroy bacteria. These rays are called **ionizing radiations** as they eject electrons out of organic molecules, thus creating ions.

Ultrasonic vibrations

These are high-frequency sound waves. If propagated in fluids, sound waves cause microscopic bubbles to form, and the water appears to boil (also called **cold boiling**). The bubbles rapidly collapse, giving tiny cavities and sending out shock waves. Microorganisms in the fluid are rapidly disintegrated by the external pressures. A device called **cavitron** is used by dentists to clean teeth. There are ultrasonic machines to clean dental plates, coins, jewellery etc.

Preservation methods

A number of physical methods are used for controlling microorganisms in foods. It checks the food spoilage. These are as follows:

Drying. It is used for preservation of meat, cereal, fish and other foods.

Salting. It is based on principle of osmotic concentration. In presence of salt, water comes out of cells due to exosmosis, causing ultimate death. This method is used in preservation of syrups, jams, jellies, etc. from bacterial contamination.

Low temperature. In the refrigerator and freezer, low temperatures retard spoilage by reducing the metabolic rates of microorganisms.

CHEMICAL CONTROL

Chemical methods are expected to remove the pathogenic organisms from an object (or body). These methods rarely achieve sterilization as in physical methods. The process of removal is called **disinfection**. If the object is non-living, the chemical is known as disinfectant. However, if the object is living, as a tissue of human body, then the chemical is an **antiseptic**. Antiseptics and disinfectants may kill the microbes or temporarily prevent their further growth and multiplication.

The effectiveness of a chemical agent is based on **phenol coefficient (PC)**. This number indicates the ability of a particular antiseptic or disinfectant as compared to phenol under identical conditions. A PC higher than one indicates that the chemical is more effective than phenol. Phenol coefficients of some common antiseptics and disinfectants are shown in Table 18.1.

TABLE 18.1 Phenol coefficients of some chemicals

Chemical agent	**Staphylococcus aureus*	**Salmonella typhi*
Phenol	1.0	1.0
Chloramine	133.0	100.0
Tincture of iodine	6.3	5.8
Lysol	5.0	3.2
Mercury chloride	100.0	143.0
Ethyl alcohol	6.3	6.3
Formalin	0.3	0.7
Hydrogen peroxide	–	0.01

*Used as standards in the laboratory

Following are some of the commonly used chemicals to control microorganisms.

Halogens

Two halogens, chlorine and iodine are most commonly used. The halogens are highly reactive elements. **Chlorine** is available either as gas or as organic or inorganic compounds. It is used in municipal water supplies, keeping bacterial population at low levels. The residue used is about 0.2-1 ppm of free chlorine. One ppm is equivalent to 0.0001 per cent. Chlorine is also useful as sodium hypochlorite (NaOCl) or as calcium hypochlorite [$Ca(oCl_2)$]. Other forms of chlorine are the chloramines. **Iodine** is used generally in the form of tincture of iodine, as antiseptic for wounds. It has 2% iodine plus sodium iodide in ethanol. **Iodophors** are complexes of iodine and detergents that release the iodine over a long period of time.

Phenolic Compounds

Phenol is the standard disinfectant, which cogulates the proteins, particularly cell membrane enzymes. Phenol is especially useful against Gram-positive bacteria. An alternative of phenol, **cresol** has become more popular in modern medicine as it is cheaper than phenol. Cresols are used as wood preservatives. Bisphenols—a combination of two phenol molecules are also prominent in modern disinfection and antisepsis. For example, orthophenylphenol is used in lysol, osyl, staphene etc. Another bisphenol, hexachlorophene was used extensively in 1950's and 1960's in toothpastes, underarm deodorants, and bath soaps. A phenol derivative, hexylresorcinol is used in a mouthwash, topical antiseptics and in throat lozenges.

Heavy Metals

The activity of heavy metals on microorganisms is termed **oligodynamic action**. Metals as silver, mercury and copper are used. Mercury is an older antiseptic, used as mercuric chloride ($HgCl_2$). In products like mercurochrome, merthiolate and metaphen, mercury is combined with organic carrier compounds, that reduces its toxicity to skin. Copper is particularly active against algae. It is used as copper sulphate in swimming pools and municipal water supplies. Silver, as silver nitrate is used as an antiseptic and disinfectant. One drop of 1% $AgNO_3$ solution is placed in eyes of newborn to protect against infection by the gonococcus, *Neisseria gonorrhoeae*. Silver nitrate can also be combined with an antimicrobial drug for use in treatment of burns.

Alcohol

It is an effective antiseptic, applied to skin. The most common is ethyl alcohol, though propyl, butyl and pentyl alcohols have a greater germicidal ability. But they are more expensive and do not easily mix with water. Methanol is toxic to tissues. Ethanol acts particularly on vegetative bacterial cells. It is strong dehydrating agent. Ethyl alcohol (70%) is mostly used.

Alkylating Agents

Formaldehyde is most common. This compound is gas at high temperature but a solid-paraformaldehyde at room temperature. Formalin is prepared by suspending 40g of this solid in water. This is used in anatomical specimens. It is also used in inactivation of viruses in vaccine preparations and in production of toxoids from toxins. Ethylene oxide (Eto) is used in sterilisation of plastic materials used in laboratory. It is used in combination with freon in a ratio of 12 : 88, available as cryoxide or steroxide. Carboxide contains Eto and carbondioxide. The gas is released into a tightly sealed chamber. Other alkalyting agents are, beta-propiolactone, and glutaraldehyde.

Hydrogen Peroxide

It is used as a rinse in wounds, scrapes or abrasions. New forms of stable H_2O_2, like Super D hydrogen peroxide have also appeared.

Soaps and Detergents

Since the pH of soaps is about 8.0, it destructs microorganisms to some extent due to alkalinity. Soap is used for mechanical washing of the skin surface. Soaps are wetting agents, emulsifying and solubilising particles that cling to a surface. Detergents are synthetic chemicals developed for their ability to be strong wetting agents and surface tension reducers.

Dyes

A group of dyes-triphenyl methane dyes is useful as antiseptics for *Bacillus* spp. and *Staphylococcus* spp., and in higher concentration to typhoid bacilli. This group includes malachite green and crystal violet (traditionally used as gentian violet for trench mouth and *Candida* infections). A second group of dyes, the acridine dyes are good antiseptics for gonococcal and staphylococcal infections. Generally dyes are more valuable for Gram-positive than Gram-negative forms. They act by combining directly with DNA and halt RNA synthesis.

Acids

Some acids are good disinfectants and antiseptics. The common ones are benzoic, salicylic and undecylnic acids for tinea infections of skin. Organic acids as lactic acid and acetic acid are good food preservatives.

CHEMOTHERAPEUTIC AGENTS

An account of this group of chemicals may appear out of place in the context of this chapter. The objective to include these chemicals is to examine their role in control of diseases of the suffering patients which could have otherwise been source of infectious microbes in the environment. Control of diseases of such suffering persons would help minimise the population of pathogenic microbes in environment.

It could become possible only since 1940's that a successful treatment of fatal disease was achieved. Early efforts of microbiologists towards control of diseases were mainly centered at enhancing the role of immune system. Vaccines for rabies, diphtheria and tetanus were developed. Among the pioneers in this field were Emil von Behring, Elic Metchnikoff, and Paul Ehrlich, all Nobel Prize winners in Physiology or Medicine. Ehrlich, is credited for the development of the first chemotherapeutic agent, who conceived the antibody molecules as "magic bullets". In the early 1900's his attention turned to magic bullets of a purely chemical nature, the chemotherapeutic agents. Ehrlich and his associates had synthesised hundreds of arsenic-phenol derivatives. These were tested by one of his associates, Sahachiro Hata, against the syphilis organism. Later, however, their attention focused on a single chemical, compound 606. This chemical, after trials in animals was made available to physicians under the trade name **Salvarsan** that soon became the first useful chemotherapeutic durg. The technical or generic name was **arsphenamine**. Ehrlich's death in 1915 and the emerging World War eroded the enthusiasm for chemotherapy. Further progress in this area was made only after about 20 years. German chemists developed some industrial dyes, that exhibited antimicrobial qualities also. A red dye, **prontosil** was synthesised in 1932, that showed activity against some Gram-positive bacteria such as streptococci and staphyococci. In 1935, a French group headed by Jacques and Thevese Trefouel announced that **sulfanilamide** was the active component of prontosil. Gerhard Domagk was awarded Nobel Prize in 1939 for successful treatment of war-related infections with sulfanilamide.

Following are the common groups of chemotherapeutic agents used in control of infectious microorganisms:

Sulfanilamide and Other Sulfonamides

Sulfanilamide was the first of a group of chemotherapeutic agents known as sulfonamides. They interfere with metabolism of bacteria through the mechanism known as **competitive inhibition**. Modern sulfonamides are typified by **sulfamethoxazole**, prescribed for urinary tract infections due to Gram-negative rods. The drug is combined with **trimethoprim**, available commercially under the trade name **Bactrim**. There are two kinds of name for each chemical, generic name and trade name. Table 18.2 shows these names for some common chemotherapeutic agents.

TABLE 18.2 Generic and trade names of some common chemotherapeutics

Generic name	*Trade name*
Sulfamethoxazole	Bactrim, Gantanol
Trimethoprim	Bactrim, Septran
Metronidazole	Flagyl
Ampicillin	Polycillin, Principen
Gentamicin	Garamycin
Neomycin	Neo-polycin
Chloramphenicol	Chloromycetin
Chlorotetracycline	Aureomycin
Oxytetracycline	Terramycin
Tetracycline	Achromycin
Erythromycin	Pediamycin
Griseofulvin	Fulvicin, Fulcin

Other common sulfonamides are, sulfacetamide, sulfabenzamide and sulfathiazole, commercially available as Triple Sulfa. This is used for vaginal infections due to *Haemophilus* sp. Another sulfonamide, sulfisoxazole is marketed as a cream for vaginal infections.

Other Chemotherapeutic Agents

There are many other agents, which became common these days. They are not related to sulfonamides. Some common ones include isonicotinic acid hydrazide, effective for tuberculosis; trimethoprim used in urinary tract infections; nalidixic acid for some Gram-negative bacteria of urinary tract infections and nitrofurantoin, also used for such infections.

Metronidazole (Flagyl) is effective against *Trichomonas vaginalis* infections and amoebiasis. However, there is evidence that it causes tumors in mice. Primaquine destroys the malarial parasite. Dapsone is used in treatment of leprosy.

Antibiotics

Alexander Fleming was a student of Almroth Wright, the British investigator who described opsonins. Fleming described the nonspecific enzyme, **lysozyme**. In 1928, he observed that a Petri dish culture of staphylococci had become contaminated with a green mold, and that the bacteria were disappearing as the mold grew over the plate. The mold isolated, was identified as *Penicillium,* and he found that the broth contained an active principle with antibacterial characteristics. Although, he failed to isolate the substance, he called it **penicillin**. He recognised the mold's potential for the treatment of human diseases, trying filtrates on infected wounds. Rene Dubos in 1939 indicated that soil bacteria could produce anti-bacterial chemicals like Fleming's penicillin. A group at Oxford University, led by the British pathologist, Howard Florey and the German biochemist, Ernst Boris Chain, reisolated penicillin and carried out careful trials with highly purified samples. In 1940, their successful attempts were published. American pharmaceutical companies developed technology for large-scale production of penicillin. Fleming, Florey and Chain received Nobel prize in 1945 for the discovery and development of penicillin. Since this was a naturally occurring product, a term **antibiotic** was introduced in medicine.

Following are some of the most commonly used antibiotics:

Penicillin

A large group of penicillin derivatives are available. Penicillin G, or benzylpenicillin is the most popular of the penicillins. Other types are penicillin F or penicillin V. All have the basic same structure with a beta-lactam nucleus and several attached groups as shown in Fig. 18.6.

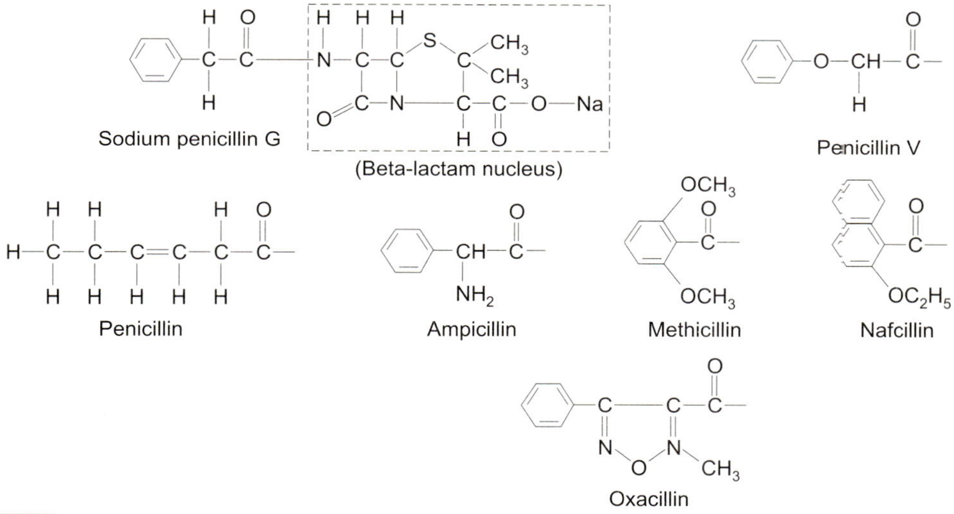

Sodium penicillin G

(Beta-lactam nucleus)

Penicillin V

Penicillin

Ampicillin

Methicillin

Nafcillin

Oxacillin

FIG. 18.6

Some members of the penicillin group of antibiotics. The beta-lactam nucleus is common to all the penicillins. Note that different penicillins are formed by varying the side group on the molecule

Penicillins are active against a variety of Gram-positive bacteria, including staphylococci and streptococci. Penicillin functions during the synthesis of bacterial cell wall. The molecule blocks the cross-linking of hexoses in the peptidogylycan layer during wall formation, causing bursting of cell. There are, however, two major drawbacks to the use of penicillin. One, the anaphylactic reaction in those who are allergic to penicillin. It causes swelling about the eyes or wrists, itchy skin etc. Second, the evolution of some penicillin-resistant bacteria, that produce an enzyme, penicillinase. This converts penicillin into harmless penicilloic acid. Modern penicillins are mostly produced from *Penicillium notatum* and *P. chrysogenum*.

Semisynthetic penicillins

In the late 1950s, the betalactum nucleus of the penicillin molecule was identified and synthesised. Various groups then could be attached to this nucleus, creating a number of new penicillins. At present thousands of penicillins are prepared by this semi-synthetic process. Ampicillin is less effective than Penicillin G against Gram-positive cocci, but valuable against some Gram-negative rods. It can be taken orally and absorbed from the intestine. Amoxicillin is a similar penicillin. Both are useful to treat urinary tract infections. Another semisynthetic penicillin, carbenicillin is used for *Pseudomonas* and *Proteus* infections of urinary tract. Others are, methicillin, nafcillin and oxacillin, which are resistant to penicillinase.

Cephalosporins

They were developed in 1960s. Cephalosporin C was isolated from the blue mold, *Cephalosporium*. A number of related semi-synthetic drugs developed from it are, cephalexin, cephalothin, cefazolin and cephaloridine. Cephalosporins are alternatives to penicillin, and are effective for staphylococcal boils or wounds, streptococci, and bacterial pneumonia, and urinary tract infections by Gram-negative bacteria.

Streptomycin (an aminoglycoside)

This was discovered by Selman A. Waksman. Several moldlike soil bacteria-actinomycetes produced antimicrobial chemicals. Of these, *Streptomyces griseus* was most effective producer. The substance produced by it was named **streptomycin** by Waksman, Elizabeth Bugie and Albert Shatz. This drug was made available in 1947 and Waksman recieved Nobel Prize in 1952 in Physiology or Medicine. Streptomycin in combination with isoniazid is important for treatment of tuberculosis. Gram-negative infections as plague, brucellosis are also treated.

Other aminoglycoside antibiotics

Gentamicin is the first drug to be given for infections by Gram-negative bacteria. It is combined with carbenicillin for *Pseudomonoas* infections, with ampicillin for strepytococcal infections of intestine, and with cephalosporin for staphlococcal disorders. Neomycin is now used as topical antibiotic for eye bacterial conjunctivitis or other Gram-negative infections. Commercially, it is available as neosporin when combined with polymyxin, and as cortisporin when combined with cortisone, bacitracin, and polymyxin. The mixtures are useful for a variety of mild skin infections due to Gram-negative or Gram-positive bacteria. All aminoglycosides are derived from the species of *Streptomyces*.

Chloramphenicol

This is the first broad-spectrum antibiotic discovered. It was isolated in 1947 by Ehrlich, Burkholder and Gotlieb. It inhibits a wide variety of Gram-positive and

Gram-negative bacteria, as well as several rickettsiae and fungi. Chloramphenicol was originally isolated from the metabolites of *Streptomyces venezuelae*. However, there are two main disadvantages during its use. In bone marrow it prevents hemaglobin incorporation into the red blood cells—**aplastic anemia**. Moreover, due to its accumulation in blood of newborn child, it causes a toxic reaction and sudden breakdown of cardiovascular system—**gray syndrome**.

Tetracyclines

They are also a abroad-spectrum antibiotics with range of activity similar to chloramphenicol. They include naturally occurring chlorotetracycline and oxytetracycline, isolated from species of *Streptomyces*. They may be taken orally. Though they have side effects problems, they remain the drugs of choice for most rickettsial and chlamydial diseases. They are used in Gram-negative infections as brucellosis, plague, cholera, for primary atypical pneumonia, as substitutes for penicillin in syphilis, anthrax, gonorrhea and pneumonia; and therapy of some protozoan infections as amoebiasis.

Other antibiotics

Erythromycin, obtained from *Streptomyces* is useful for primary atypical pneumonia, staphylococcal and streptococcal infections and syphilis. Vancomycin, also a product of *Streptomyces* is given intravenously against Gram-positive infections. Other antibiotics are rifampin for leprosy and tuberculosis, clindamycin and lincomycin, active against streptococci, staphylococci and other Gram-positive organisms.

Bacitracin and polymyxin, obtained from *Bacillus* species are used topically. The former is used as ointment for staphylococci, and latter for Gram-negative bacilli. Spectinomycin, a product of *Streptomyces* became popular in late 1970s as substitute for penicillin in case of gonorrhea caused by PPNG. It is given intramuscularly.

Antifungal antibiotics

Nystatin, a product of *Streptomyces* is used as cream or ointment or in suppository form, for infection of oral cavity, vagina or intestine due to *Candida albicans*. Griseofulvin is used for fungal infections of skin, hair and nails. It is effective against ringworm and eczema. This is a product of *Penicillium*. For serious systemic fungal infections, amphotericin B is used. This is effective for organisms of histoplasmosis, blastomycosis, cryptococcosis etc.

SITES OF ACTION OF ANTIBACTERIAL AGENTS

A knowledge of the sites of action of antibiotics should help designing new and more effective chemotherapeutic agents. The sites of action of some important antibiotics are indicated in Table 18.3.

TABLE 18.3 Sites of action of some antibiotics

Antibiotic	*Site of action*	*Effect*
Penicillin, Cephalosporin	Cross-linking of peptidoglycan	Inhibits bacterial cell wall synthesis
Streptomycin	30s ribosome subunit	Inhibits bacterial protein synthesis
Chloramphenicol	Translation of mRNA on 50s ribosome	As above
Erythromycin	50s ribosome	As above
Rifampicin	RNA polymerase	Inhibits mRNA synthesis
Nalidixic acid	DNA gyrase	Inhibits DNA replication
Polymixin	Cell membrane	Destroys membrane functions

It may be seen that antibacterials act at different sites of the pathogenic bacterium to which they are applied. They may act accordingly as follows:

Cell Wall Inhibitors

These include two widely used classes of antibiotics, the **penicillins** and **cephalosporins**. Both contain a β-lactam ring (Fig. 18.7). They act on various Gram-positive and Gram-negative rods and cocci, responsible for various diseases. They inhibit the formation of peptide cross-linkages within the peptidoglycan backbone of the cell wall. Other cell wall inhibitors are **vancomycin, bacitracin** and **cycloserine**.

Protein Synthesis Inhibitors

Such antibiotics are streptomycin, gentamicin, neomycin, kanamycin, tobramycin and amikacin. These are called **aminoglycoside antibiotics**, and used for Gram-negative bacteria. They bind to the 30s ribosomal subunit of the 70s prokaryotic ribosome. In addition to these, a number of other antibiotics inhibit protein synthesis. These are tetracyclines, chloramphenicol, erythromycin, lincomycin, clindamycin and spectinomycin. Chloramphenicol, unlike others acts primarily by binding to the 50s ribosomal subunit, preventing the binding of tRNA molecules to both the amino-acyl and peptidyl binding sites of the ribosome. Erythromycin also binds to 50s ribosomal subunits.

Membrane Transport Inhibitors

The polymyxins, such as polymyxin B changes structure of cell membrane and causes leakage of cell contents of Gram-negative bacteria. *Pseudomonas* spp and other Gram-negative bacteria, resistant to penicillins and aminoglycoside antibiotics are controlled by polymyxins B and E.

DNA Inhibitors

They block DNA replication. In particular **quinolones** interfere with DNA gyrase, preventing the establishment of replication fork. The quinolones include nalidixic

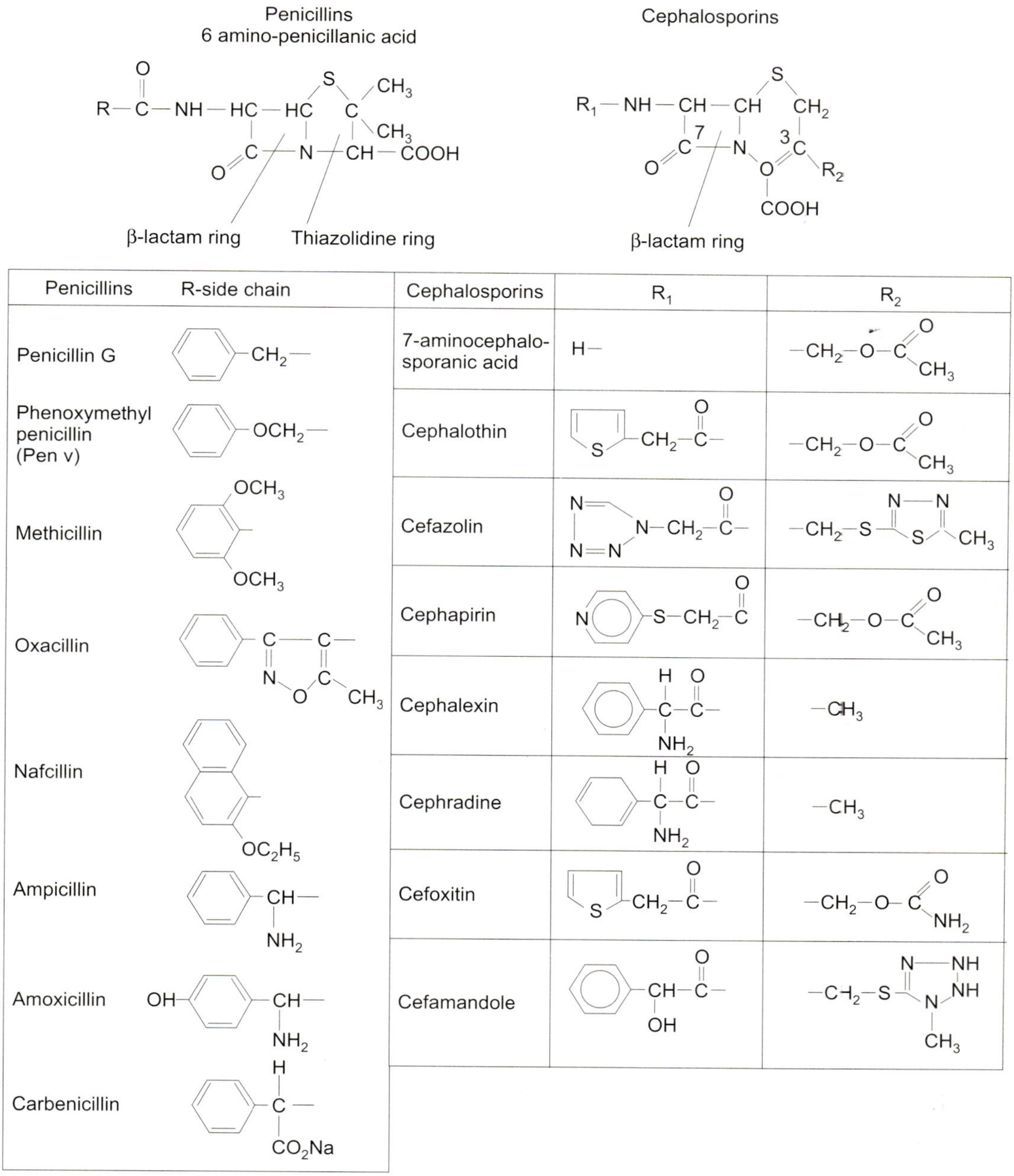

Penicillins
6 amino-penicillanic acid

R—C(=O)—NH—HC—HC—S—C(CH_3)(CH_3)—CH—COOH; C(=O)—N

β-lactam ring   Thiazolidine ring

Cephalosporins

R_1—NH—CH—CH—S—CH_2—C(3)—R_2; 7 C(=O)—N—C—COOH

β-lactam ring

Penicillins	R-side chain
Penicillin G	C_6H_5—CH_2—
Phenoxymethyl penicillin (Pen v)	C_6H_5—OCH_2—
Methicillin	$C_6H_3(OCH_3)_2$—
Oxacillin	C_6H_5—C—C—; N—O—C—CH_3
Nafcillin	$C_{10}H_6(OC_2H_5)$—
Ampicillin	C_6H_5—CH(NH_2)—
Amoxicillin	OH—C_6H_4—CH(NH_2)—
Carbenicillin	C_6H_5—C(H)(CO_2Na)—

Cephalosporins	R_1	R_2
7-aminocephalo-sporanic acid	H—	—CH_2—O—C(=O)—CH_3
Cephalothin	thienyl—CH_2—C(=O)—	—CH_2—O—C(=O)—CH_3
Cefazolin	tetrazolyl (N=, N=N) N—CH_2—C(=O)—	—CH_2—S—thiadiazolyl (N—N, S)—CH_3
Cephapirin	pyridyl (N)—S—CH_2—C(=O)	—CH_2—O—C(=O)—CH_3
Cephalexin	C_6H_5—C(H)(NH_2)—C(=O)—	—CH_3
Cephradine	C_6H_7—C(H)(NH_2)—C(=O)—	—CH_3
Cefoxitin	thienyl—CH_2—C(=O)—	—CH_2—O—C(=O)—NH_2
Cefamandole	C_6H_5—CH(OH)—C(=O)—	—CH_2—S—tetrazolyl (N—NH, NH, N—CH_3)

FIG. 18.7

Biochemical structure of penicillins and cephalosporins, all of which contain a β-lactam ring

acid, ciprofloxicin, norfloxicin, amifloxicin and enoxicin, effective against a range of Gram-positive and Gram-negative bacteria.

Other Sites of Inhibition

Sulphonamides, sulphones and para- aminosalicylic acid are structural analogues of the vitamin para- aminobenzoic acid. These analogues are effective competitors with the natural substrate for the enzymes involved in synthesis of folic acid, as such, and thus inhibit the formation of this required co-enzyme (folic acid) causing a bacteriostatic effect. Trimethoprim is an inhibitor of dihydropholate reductase. Dihydrofolic acid is a co-enzyme required for 1-carbon transfers, as those required in synthesis of thymidine and purines. This antibiotic is broad-spectrum one used to treat bacterial infections of urinary and intestinal tracts.

ANTIBIOTIC SENSITIVITY ASSAYS

These assays are used to study the inhibition of a test organism by one or more antibiotics or other chemotherapeutic agents. Two general methods are commonly used:

Tube Dilution Method

This is often used to determine the smallest amount of antibiotic necessary to inhibit a test organism. This amount is known as the **minimum inhibitory concentration** (MIC). A set of tubes with different concentrations of a particular antibiotic are prepared. The tubes are inoculated with the test organism, incubated, and examined for growth of bacteria. Growth is seen to diminish as the concentration of antibiotic increases, and eventually an antibiotic concentration may be observed at which growth fails to occur. This is the MIC.

Agar Diffusion Method

The principle used here is that antibiotic will diffuse from a paper disc or small cylinder into an agar medium that contains test organisms. Inhibition is observed as a failure of the organism to grow in the region of the antibiotic. A common application of this method is the **Kirby-Bauer test**, developed in the 1960s. The procedure is used to determine the sensitivity of an organism isolated from a patient to a series of antibiotics. The results serve a guide to physician to prescribe a drug.

An agar medium such as Mueller-Hinton medium is inoculated with the organism and poured to the plate. Paper discs containing known concentrations of antibiotics are applied to the surface, and the plate is incubated. The appearance of a zone of inhibition surrounding the disc is indicative of sensitivity. By comparing the diameter of the zones to a standard table, one may determine if the test organism is susceptible or resistant to the antibiotic. If the organism is susceptible, it is likely to be killed in the blood stream of the patient if that concentration of the drug is reached. Resistance indicates that the antibiotic will not be effective at that concentration in the blood stream.

19

MICROBIAL SAFETY AND RISK ASSESSMENT

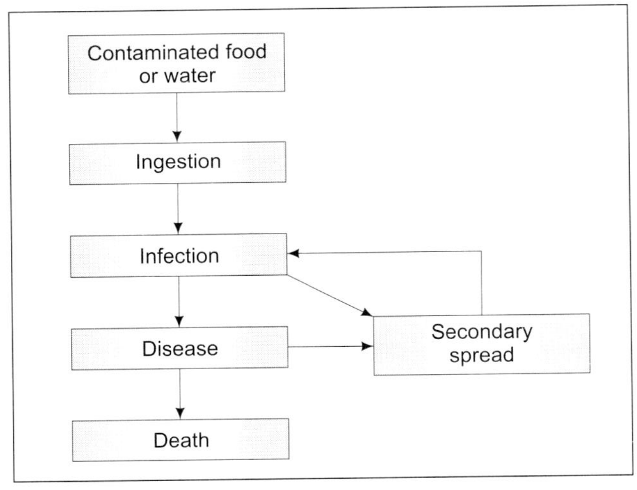

Chapter Outline

In recent years much attention has been paid towards assessment of risks/hazards associated with concentrations of pathogenic microorganisms in environment (water, food, soil, air), release of genetically modified organisms and consequences of microbial bioremediation of contaminated sites. There is need of assessing microbial hazards and deciding whether a particular microbe presents a reasonable or unreasonable risk to human health and environment. The task of interpreting data on the occurrence of pathogens in the environment is often critical in making decisions about potential health risks and corrective measures. The objective of this chapter is to introduce to the reader basic concepts of microbial safety and risk assessment.

WHAT IS RISK AND RISK ASSESSMENT?

Risk, which is common to all life, is an inherent property of everyday human existence. It is therefore a key factor in all decision making. Risk assessment means different things to different persons. To financial analysts it may be financial risk, to drivers it may be fatality risk or to factory worker it may be risk of cancer incidence due to exposure. But all these have one thing common—the measurable phenomenon called risk that can be expressed in terms of probability. The terms **hazard** and **risk** may be used interchangeably. Both are synonyms of the word danger, and each implies an element of chance that an adverse effect will occur under a particular set of conditions. **Risk** can be defined as "a measure of the probability that a certain adverse effect may happen". Accordingly **risk assessment** can be defined as "the process of estimating both the probability that an event will occur and the probable magnitude of its adverse effects—economic, health or safety related, or ecological-over a specific time period". For instance, one might calculate the health risks associated with the presence of pathogens in drinking water or in food or in air. Risk assessment as a formal discipline emerged in the 1940s and 1950s, paralleling the risk of nuclear industries.

Safety may be defined as "a value judgement of the acceptability of risk." A pathogen in food or water may be regarded safe if its associated risks are judged to be acceptable. Such a definition implies that two different activities are required for determining safety. (i) a risk assessment to evaluate potential harm to organisms, and (ii) a value judgement regarding the acceptability of that risk.

ELEMENTS OF RISK ASSESSMENT

Chemical risk assessment has been used to judge the safety of our food and water supply. Such assessments are important in setting standards for chemical contaminants in the environment. Whether chemical or microbial, risk assessment consists of the following **four** basic steps:

Hazard Identification

This is the first step where we determine the nature of hazard. For instance, in pollution problems, the hazard in question, besides a specific physical or chemical agent, may be a microorganism, which is to be identified with a specific disease/illness. Its toxic effects on human are to be documented.

As shown in Fig. 19.1, clinical studies of disease can be used to identify very large risks (between 1:10 and 1:100), most epidemiological studies can detect risks down to 1:1,000, and very large epidemiological studies can examine risks in the 1:10,000 range.

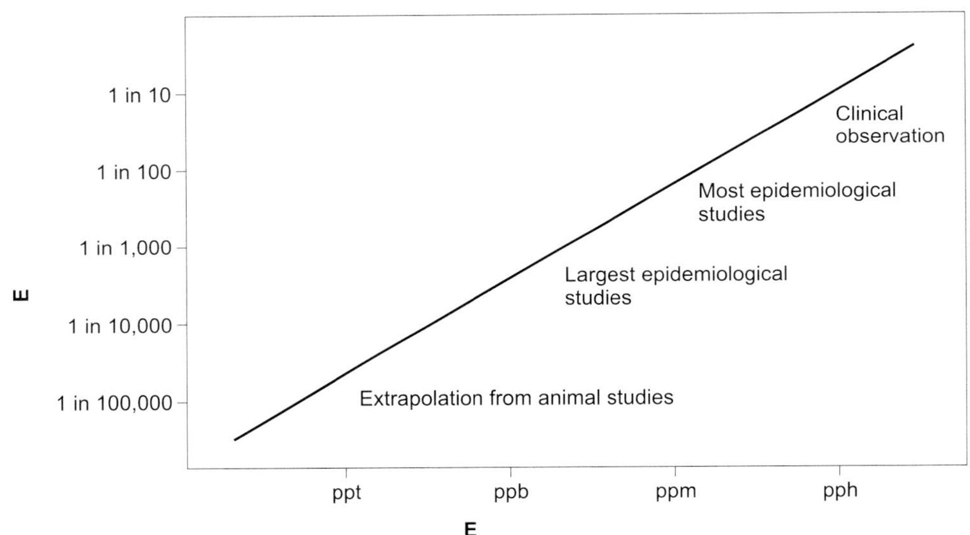

FIG. 19.1

Sensitivity of epidemiology in detecting risks of regulatory concern. The generalized units of ppt (parts per thousand), ppb (parts per billion), ppm (parts per million), and pph (parts per hundred) are used here for comparative purposes). (Modified from National Research Council, 1993)

Exposure Assessment

Here we determine the concentration of a contaminating agent in the environment and estimate its rate of intake. For example, to determine the concentration of *Salmonella* in a meat product and determine the average dose a person would digest.

Thus exposure assessment is the process of measuring or estimating the intensity, frequency, and duration of human exposures to an environmental agent. Exposure can occur via inhalation, ingestion of contaminated water or food, or via the skin.

Dose-Response Assessment

It is the quantification of adverse effects arising from exposure to a hazardous agent based on the degree of exposure. This assessment is usually expressed mathematically as a plot showing the response in living organisms to increasing doses of the agent (e.g. a virus). Chemical or microbial contaminants are not equal in their capacity to cause adverse effects. In order to determine such a capacity of agents to cause harm, we need quantitative toxicity or infectivity data. These data can be derived from occupational, clinical and epidemiological studies. Most toxicity data are, however, obtained from experimental animals (mice, rats) in laboratory conditions. Animals are exposed to increasing doses of chemicals and adverse effects are observed. The results are expressed as **dose-response-relationship**—a quantitative relationship that indicates the agent's degree of toxicity to exposed species.

Responses can vary widely from no observable effect to death. The objective of a dose-response assessment is to obtain a mathematical relationship between the amount of a toxicant or microorganism to which a human is exposed and the risk of an adverse outcome from that dose. The data obtained from such experiments are presented as a dose-response curve (Fig. 19.2). In case of pathogen, the ordinate may represent the risk of infection and not necessarily illness, whereas the abscissa represents the dose (concentration of pathogen). However, such curves obtained from animal studies must be interpreted with care. Results of such curves may differ with the method employed. In microbiological literature, the term **minimum infectious**

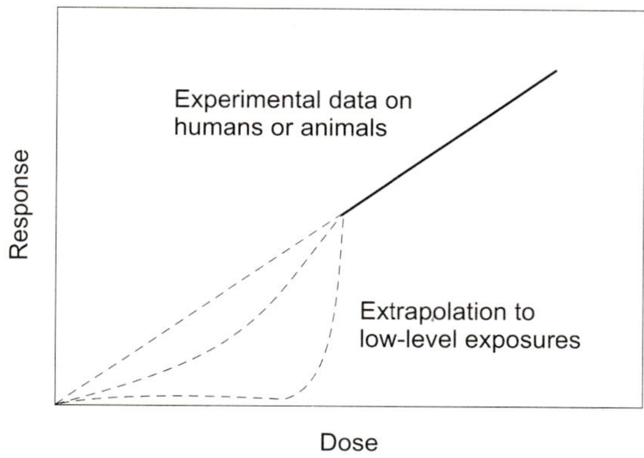

FIG. 19.2

Extrapolation of dose-response curves. (Adapted from U.S. EPA, 1990)

dose is used frequently, implying that a threshold dose exists for microbes. In reality, the term used usually refers to ID 50 or the dose at which 50% of the animals or humans exposed became infected or exhibit any symptoms of a disease/illness.

Risk Characterisation

This includes estimation of the potential impact (e.g. human illness or death) of a microorganism or chemical based on the severity of its effects and the amount of exposure. In this final phase, exposure and dose-response assessments are integrated to obtain probabilities of effects in humans under specific exposure conditions. Quantitative risks are calculated for appropriate media and pathways. The information is then used by risk managers to develop standards for specific toxic chemicals or infectious microorganisms in different media like drinking water, food supply etc.

MICROBIAL RISK ASSESSMENT

In area with severely contaminated water supplies, there occur sudden outbreaks of waterborne diseases caused by enteric pathogens. In such a situation with high-level doses, cause and effects can be easily determined. However, under low-level doses (1:1000), such determinations become difficult. Microbial risk assessment allows us to estimate responses in terms of **risk of infection** in a quantitative fashion. Although no accepted formal protocol for microbial risk assessment exists, it usually follows the steps used in health-based risk assessments—hazard identification, exposure assessment, dose-response, and risk characterisation. Hazard identification in case of pathogens is complicated because several outcomes-from asymptomatic infection to death (Fig. 19.3) are possible, and these outcomes depend on the complex interactions between the host and the pathogen.

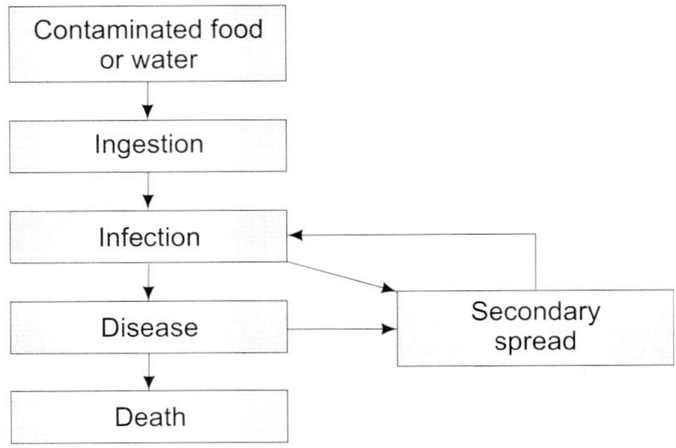

FIG. 19.3
Outcomes of enteric pathogen exposure

20

LABORATORY EXERCISES

EXERCISE 1

Objective: *Calibration of microscope and measurement of microorganisms.*

Theory. Microorganisms vary in size and shape. Methods have been developed to record the dimensions of their size in terms of diameter, length, breadth etc. Since microbes can be viewed only under a microscope with proper magnification, the scale to be used for their measurements is to be put somewhere in the microscope itself.

The scale put in the ocular (eye piece) lens of the microscope needs to be calibrated with a known scale of measurement, the stage micrometer. Measurements of the microorganisms can be easily made after calibrating the ocular micrometer.

Materials. Prepared slides of yeast protozoa, bacterial cocci and bacilli and fungal spore; Ocular micrometer; Stage micrometer; Microscope; Immersion oil; Xylene; Lens paper.

Procedure. 1. Remove the ocular (eye piece) lens and insert the ocular micrometer on the circular shelf (metal diaphragm). Replace the ocular lens and mount in the microscope and observe. There will be seen scale lines of ocular micrometer in sharp focus (Fig. 20.1c). The lines and distances will remain unchanged under different objectives.

2. Mount the stage micrometer on the microscope stage and bring its scale in the centre of microscope field under a sharp focus. This is done first with low-power objective (Fig. 20.1d) and thereafter also with high-power and oil-immersion objectives (Fig. 20.1e, f).

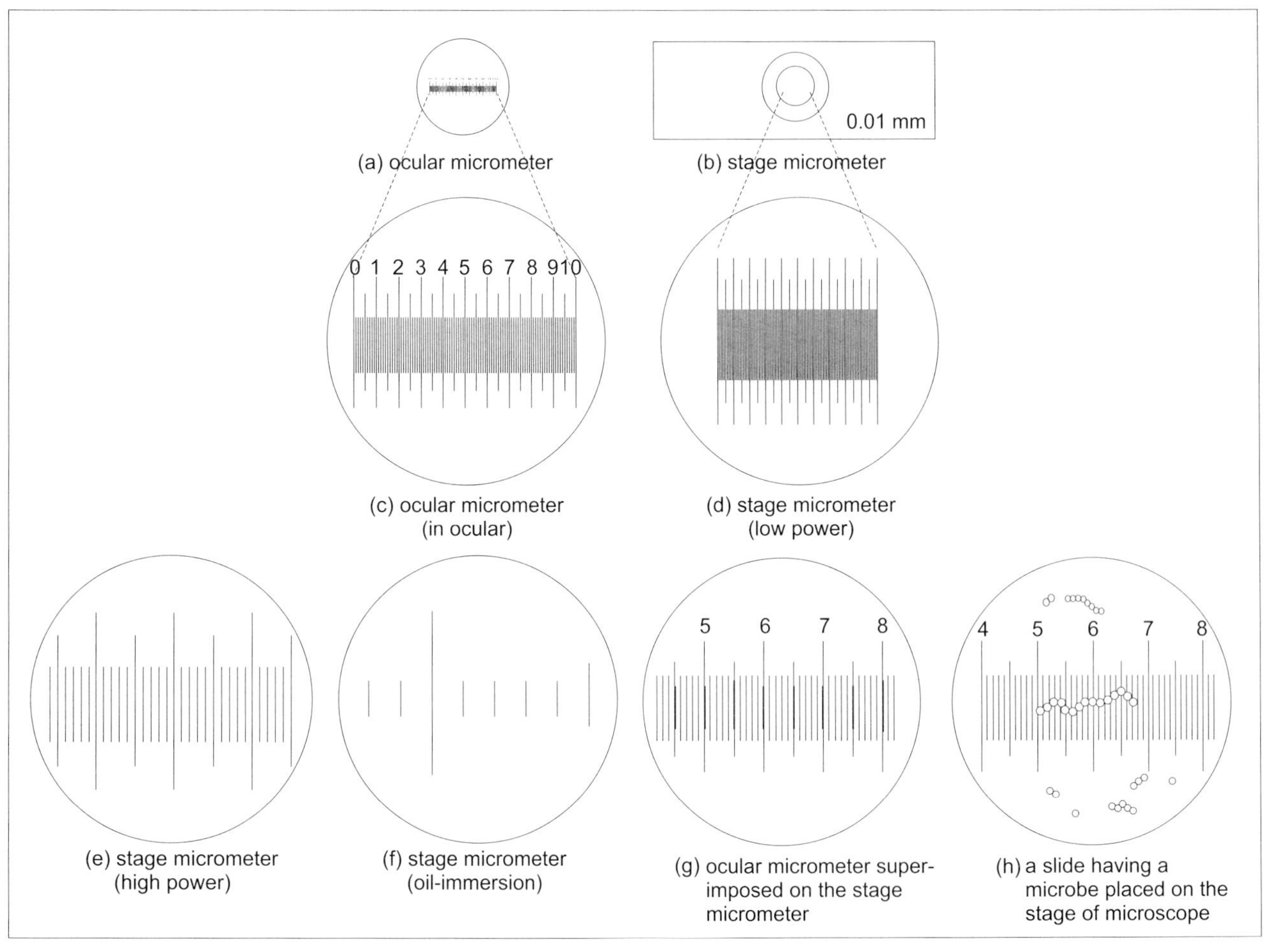

FIG. 20.1

Standardization (calibration) of the ocular micrometer and its use in micrometry

3. Adjust the scales of ocular micrometer and stage micrometer in such a way that the lines of the former superimpose upon those of the latter. If necessary the ocular micrometer may be rotated or stage micrometer be moved on the stage so that lines of ocular micrometer superimpose upon the stage micrometer. The scales of both micrometers are to be adjusted in a way that lines of the two coincide at one end of the microscope field. This is shown in Fig. 20.1g. Now count the spaces of each micrometer to a point where the lines of the two micrometers coincide again. In this way we can find out how many divisions of ocular micrometer (unknown scale) are equal to how many divisions of stage micrometer (known scale). As shown in Fig. 20.1g, five divisions of ocular micrometer equal to one division of stage micrometer or 10 divisions of the former equal to two divisions of the latter. This is to be found out for (i) low power, (ii) high power, and (iii) oil-immersion objectives. For each objective, at least five readings are taken at random to obtain mean valve of one division of ocular micrometer.
4. Calculate the mean value of one division of ocular micrometer for different objectives in the following way:

   We know that:

   100 divisions of stage micrometer = 1 mm = 1000 μm or

   1 division of stage micrometer = 0.01 mm = 10 μm (known)

   Now for (i) low power objective:

   Suppose, 40 divisions of ocular micrometer = 60 divisions of stage micrometer

   i.e. 60×10 μm = 600 μm

   1 division of ocular micrometer = 15 μm

   (ii) high power objective:

   Suppose, 15 divisions of ocular micrometer = 6 divisions of stage micrometer
   i.e. 6×10 μm = 60 μm

   1 division of ocular micrometer = 4 μm

   (iii) oil-immersion objective:

   Suppose, 10 divisions of ocular micrometer = 2 divisions of stage micrometer

   i.e. 2×10 μm = 20 μm

   1 division of ocular micrometer = 2 μm.
5. Replace the stage micrometer with the slide bearing the microorganism whose measurements are to be taken. For instance, total length of the chain of a bacterium equals 18 divisions of ocular micrometer, when examined under oil-immersion (Fig. 20.1h). Thus length of the chain would be 18×2 μm = 36 μm.

Observations. Record the observations as follows:

Organisms	*Size of the microbes (μm)*					
	Length			*Width/Diameter*		
	No. of ocular divisions	*Calibration factor*	*Size*	*No. of ocular divisions*	*Calibration factor*	*Size*
Yeast	8	3.75	30	6	3.75	22.5
Protozoa	5	15	75	2	15	30
Cocci	2	2	4	2	2	4
Bacilli	4	2	8	2	2	4
Fungal spore	10	3.75	37.5	4	3.75	15

EXERCISE 2

Objective: ***Microscopic examination of live bacteria by hanging-drop procedure***

Theory. Bacteria vary in shape and size and their cells may bear flagella. The students should make themselves familiar with various forms of bacteria and the way of their motion by flagella.

Materials. Broth culture of the bacterium (24-hour old); Bunsen burner; Inoculating loop; Depression slide; Coverslip; Microscope; Petroleum jelly.

Procedure. Apply a ring of petroleum jelly around the concavity of the depression slide. Place a loopful of the culture in the centre of clean coverslip under sterile conditions. Place the depression slide with the concave surface facing down over the coverslip so that depression covers the drop of the culture. Press the slide gently to form a seal between the slide and the coverslip. Quickly turn the right side up so that the drop continues to adhere to the inner surface of the coverslip. First focus the slide under low-power and then adjust under oil-immersion objective for detailed observation. Observe the size, shape and motility of the bacteria present.

Observations. Record the observations as follows:

Organism	*Size, shape, true motality or Brownian movement*
Pseudomonas sp.	
Bacillus sp.	
Staphylococcus sp.	

EXERCISE 3

Objective: *To isolate, observe, and quantify the filamentous soil fungi by serial dilution and pour plating techniques*

Theory. Fungi are abundant in surface soils and as decomposers play important role in nutrient cycling. Since soils contain high population of fungi, samples are to be adequately diluted to obtain discrete colonies on culture media used for their isolation. Inoculated plates are incubated at appropriate temperature and fungal colonies are counted. Since it is assumed that one fungal colony has developed from one organism (a viable spore, hypha, hyphal fragment) the term colony forming units (CFUs) is used to express final results of calculation in terms of CFUs g^{-1} oven dry soil.

Materials. Moist soil sample (approx. 20g); Deionised water; Pipettes (25 ml, 1ml); Bench top balance; Weighing dishes; Sterile 95 ml water blank; Sterile 9 ml water blanks; Rose Bengal agar (150 ml); Filter-sterilised streptomycin solution (30 μg ml^{-1}); Sterile Petri dishes (9); Vortex; Water bath at 45°C; Lactophenol mounting fluid; Microscope; Microscope slides.

Procedure. Soil samples are weighed before putting 10g of soil to bottle A to find out the weight loss due to evaporation of moisture. Thus actual soil moisture and soil weight (real) are calculated at the time of preparation of dilution series and subsequent plating.

Beginning at step 1 of Fig. 20.2, prepare a 10-fold dilution series of the soil sample. Ten gram of moist soil is added to 95 ml (solution A) of deionised water and shaken well to disperse the organisms. The reason of using 10g of moist soil is that 10g of moist soil occupies roughly 5 ml of water. Thus we have 10 g of soil in 5 ml + 95 ml = 100 ml total volume (i.e. a 1:10 w/v dilution). Next, 1.0 ml of suspension is removed from bottle A and added to the tube B containing 9 ml of the same dispession solution as used in A. The tube is capped and vortexed. The dilution series is continued to the progressively higher dilutions (tube C, D and E). The three most diluted suspensions are plated. The three dilutions (C, D, E) are plated so as to obtain discrete, countable number of colonies (Fig. 20.3).

Pour plates are prepared using aliquots of spore suspensions of diluted series. The dilution is vortexed and 1.0 ml of suspension is added to each of two sterile Petri dishes (step 2, Fig 20.2). A suitable volume of molten agar is poured into the plate before the soil partides in the suspension settle down (step 3 a, Fig 20.2). The plate is gently swirled (step 3 b, Fig 20.2) to distribute the agar and inoculum across the bottom of the plate. Allow the agar to solidify and incubate the plates upside down (to prevent condensation from falling on the growing surface of agar). Colonies are counted after incubating the plates for 5-7 days.

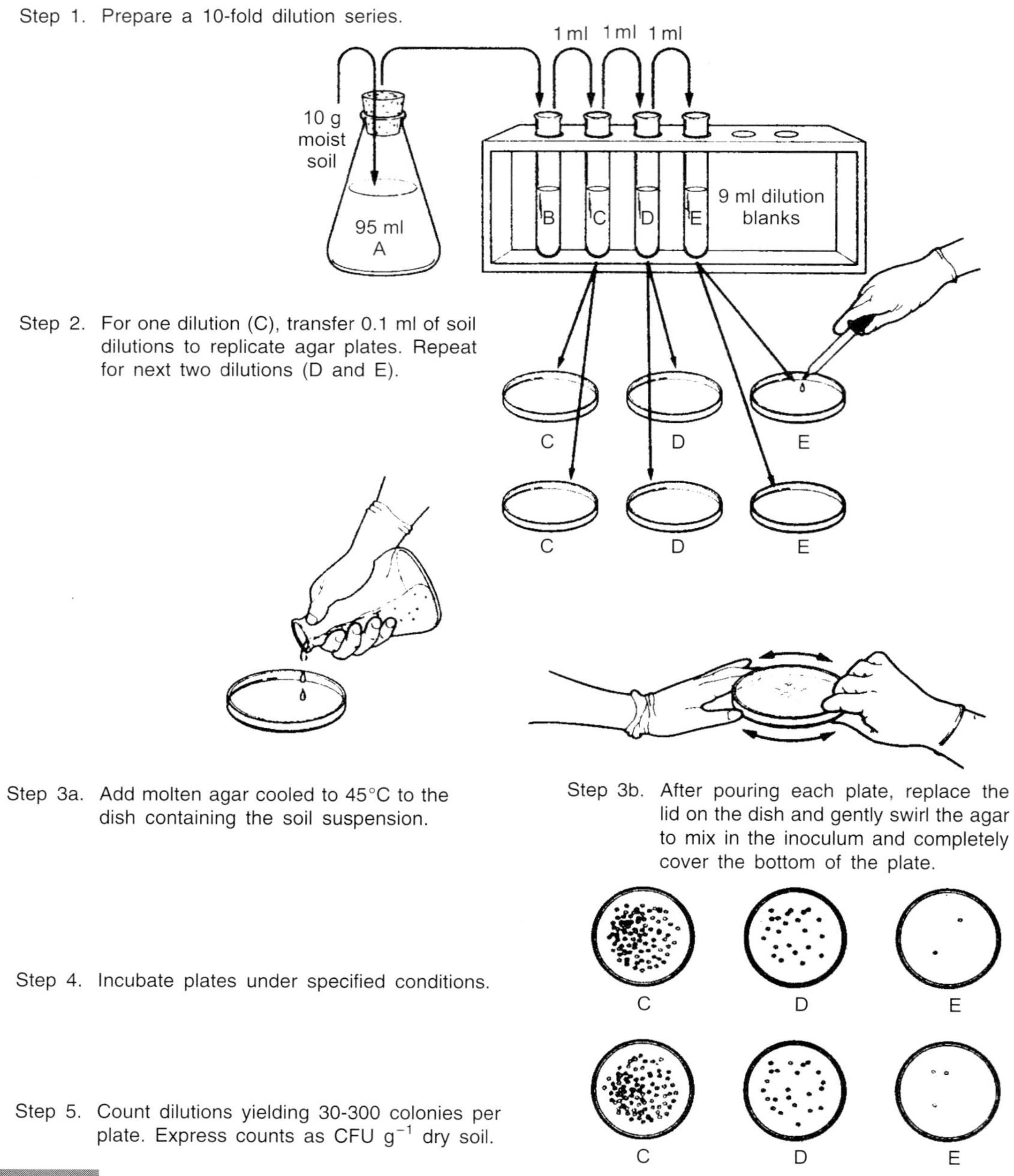

Step 1. Prepare a 10-fold dilution series.

1 ml 1 ml 1 ml

10 g moist soil

95 ml A

B C D E

9 ml dilution blanks

Step 2. For one dilution (C), transfer 0.1 ml of soil dilutions to replicate agar plates. Repeat for next two dilutions (D and E).

C D E

C D E

Step 3a. Add molten agar cooled to 45°C to the dish containing the soil suspension.

Step 3b. After pouring each plate, replace the lid on the dish and gently swirl the agar to mix in the inoculum and completely cover the bottom of the plate.

Step 4. Incubate plates under specified conditions.

C D E

Step 5. Count dilutions yielding 30-300 colonies per plate. Express counts as CFU g^{-1} dry soil.

C D E

FIG. 20.2

Schematic showing different steps of culture and plate counts of filamentous fungi

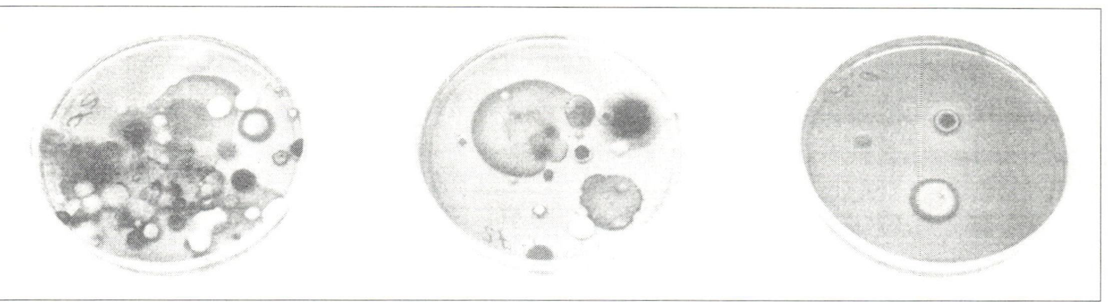

FIG. 20.3
Discrete fungal colonies appearing on plates incubated after dilution and plating of soil

Observations and calculations. Examine the incubated plates for each dilution, 10^{-3}, 10^{-4} and 10^{-5} soil per plate (in this case). Make suitable preparations with lactophenol, identify the genera of fungi and count the colonies. Examine the fungi microscopically. Identify at least three different genera.

The plates to be selected for count should have discrete countable colonies. Overgrown plates, and plates with <10 colonies should be discorded.

Tabulate all results as below including individual plate counts and mean counts. Calculate the total number of filamentors fungi (as CFUs) per gram dry weight of soil. Compute the sample mean, standard deviation and coefficient of variation.

Example. A 10 gram soil sample with a moisture content of 20% on a dry wt. basis is analysed for CFUs as follows:

10g soil → 95 ml saline (solution A) 10^{-1} (w/v) dilution. Subsequent dilutions (v/v) were obtained as B – 10^{-2}; C – 10^{-3}; D – 10^{-4} and E – 10^{-5} (Fig. 20.2).

1 ml of solution E is pour plated onto medium, and it yielded, say 200 colonies.

$$\text{No. of CFUs} = \frac{1}{\text{dilution factor}} \times \text{no. of colonies}$$

$$= \frac{1}{10^{-5}} \times 200 \text{ CFUs g}^{-1} \text{ moist soil}$$

$$= 2.00 \times 10^{-7} \text{ CFUs g}^{-1} \text{ moist soil}$$

But, for 10g of moist soil:

$$\text{Moisture content} = \frac{\text{Moist weight} - \text{Dry weight (D)}}{\text{Dry weight (D)}}$$

$$\text{i.e. } 0.20 = \frac{10 - D}{D} \quad \text{i.e. } D = 8.33 \text{ g}$$

(20% moisture content)

$$\text{Number of CFUs per g dry soil} = 2.00 \times 10^7 \times \frac{1}{8.33} = 2.4 \times 10^7$$

EXERCISE 4

Objective: ***To isolate, observe and quantify the soil bacteria and actinomycetes by serial dilution and spread plating techniques***

Theory. Soil bacteria are the most abundant microbes found in surface soils and are diverse in terms of metabolic and nutritional relations. Like fungi, heterotrophic culturable bacteria and actinomycetes also play important role in nutrient cycling. As in case of fungi, the soil sample is diluted serially and plated on agar to obtain countable number of colonies on the plates (Fig. 20.4).

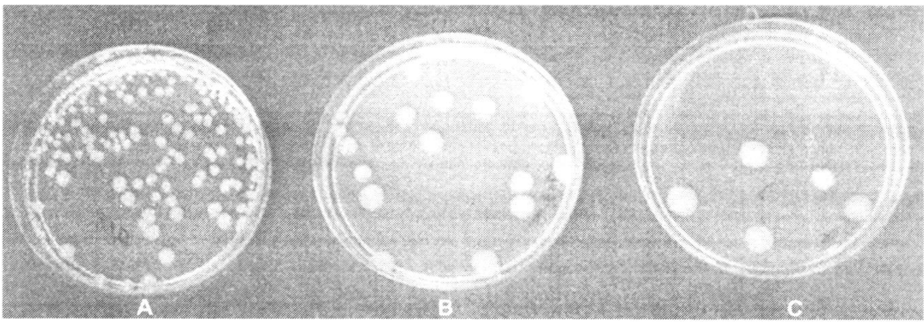

FIG. 20.4

Results of a successive bacterial plate count assay. The plate (A) has the most bacterial colonies (lowest soil dilution). The other two plates (B, C) result from subsequent dilutions. From these three plates, plate A should be counted since it has between 30 and 200 colonies

Materials. Moist soil sample (approx. 20g); Deionised water; Benchtop balance; Weighing dishes; Sterile 95ml water blank; Sterile 9 ml water blanks; Pipettes (1 ml); Peptone-yeast agar plates (for bacteria); Glycerol-casein agar plates amended with cycloheximide (for actinomycetes); Glass spreader; Ethyl alcohol; Vortex.

Procedure. Weigh the soil sample before proceeding for dilution and plating to find out the weight loss due to evaporation of moisture as done for fungi. Find out the moisture content and actual soil weight to be put to bottle A for dilution at 10^{-1}. As with fungi, prepare a dilution series of the soil sample (Fig. 20.5). Put 10g of soil in a 95 ml water blank and vortex well. Before the soil settles in bottle A, remove 1ml of the suspension with pipette and add it to a 9 ml water blank. Vortex well and repeat the previous step thrice, each time with a fresh 9 ml water blank. Vortex well, each time. In this way obtain the dilutions ca. 10^{-1}, 10^{-2}, 10^{-3}, 10^{-4} and 10^{-5}g soil ml^{-1} (bottle A through tubes B to E).

Prepare two or three spread plates for each dilution 10^{-3}, 10^{-4}, 10^{-5}. After vortexing, place a 0.1 ml drop of each dilution to three separate, labelled peptone-yeast agar plates (for bacteria). Spread the inoculum on agar surface using glass hockey spreader. Replace the lid and incubate inverted plates at room temperature for one week.

1. Prepare a 10–fold dilution series:

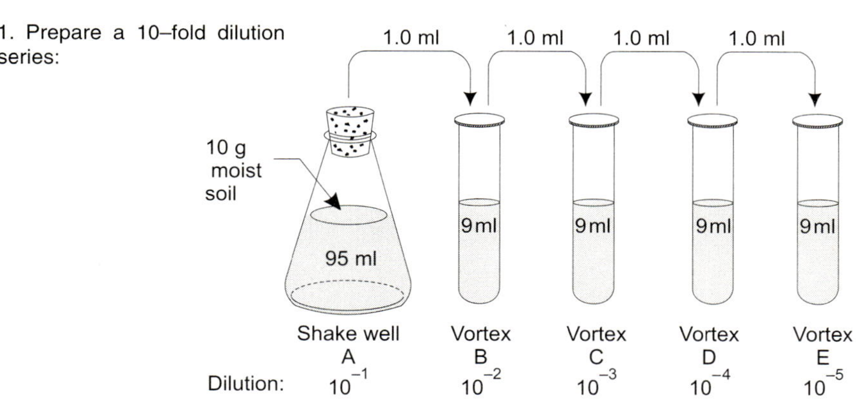

2. For one dilution (C), transfer 0.1 ml of suspension to each plate. After inoculating all replicate plates in one dilution, go to 3. Repeat for next two dilutions (D and E).

3. For each plate, sterilize a glass hockey stick spreader in a flame after dipping it in ethanol. Let the spreader cool briefly. Go to 4.

4. Briefly touch the spreader to the agar of an inoculated plate to cool, away from the inoculum. Then, spread the inoculum by moving the spreader in an arc on the surface of the agar while rotating the plate.

5. Repeat steps 2, 3 and 4 for each dilution. Let the agar solidify, tape the plates together, and incubate them upside down for one week.

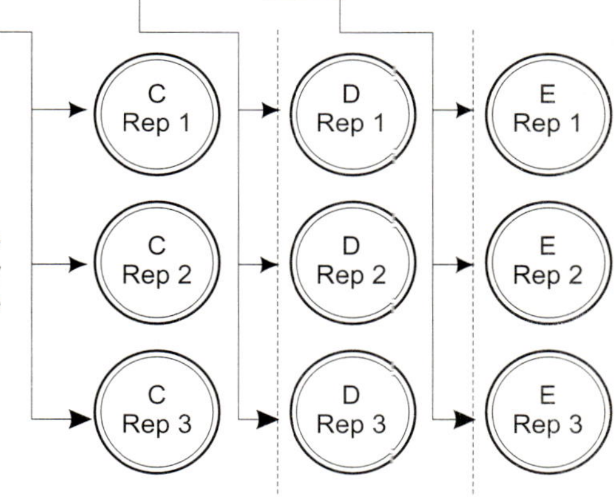

FIG. 20.5

Schematic showing the steps of the procedure for viable plate counts of bacteria and actinomycetes. Use tubes B, C, and D for the actinomycetes also plating 0.1 ml of soil suspension

For actinomycetes, use dilutions 10^{-2}, 10^{-3} and 10^{-4} from the above series. Spread plate 0.1 ml of vortexed suspension on labelled glycerol-casein plates as above. Make three replicates for each dilution. Incubate the inverted plates at room temperature for two weeks.

Observations and calculations. Examine all of the bacteria plates incubated for one week. Count the number of CFUs for each plate, including any actinomycetes. Average the total for each dilution. Count only those plates that have countable colonies (30 – 200 colonies per plate). Calculate the sample mean of CFUs per gram of dry soil in the same way as done for soil fungi, **except** that the effective dilution is increased by one order of magnitude since only 0.1 ml of inoculum was used for spread plate as opposed to 1 ml in the pour plates for fungi. Calculate also the standard deviation and coefficient of variation for the sample.

Actinomycte plates are examined and counts are made as for bacteria for CFUs for each plate, subtracting any bacteria. Average the totals and express CFUs of actinomycetes per gram of dry soil.

EXERCISE 5

Objective: *To isolate microorganisms in pure cultures*

Theory. Microorganisms are to be isolated as pure cultures for detailed studies on their morphological and physiological characteristics. Accordingly techniques have been devised to isolate bacteria, actinomycetes and fungi in pure culture.

Materials. Mixed cultures of microbes; Petri dishes; Nutrient agar plates; Needles; Gas burner; Alcohol.

Procedure. This is done in the following two steps.

Step 1. Isolation of discrete colonies from a mixed culture

Three techniques are used for the separation of a mixed culture so that discrete colonies can be isolated: (i) streak plate method, (ii) spread plate method, and (iii) pour plate method.

1. ***Streak plate method.*** A four-way or quadrant streak is done as shown in Fig. 20.6. Place a loopfull of culture on the agar surface in area 1. Flame and cool the loop, and drag it rapidly several times across the surface of area 1. Reflame and cool the loop and turn the Petri dish at 90 degrees. Touch the loop to a corner of the culture in area 1, and drag it several times across the agar in area 2. The loop should never again enter the area 1. Reflame and cool the loop, and again turn the dish at 90 degrees. Streak area 3 in the same manner as area 2. Without reflaming the loop, again turn the dish at 90 degrees, and now drag

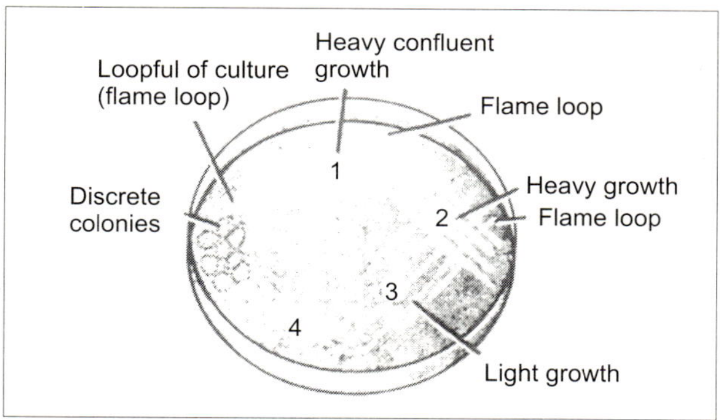

FIG. 20.6
Four-way streak plate inoculation

the culture from a corner of area 3 across area 4, using a wider streak. The loop must not touch any of the previously streaked area.

2. ***Spread plate method.*** It requires the use of a previously diluted mixture of microbes. During inoculation, the cells are spread over the surface of solid agar medium with a sterile L-shaped bent rod (hockey spreader) while the dish is spun on a turntable. A loopful of the culture is placed with a sterile loop in the centre of the agar plate placed on turntable. While the turntable is spinning, the sterile bent rod is lightly touched to the agar surface and moved back and forth to spread the culture.
3. ***Pour plate method.*** It requires a serial dilution of the mixed culture by means of a loop or pipette. The diluted inoculum is then added to molten agar medium in a Petri dish, mixed and allowed to solidify (Fig. 20.7).

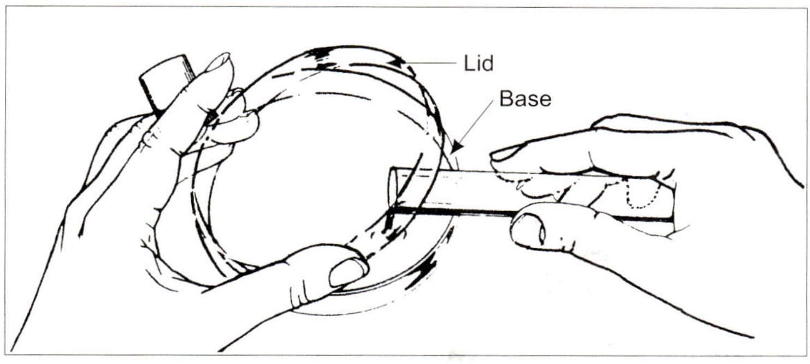

FIG. 20.7
Pour-plate technique

Step 2. Isolation of pure cultures from a streak plate or spread plate or pour plate preparation

Flame the needle, touch the cool needle to the surface of the selected colony growing in plate. Uncap the agar slant and pass the neck of the tube over flame once. Inoculate the slant by drawing the needle upward in a zigzag way along the surface of the agar. Do not dig into the agar.

Flame the neck of the tube and recap. Flame the needle. The procedure is shown in Fig. 20.8.

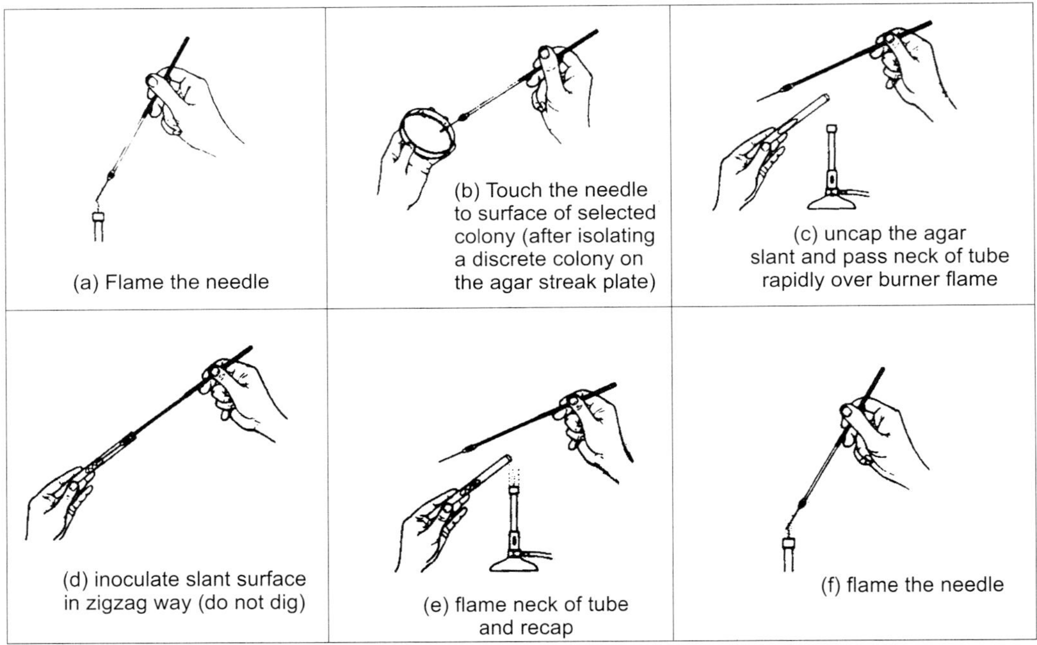

FIG. 20.8

Procedure for the preparation of a pure culture

EXERCISE 6

Objective: *Transfer of microorganisms for sub-culturing (Culture transfer technique).*

Theory. Microorganisms can not be maintained in pure cultures indefinitely. These are to be, therefore, transferred from pure culture to fresh culture media (slants) at regular intervals.

Materials. 24-hr. nutrient broth and nutrient agar slant cultures; Nutrient broth; Nutrient agar slant; Burner; Inoculating needle or loop.

Procedure. Place the tubes in the palm of your hand, secure with thumb and separate to form V. Microorganisms are transferred from one culture to fresh tube or Petri dish by a transfer needle or loop (Fig. 20.9). Flame the needle or loop until it becomes red. With sterile needle or loop in hand, uncap the tubes. Flame the necks of tubes by rapidly passing them through the flame once. Withdraw the inoculum from stock culture with needle or loop and transfer to the tube/plate in which sub-culture is done. Flame the necks of tubes. Recap the tubes. Reflame the needle or loop. The procedure is shown in Fig. 20.10.

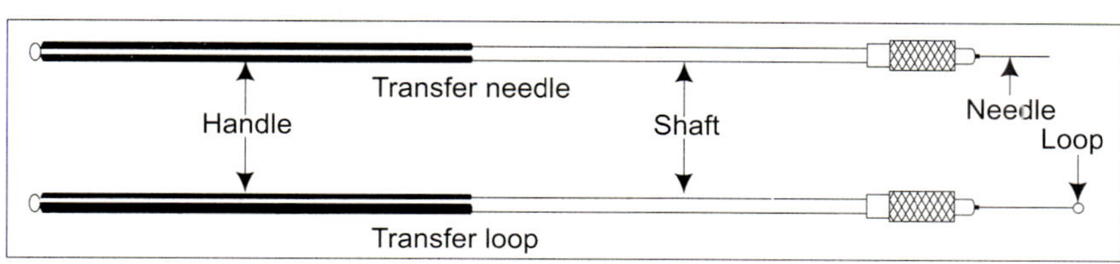

FIG. 20.9
Transfer needle and loop

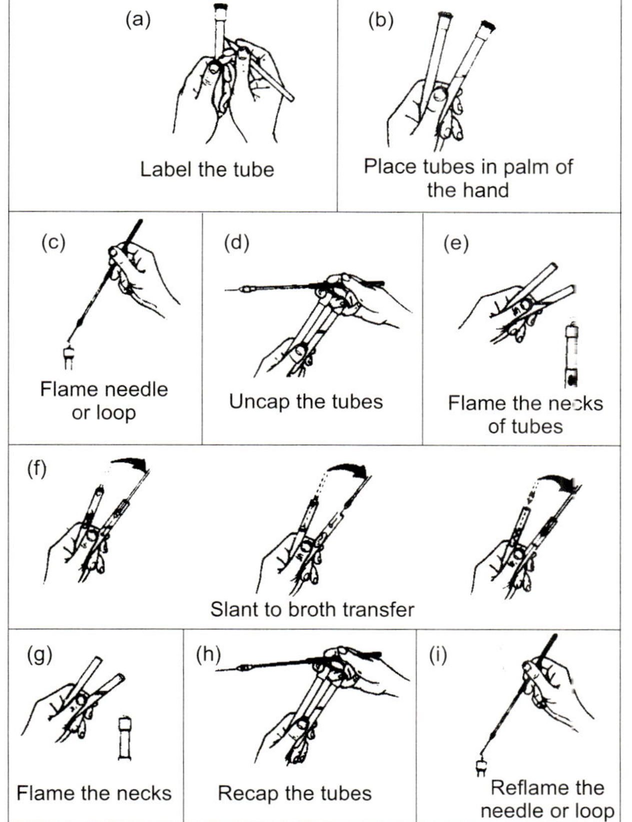

FIG. 20.10
Sub-culturing procedure

EXERCISE 7

Objective: ***Staining of bacteria with Gram stain***

Theory. A Danish scholar, Christian Gram in 1884 devised a differential staining procedure which differentiates between two groups of bacteria: Gram-positive and Gram-negative. This procedure is called Gram stain technique.

Materials. 24-hour agar slant cultures of *Escherichia coli, Staphylococcus aureus* and *Bacillus cereus;* Crystal violet, Gram's iodine; Decolouriser-ethyl alcohol (95%) or 1:3 (v/v) acetone: isopropyl alcohol; Safranin, Bunsen burner, Inoculating needle or loop; Staining tray; Glass slides; Blotting paper; Lens paper; Microscope.

Procedure. 1. Clean four glass slides.

2. Under sterile conditions, prepare a smear of each of the three microbes and on the remaining slide prepare a smear consisting of a mixture of *S. aureus* and *E. coli.* This is done by placing a drop of water on the slide and then transferring each microbe separately to the drop of water on the slide with a sterile, cooled needle. Both microbes are then mixed and spread by means of a circular motion of the inoculating needle.
3. Allow smears to air dry, then heat fix, and proceed for staining through steps 4 to 12, shown in Fig. 20.11.

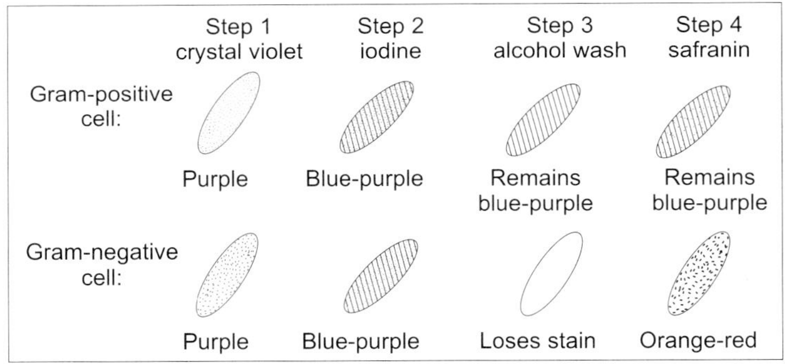

FIG. 20.11

Different steps of Gram-staining. Reactions of two kinds of bacterial cells are also shown

4. Flood the smears with crystal violet and allow to stand for one minute.
5. Wash with tap water.
6. Flood smears with the Gram's iodine mordant and allow to stand for one minute.

7. Wash with tap water.
8. Decolourise with decolouriser. Do not overdecolourise. Add reagent drop by drop until crystal violet fails to wash from smear.
9. Wash with tap water.
10. Counterstain with safranin for 45 seconds.
11. Wash with tap water.
12. Blot dry and examine under oil immersion.

Observations. Record observations for *E. coli, B. cereus* and *S. aureus* for cell morphology, shape arrangement, cell colour and Gram reaction.

EXERCISE 8

Objective: *To determine the algal population of soil by most probable number (MPN) method*

Theory. Since algae are photoautotrophs, most algal cells would be found at the soil surface. They are important colonisers of developing soils, poor in organic matter. They also help in soil aggregation through secretion of extracellular polysaccharides. The media used for their isolation contain essential macro-and micronutrients. MPN technique is a statistical method based on dilution of organisms to extinction. It is simple and involves numerical values of the tubes showing positive sign of algal growth and using the data for reference to a MPN table.

Materials. Fresh soil (25g); Plastic container; Weighing dishes; Benchtop balance; Deionised water; Dissecting probe; Vortex mixer; Water blank (95 ml); Blanks of modified Bristol's solution, each with 9 ml (30); Pipettes, 1 ml (10).

Procedure and calculations. Prepare a 10-fold dilution series, 10^{-1} to 10^{-6} g soil ml^{-1} using 10g of the soil, one 95 ml and five 9 ml blanks containing a modified Bristol's solution (Fig. 20.12). For each of the dilutions, 10^{-2} to 10^{-6}, inoculate five 9 ml blanks (replicates) of modified Bristol's solution by aseptically transferring, for example, 1ml of the contents of tube B to each of the five blanks, as shown in step 2 of Fig. 20.12. Incubate the capped tubes for four weeks in a greenhouse or growth chamber.

Examine the incubated tubes to record the number of tubes showing a positive sign of growth for each dilution. Calculate the number of algal cells on a dry weight soil basis as follows:

Assume dry weight of the 10g of fresh soil taken as 8.2g. Therefore, amounts of dry soil in each tube range from 8.2×10^{-3} (tube B) to 8.2×10^{-7} (tube F). The appearance of tubes after incubation is shown in Fig. 20.13. Shaded tubes indicate the

Step 1. Prepare a 10-fold dilution series

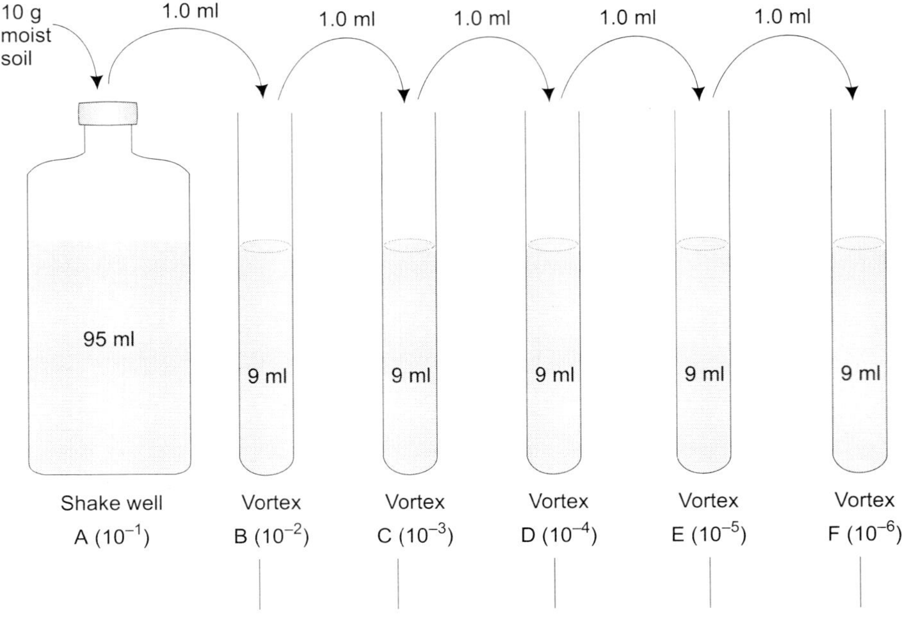

Step 2. Label all the tubes below with the soil and dilution. Transfer 1.0 ml from each tube into 5 replicate tubes. Vortex the above tubes often. Go to 3.

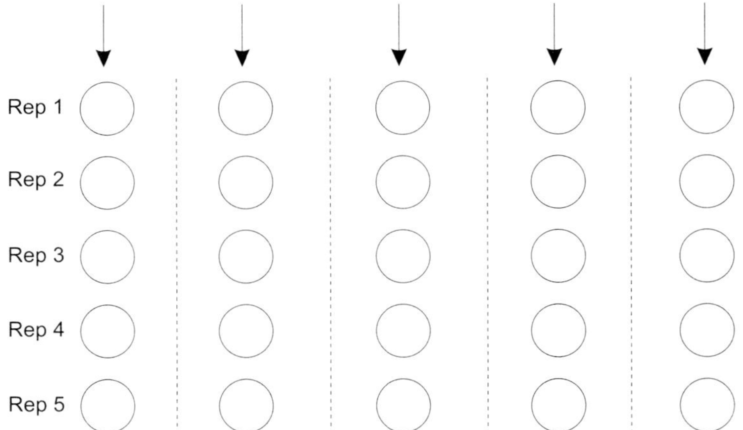

Step 3. After making sure that the contents of all the dilution tubes have been well suspended by votexing, incubate the tubes for four weeks in light and warm area.

FIG. 20.12

Schematic showing different steps of the procedure for determining soil algal counts by MPN method

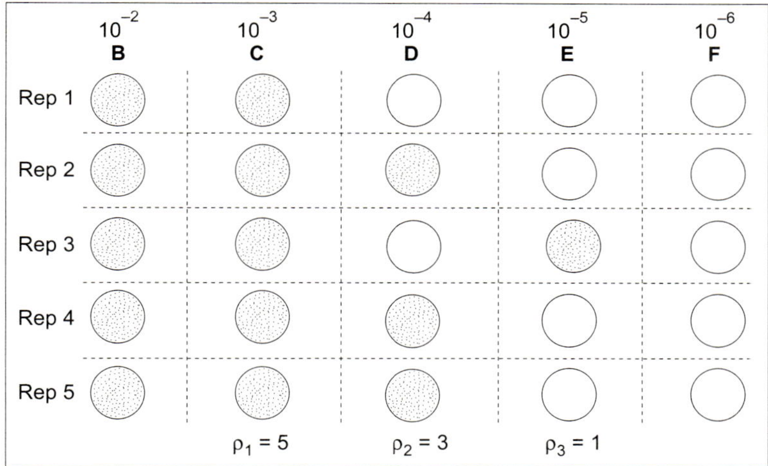

FIG. 20.13

Hypothetical outcomes of an algal enumeration assay by MPN discussed as an example in the calculations section. Shaded circles represent tubes with a positive sign of growth, whereas the empty ones represent tubes with no sign of growth and are, therefore negative

positive sign and empty tubes negative sign of growth. p_1 is to be choosen as the number of replicate tubes of the highest dilution that has the highest number of positive tubes. Here, the replicates from tube B do not count because those of tube C are from a higher dilution. In contrast, the number of tubes from tube D that show a positive sign of growth is less than those from tube C. So, $p_1 = 5$. Choose p_2 and p_3 to be the number of tubes in the next two higher dilutions that show a positive sign of growth. Thus $p_2 = 3$ and $p_3 = 1$. Look down the first column in Table 20.1 to find your value for p_1. Do the same in the p_2 column. Then find where your value of p_3 (across the top) intersects the row defined by your values of p_1 and p_2. In this case, the value is 1.1 organisms ml^{-1}. Divide the tabular value by the concentration of soil in the dilution to which you assigned Δ_2 (Tube D) to obtain the number of organisms per gram of dry soil.

$$\text{Therefore, } \frac{1.1 \text{ organisms}}{\text{ml}} \times \frac{\text{ml}}{8.2 \times 10^{-4}} = 1.34 \times 10^4 \text{ algal cells g}^{-1} \text{ dry soil.}$$

The upper and lower 95% confidence limits are calculated from Table 20.2. In this experiment, we used a 10-fold serial dilution with five tubes per dilution. From Table 20.2 find the dilution ratio used (10) in the top row and then find where it intersects 5 tube per dilution (at 3.30). The upper confidence limit is $1.34 \times 10^4 \times 3.30 = 4.42 \times 10^4$ algal cells g^{-1} dry soil.

$$\text{The lower confidence limit is } \frac{1.34 \times 10^4}{3.30} = 4060 \text{ algal cells g}^{-1} \text{ dry soil.}$$

TABLE 20.1 Most probable numbers for use with the experimental design in this exercise. (Alexander, 1982). Value in bold face is explained in the calculation section

		Most probable number for indicated values of ρ_3					
ρ_1	ρ_2	***0***	***1***	***2***	***3***	***4***	***5***
0	0	-	0.018	0.036	0.054	0.072	0.090
0	1	0.018	0.036	0.055	0.073	0.091	0.11
0	2	0.037	0.055	0.074	0.092	0.11	0.13
0	3	0.056	0.074	0.093	0.11	0.13	0.15
0	4	0.075	0.094	0.11	0.13	0.15	0.17
0	5	0.094	0.11	0.13	0.15	0.17	0.19
1	0	0.020	0.040	0.060	0.080	0.10	0.12
1	1	0.040	0.061	0.081	0.10	0.12	0.14
1	2	0.061	0.082	0.10	0.12	0.15	0.17
1	3	0.083	0.1	0.13	0.15	0.17	0.19
1	4	0.11	0.13	0.15	0.17	0.19	0.22
1	5	0.13	0.16	0.17	0.19	0.22	0.24
2	0	0.045	0.068	0.091	0.12	0.14	0.16
2	1	0.068	0.092	0.12	0.14	0.17	0.19
2	2	0.093	0.12	0.14	0.17	0.19	0.22
2	3	0.12	0.14	0.17	0.20	0.22	0.25
2	4	0.15	0.17	0.20	0.23	0.25	0.28
2	5	0.17	0.20	0.23	0.26	0.29	0.32
3	0	0.078	0.11	0.13	0.16	0.20	0.23
3	1	0.1	0.14	0.17	0.20	0.23	0.27
3	2	0.14	0.17	0.20	0.24	0.27	0.31
3	3	0.17	0.21	0.24	0.28	0.31	0.35
3	4	0.21	0.24	0.28	0.32	0.36	0.40
3	5	0.25	0.29	0.32	0.37	0.41	0.45
4	0	0.13	0.17	0.21	0.25	0.30	0.36
4	1	0.17	0.21	0.25	0.31	0.36	0.42
4	2	0.22	0.26	0.32	0.38	0.44	0.5
4	3	0.27	0.33	0.39	0.45	0.52	0.59
4	4	0.34	0.40	0.47	0.54	0.62	0.69
4	5	0.41	0.48	0.56	0.65	0.72	0.81
5	0	0.23	0.31	0.43	0.58	0.76	0.95
5	1	0.33	0.46	0.64	0.84	1.1	1.3
5	2	0.49	0.7	0.95	1.2	1.5	1.8
5	3	0.79	***1.1***	1.4	1.8	2.1	2.5
5	4	1.3	1.7	2.2	2.8	3.5	4.3
5	5	2.4	3.5	5.4	9.2	16	-

TABLE 20.2 Factors for calculating the upper and lower 95% confidence intervals for an MPN analysis of the design presented in this laboratory exercise (Alexander, 1982)

Number of replicate tubes	*Dilution ratio*			
P1	*2*	*4*	*5*	*10*
1	4.00	7.14	8.32	14.45
2	2.67	4.00	4.47	6.61
3	2.23	3.10	3.39	4.68
4	2.00	2.68	2.88	3.80
5	1.86	2.41	2.58	***3.30***
6	1.76	2.23	2.38	2.98
7	1.69	2.10	2.23	2.74
8	1.64	2.00	2.12	2.57
9	1.58	1.92	2.02	2.43
10	1.55	1.86	1.95	2.32

EXERCISE 9

Objective: ***To demonstrate the dehydrogenase activity of soil microorganisms***

Theory. Dehydrogenases are found in all living organisms and they take part in many reactions involving transfer of pairs of electrons. As dehydrogenases also take part in the electron transfer system of aerobic organisms, the activity of these enzymes is a measure of respiration along with general metabolic activity. By far the most commonly used version of dehydrogenase assay is that of Casida *et al* (1977), which involves incubation of soil mixed with TTC in air-tight conditions to exclude oxygen. After incubation the soils are extracted with a solvent to remove TPF, using ethanol. The TPF concentration is determined by absorption spectophotometry at $\lambda = 485$ nm using a standard curve.

Materials and procedure. ***First part.*** Dry-wt basis, 24g of soil for each soil type; Plastic vial with tight lid; (blank); Plastic vials(4) with tight lids for each soil type; Analytical balance; Benchtop balance; Weighing dishes; Spatula; Water (deionised); Pipette (1 ml); 3% (w/v) TTC, 1 ml for each vial.

Weigh 6 g (dry wt.) for each soil in each of four plastic vials. Add 0.5% glucose (dry wt.) to two duplicate vials. Two samples will have no glucose. Prepare one blank (no soil). Thus, there will be four vials with soil for each type, and 1 blank.

Add 1 ml of 3% TTC and 2.5 ml deionised water to each of the five vials including blank. Mix the soil and liquid with stirring rod and put caps on vials and seal them. Incubate the vials for one week.

Second part. Incubated vials; Methanol; Graduated Cylinder; Filtration funnel; Whatman 42 filter paper; Fume hood; Volumetric flask (50 ml); Pipette (5 ml); Cuvetts; Spectrophotometer at λ = 485 nm.

Add 10 ml of methanol to each vial, stir and transfer the suspension to a funnel fitted with filter paper. Collect the filtrate in 50 ml flask. Wash the vial and funnel containing the sediment twice with 10ml methanol until the filtrate is clear of red colour. Add more methanol to make the total volume to 50 ml.

Transfer 5 ml of each sample to a cuvette. Read the absorbance at 485 nm on spectrophotometer using the blank as the zero. In case, readings reach outside the calibration range (Table 20.3, standard curve) dilute with methanol. Rinse the cuvettes with 1-2 ml of methanol between samples. Express the results as GTPF g^{-1} dry soil. Report the average of two duplicates.

Example:

Soil A	*Absorbance value*
Container *x*	0.227
Container *y*	0.271
Average (A)	0.249

Using the Table 20.3:

$$x = \frac{A - 0.00629}{0.0415}$$

$$\text{ie. } x = \frac{0.249 - 0.00629}{0.0415}$$

$$\text{ie. } x = 5.85\ \mu\text{g/ml of extractant, ie. } x = \frac{5.58 \times 50}{6} g^{-1} \text{ soil}$$

$$\text{ie. } x = 48.7\ \mu\text{g TPF } g^{-1} \text{ soil}$$

$$x = \frac{A - 0.00629}{0.0415}$$

$$r^2 = 1.000$$

TABLE 20.3 Standard curve data and equation for TPF using visible absorption spectrophotometry at λ = 485 nm. *A* is the absorbance value; *X* = concentration of TPF analysed in the solution (µg ml^{-1})

X (TPF) in µg ml^{-1}	*A*
0.00	0.000
5.00	0.214
10.00	0.423
15.00	0.642
20.00	0.833
25.00	1.051
30.00	1.244

EXERCISE 10

Objective: *Isolation of 2,4-Dichlorophenoxyacetic acid (2,4-D)-degrading bacteria by substrate enrichment method*

Theory. The method is based on the concept of natural selection. The organism capable of exploiting a particular (specific) substrate most under other environmental conditions (temp, pH, O_2 etc.) is the one likely to be selected. The soil is amended with a particular substrate for the rapid isolation of an organism.

The bacteria capable of degrading the herbicide, 2,4-D can be isolated by enrichment of the soil with the herbicide, and incubating the soil for several weeks. This is followed by dilution and plating on a selective, differential enrichment medium, the eosin-methylene blue (EMB)-2,4-D agar.

Materials. Soil; Microcosm jars (600 ml polypropylene); 2,4-D indicator plates (EMB 2,4-D agar); Dilution water blanks (9 ml and 95 ml); Pipettes (1 ml); Stick spreader; Vortex mixer; Gas burner; Balance; Spatula; Ethyl alcohol; Reagent grade 2,4-D; 2,4-D, stock solution (1%); 2,4-D screening broth; Sterile screw cap tubes, each containing 5 ml of 2,4-D screening broth.

Procedure. Put 150g of soil into each of two microcosm jars. Label one as control and another as 2,4-D enrichment. Bring the moisture content of control soil to field capacity level. Add same amount of water to 2,4-D enrichment also, but include in this enough 1% 2,4-D stock solution to have a final soil concentration of 500 µg 2,4-D g^{-1} dry soil. Mix thoroughly. Cover the jars with lids (not tightly, to allow aeration). Weigh the 2,4-D enriched jars and incubate at 25° for one week.

Prepare a dilution series for the control soil for plating on 2,4-D indicator plates. Specifically use 10g of soil and dilute in a 95 ml water blank to obtain 10^{-1} dilution. Re-weigh the control jar after removing the soil. Incubate at 25°C for one week.

Transfer a 1ml aliquot to a 9 ml dilution blank to have 10^{-2} dilution. Spread plate 0.1 ml of each dilution onto 2,4-D indicator plates to give final dilutions of 10^{-2} and 10^{-3} respectively. Use duplicate plates for each dilution. Incubate plates at 25°C for one week (1st lot). These plates will provide an estimate of the original population of bacteria in the soil capable of degrading 2,4-D. Normally this number will be low.

Reweigh microcosms and add water to have the moisture content at original level. Take 0.5g soil sample from each microcosm and put each to a 5 ml screening broth, containing 2,4-D and an indicator dye, in the broth tube. Place each tube on a shaker and observe the growth over the next 2-3 weeks. A change in colour from green to yellow will indicate degradation of 2,4-D and the presence of degrading bacteria in the sample. Prepare a dilution series for each soil for plating on 2,4-D indicator plates. Prepare a 10^{-1} dilution with 10 g of soil, diluting it in 95 ml of water blank. Transfer 1ml aliuots to 9 ml dilution blanks to obtain dilution through 10^{-3}. Spread plate 0.1 ml aliquots of 10^{-1}, 10^{-2} and 10^{-3} dilution on 2,4-D indicator plates. Final dilutions will be 10^{-2}, 10^{-3} and 10^{-4} respectively. Incubate the plates at 25°C for one week. (2nd lot) Reweigh microcosms and incubate for another two weeks.

Examine plates of 1st lot incubated earlier. The 2,4-D degrading colonies developing on indicator plates will appear black. Count the black colonies and calculate the mean number of 2,4-D degrading bacteria g^{-1} soil.

Examine the plates of 2nd lot. Similarly there will appear black colonies on 2,4-D indicator plates. These are presumptive 2,4-D degraders. Count plates with between 30 and 200 colonies. Calculate the mean number of 2,4-D-degrading colonies g^{-1} soil.

EXERCISE 11

Objective: ***To demonstrate the ability of soil bacteria to adapt to metal stress***

Theory. Metal contamination of the environment is arising at alarming rate. Microorganisms, exposed to metal stress are able to develop resistance against metals. Metal resistance in microorganisms is being successfully applied in bioremediation of metal-contaminated sites.

Materials. Soil; Plastic cups; Balance; Weighing dishes; Dissecting probe; Stock solution of $Cd(NO_3)_2$. $4H_2O$ (0.6%); Peptone-yeast (PY) agar plates; PY plates amended with Cd (500 μg Cd ml^{-1} agar); Sterile 95 ml water blanks; Sterile 9 ml water blanks; Pipettes (1 ml); Gas burner; Stick spreader; Ethyl alcohol.

Procedure. Weigh out two soil samples (50g, dry wt. basis). To one sample, add Cd as Cd $(NO_3)_2$ to provide 500 μg of Cd g^{-1} dry soil using 0.6% stock solution. It requires addition of 1.14 ml of the stock solution to the soil. Add 0.75 ml of 6% KNO_3 stock solution to the control soil to normalise soil nitrogen.

Adjust soil moisture content of the control and Cd-amended samples. Using the one 95 ml and four 9 ml water blanks, dilute a 10 g sample of the control soil out to the 10^{-5} dilution. Cover both samples with plastic wrap to reduce moisture losses, secure with a rubber band, and record the weights. Incubate at 25°C for two weeks.

Prepare spread plates for 10^{-3}, 10^{-4} and 10^{-5} dilutions on PY agar plates, using 0.1 ml of inoculum for each dilution. The final counts will be 10^{-4}, 10^{-5} and 10^{-6}. These plates will indicate total heterotrophic counts. Similarly plate out the 10^{-1}, 10^{-2}, and 10^{-3} dilutions on PY agar plates amended with Cd (final dilutions,10^{-2}, 10^{-3} and 10^{-4}). These plates will indicate numbers of Cd-resistant bacteria initially present in the control soil.

EXERCISE 12

Objective: ***To measure the amount of biodegradable organic matter in wastewater***

Theory. A knowledge of the organic matter content of the wastewater is desirable for assessment of the effectiveness of wastewater treatment processes. BOD (biochemical or biological oxygen demand) is one of the most important methods used by treatment plants to monitor the efficacy of treatment process. BOD is the amount of dissolved oxygen in water consumed by microorganisms for the biochemical oxidation of organic and inorganic matter. The amount of oxygen being consumed can give a rough idea of how much organic matter is left in water. Conventional treatment removes up to 95% of BOD in wastewater. Domestic wastewater, effluents from treatment plants, and surface water receiving wastewater discharges are normally rich in microbial populations. These microbes use the dissolvel oxygen. The wastewater treatment decreases the BOD, indicating a decrease in organic matter content of the water.

BOD is measured as BOD_5, the oxygen demand over a 5-days incubation period. BOD in wastewaters often exceeds the dissolved oxygen (due to richness of microflora) available in water. Hence, samples must be diluted in a solution of neutral phosphate. Dissolved oxygen in diluted samples (five dilutions recommended) is measured with a membrance electrode or by the Winkler method.

Materials. Wastewater sample; Distilled water (2 litre); Hach Buffer nutrient pillows (contain phosphate buffer and nutrients); 2-chloro- 6–(trichloromethyl) pyridine (TCMP); Chlorine test kit, free and total; DPD total chlorine reagest pillows; Sodium thiosulphate (0.1%); Beaker (2l); 10 – ml pipettes (20); Beaker (500 ml); Stir plate and sir bar; BOD bottles, 60 ml (9) (Fig. 20.14); Membrane electrode; Dark incubator at 20°C.

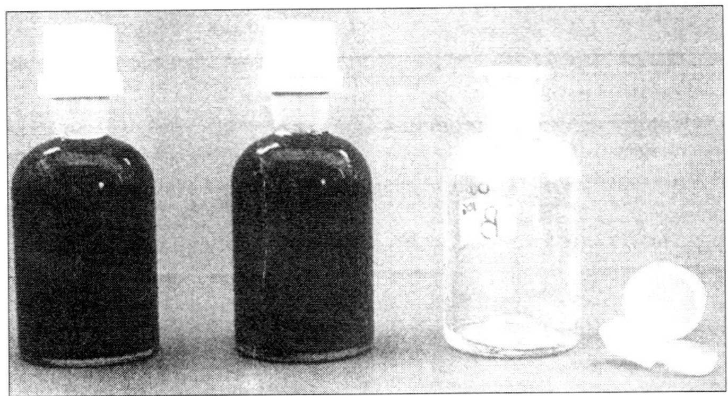

FIG. 20.14
BOD bottles

Preparation of samples. Add the correct number of Hach buffer nutrient pillows to 2 l of ditilled water to make the **dilution water**. Take 10 ml of water sample and place in the Hach chlorine test kit tube. Add a DPD reagent pillow to the tube. Follow the kit directions to determine the chlorine residual in the water. Based on the chlorine residual, determine the amount of 0.1 N sodium thiosulphate ($Na_2S_2O_3$) needed to neutralise the chlorine sample.

$$\text{ml of } Na_2S_2O_3 \text{ needed} = \frac{RS}{(35{,}450)\,(N)}$$

Where R = chlorine residual in mg/l

S = volume of water sample (ml)

N = normality of $Na_2S_2O_3$

35,450 = mol. wt. in mg/l of chlorine

Add the required volume of $Na_2S_2O_3$ to the water sample.

Procedure. Put 300 ml of dilution water is a beaker. Add 3 mg of TCMP.

To prepare dilution blanks, fill 5 BOD bottles with 54 ml each of the dilution water prepared above. Make sure there is dissolved oxygen by shaking each bottle vigorously. Measure the residual chlorine as described above and neutralize the water sample with correct amount of $Na_2S_2O_3$. To make dilutions, add 6 ml of the water sample to the first bottle. Make serial dilutions by transferring 6ml from each bottle to the next bottle (Fig. 20.15).

Measure the dissolved oxygen in each bottle with a membrane electrode within 30 minutes of making the dilution. If necessary, add additional dilution water to fill

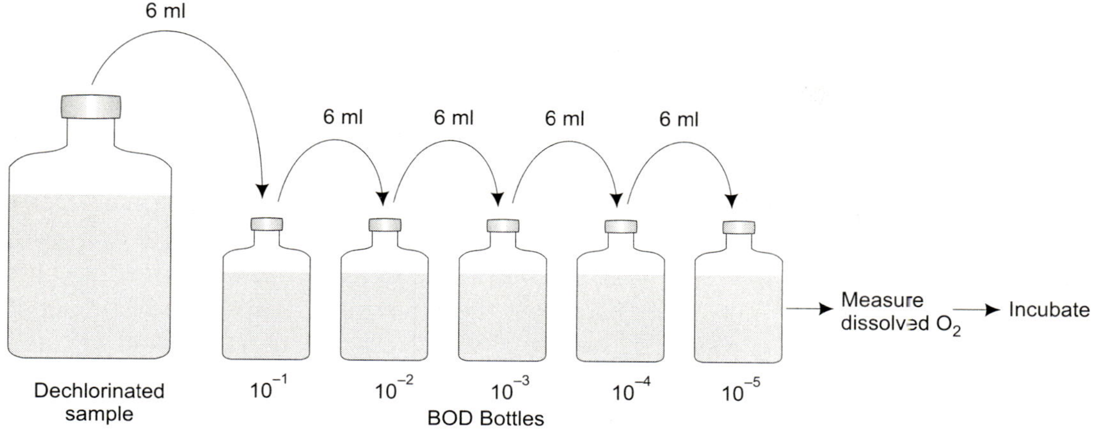

FIG. 20.15
Procedure of making dilutions of the sample

the bottle to the top (no air space). Seal the bottle tightly. Wrap bottles in aluminium foil to block light and incubate at 20°C for 5 days. Open each bottle and measure the dissolved oxygen with membrane electrode. For BOD_5 calculation use the bottles that have a final dissolved oxygen value at least 1 mg/l and a dissolved oxygen depletion of at least 2.0 mg/l over 5 days ($D_1 - D_2 > 2.0$ mg/l).

$$BOD_5 \text{ (mg/l)} = (D_1 - D_2)/P$$

Where D_1 = initial dissolved oxygen (mg/l) in the diluted sample.

D_2 = dissolved oxygen (mg/l) in the diluted sample after 5 days of incubation.

P = The decimal volumetric fraction of sample used (Ex. 10^3 dilution = $1/10^3$, or 0.001).

Example. The 1:100 dilution has the following values:

D_1 = 4.2 mg/l

D_2 = 1.3 mg/l

P = 1/100, or 0.01

Therefore, $BOD_5 = (4.2 - 1.3)/0.01 = 290$ mg/l

EXERCISE 13

Objective: ***To detect coliform bacteria in water by the most probable number (MPN) method (Bacteriological examination of water)***

Theory. The microorganisms pathagenic to humans that are transmitted by water include bacteria, viruses and protozoa. They usually grow in the intestinal tract of man and excreted in the feces. Routine examination of fecal pollution of water for the presence of intestinal pathogens would be very tedious and difficult. However, it is easier to demonstrate the presence of some of the intestinal (nonpathogenic) bacteria like *E.coli* and *Streptococcus faeacalis*. These are always found in intestines and normally not present in water. Hence, their presence in water indicates that water is contaminated with fecal material. In polluted waters, coliform bacteria (of which *E.coli* is a member) are found in densities roughly proportional to the degree of fecal pollution.

The coliform group includes all aerobic and facultatively amaerobic, Gram-negative, non-spore forming, rods which ferment lactose with gas production in media within 48 hr at 35°C. Coliform bacteria include *E.coli, Citrobactor, Enterobacter* and *Klebsiella* species.

MPN test and the membrance filter test have been most commonly used for detection of coliforms in water. The latter method can not be used for turbid waters as they clog the filters. The MPN test consists of **three** steps: a presumptive test, a confirmation test, and a completed test.

In **presumptive test,** a set of tubes of lauryl sulphate tryptose (LST) lactose broth is inoculated with water sample and incubated for 24-48 hrs at 35 ± 0.5°C (Fig. 20.16). Lauryl sulphate inhibits the growth of Gram-positive microbes while encouraging the growth of coliforms. Coliforms ferment lactose to produce acid and gas under anaerobic conditions. Gas formation in 24-48 hr is a positive test which is observed by its presence in the inverted Durham tube (Fig 20.17). Once it has been established that gas-producing lactose fermenters are absent in water, it is presumed to be safe. If there are gas-producers, the presence of Gram-negative lactose fermenters is confirmed by the next step, the confirmation test because gas formation may also be caused by noncoliforms, such as *Clostridium perfringens,* a Gram-positive bacterium.

In **confirmation test,** the growth sample from positive presumptive tubes is inoculated on media, such as Levine's eosin methylene blue (EMB) agar and Endo agar. Levine's EMB agar contains methylene blue which inhibits Gram-positive bacteria. Gram-negative coliforms that grow on this medium will produce "nucleated colonies" (dark centres). In this group, *E. coli* colonies will be small and have metallic sheen, whereas *Enterobacter aerogenes* colonies usually lack the sheen and are larger. But *E. coli* is more reliable indicator. Endo agar with fuschsin sulphite indicator also makes identification of lactose fermenters easier. Coliform colonies and the surrounding medium appear red on Endo agar. Non-coliforms are colourless and do not affect the colour of the medium.

For **completed test**, typical colonies are isolated on nutrient slants and examined for Gram reaction. Gram-negative staining without spores indicate a positive completed test. All three tests are necessary to prove that the bacteria in water sample are true coliforms. However, in actual practice, once the presumptive and confirmed tests gave the same results, then there is hardly any need to waste time in the completed test.

Presumptive Test

Materials. Water sample; Three test tubes containing Durham tubes and double strength LST lactose broth (DSLB); Six test tubes containing Durham tubes and single strength LST lactose broth (SSLB); Pipette (10 ml); Pipette (1 ml); Incubator.

Procedure. Set up three DSLB and six SSLB tubes (step 1, Fig. 20.16). Label each tube as 10 ml, 1.0 ml and 0.1 ml respectively. Mix the bottle of water sample to be tested thoroughly. Transfer 10 ml of water with a pipette to the DSLB tubes. Transfer 1.0 ml of water with pipette to each of the middle set of SSLB tubes and 0.1ml to each of the last three SSLB tubes. Incubate the tubes at 35°C for 48 hours.

Observations and calculations. Remove the tubes from incubator and examine them to record the number of tubes in each set that produce gas bubble(s) in the inverted Durham tube. Determine the MPN by referring to Table 20.4.

If you had gas in the first three tubes (10 ml ones) and gas only in one tube of the second series (1 ml one), but none in the last three tubes (0.1 ml ones), the test would be read as 3-1-0. Table 20.4 indicates that the MPN for this reading would be 43. This means that the particular water sample would have approximately 43 microbes per 100 ml with 95% probability of there being between 7 and 210 microbes. MPN of 43 is a statistical probability number. If there is no gas production in any of the tubes, it shows that the water is safe.

Confirmation Test

Materials. Incubated tubes; One Petri dish of Levine's EMB agar; One Petri dish of Endo agar.

Procedure. Select one positive lactose broth tube from the presumptive test and streak one Petri dish each of Levine's EMB agar and Endo agar (step 3, Fig. 20.16) to obtain well isolated colonies. Incubate the dishes at 35°C for 24 hour.

Observations. Typical coliform colonies would develop on both kinds of media, indicating that confirmed test is positive. If presumptive test was negative, this test would also be negative (no colonies on any of the media) and the water is considered safe to drink.

Presumptive test

Step 1. Transfer the specified volumes of sample to each tube, incubate for 24 h at 35°C. Proceed to 2.

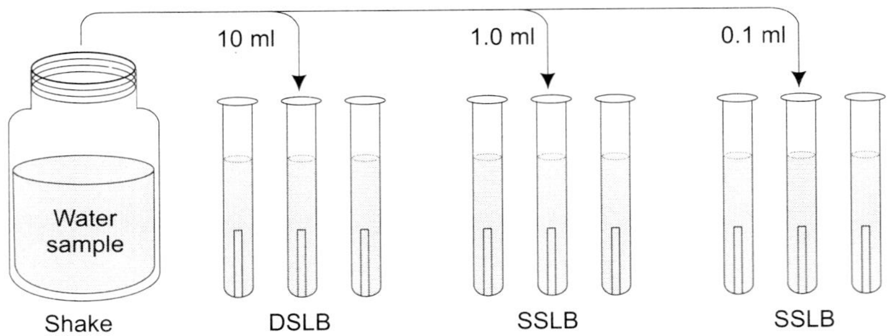

Step 2. Examine the tubes and record those having 10% gas volume or more. Proceed to 3.

Confirmation test

Step 3. Choose any positive tube from above as indicated by the presence of gas trapped in the inner tube and streak a plate of Levine's EMB agar and Endo agar. Incubate for 24 h at 35°C. Proceed to 4.

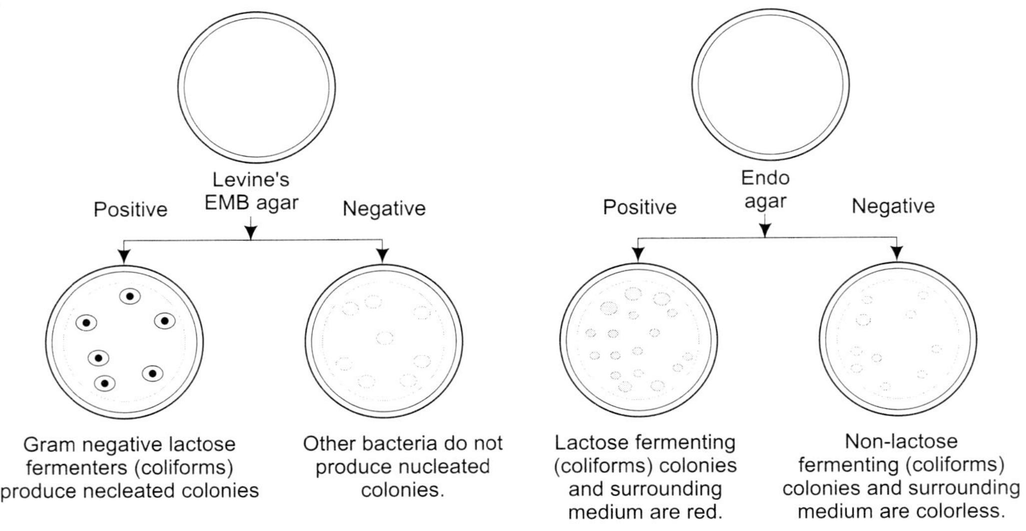

Step 4. Examine the plates for typical coliform colonies as described beneath each illustration above.

FIG. 20.16

Procedure for performing presumptive and confirmation tests of the MPN method to detect the coliforms in water samples

TABLE 20.4 Most probable number (MPN) table used for evaluation of the data in this experiment, using three tubes in each dilution. The value in bold face is referred to in the example in the calculations section

Number of positive tubes in dilutions				Number of positive tubes in dilutions			
10 ml	*1 ml*	*0.1 ml*	*MPN per 100 ml*	*10 ml*	*1 ml*	*0.1 ml*	*MPN per 100*
0	0	0	<3	2	0	0	9.1
0	1	0	3	2	0	1	14
0	0	2	6	2	0	2	20
0	0	3	9	2	0	3	26
0	1	0	3	2	1	0	15
0	1	1	6.1	2	1	1	20
0	1	2	9.2	2	1	2	27
0	1	3	12	2	1	3	34
0	2	0	6.2	2	2	0	21
0	2	1	9.3	2	2	1	28
0	2	2	12	2	2	2	35
0	2	3	16	2	2	3	42
0	3	0	9.4	2	3	0	29
0	3	1	13	2	3	1	36
0	3	2	16	2	3	2	44
0	3	3	19	2	3	3	53
1	0	0	3.6	3	0	0	23
1	0	1	7.2	3	0	1	39
1	0	2	11	3	0	2	64
1	0	3	15	3	0	3	95
1	1	0	7.3	3	1	0	**43**
1	1	1	11	3	1	1	75
1	1	2	15	3	1	2	120
1	1	3	19	3	1	3	160
1	2	0	11	3	2	0	93
1	2	1	15	3	2	1	150
1	2	2	20	3	2	2	210
1	2	3	24	3	2	3	290
1	3	0	16	3	3	0	240
1	3	1	20	3	3	1	460
1	3	2	24	3	3	2	1100
1	3	3	29	–	–	–	–

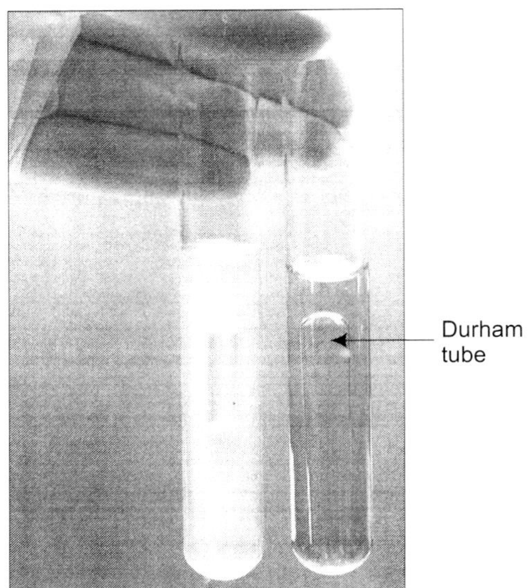

FIG. 20.17
Inverted Durham tubes to collect gas

EXERCISE 14

Objective: ***To detect coliform and fecal coliform bacteria in water by membrane filter method.***

Theory. Besides MPN method, a membrane filter is also commonly used for detecting and quantifying bacteria in water. As compared to MPN method, it is more direct and quicker (consuming hardly 18-24 hours) method of bacteriological examination of water. For this technique, the coliform group may be defined as "all aerobic and facultatively anaerobic, Gram-negative, non-spore forming, rods that develop a red colony with a metalbic green sheen within 24 hours at 35°C on an Endo-type medium containing lactose".

A known volume of water is filtered through a membrane with pore size of 0.45 μm, which traps bacteria on its surface. The membrane is then placed on selective agar or a thin absorbent pad, saturated with a medium designed to grow or permit differentiation of the microbes. For instance, a modified Endo medium, if **coliforms** are sought, and modified Entero-coccus-agar medium (KF agar) containing azide and fecal coliform medium (mFC broth) to detect **fecal streptococci**. After incubation at 44.5°C colonies are counted.

Materials. Filter unit, graduated, with top; Filter flask (1 litre); Sterile 0.45 µm pore size, 47 mm diam. membrane filters with pads (6); Sterile 50 × 12 mm Petri dishes (6); Forceps; Gas burner; Ethyl alcohol; Pipette (10 ml); mEndo broth-MF sterile (6 ml); mFC agar sterile (15 ml); Water bath at 44.5°C; Plastic bag for waterbath large enough to hold the plates; Vacuum hoses for the filter; Vacuum source; Flask with 200 ml, sterile, distilled water; Water sample to be examined (1 litre); Dissection microscope (10 – 15 X).

Procedure. Place a sterile blotter pad with flamed forcep in the bottoms of each of three Petri dishes. Pipette 2 ml of mEndo broth-MF onto each pad and replace the covers. Pipette 5 ml of molten mFC agar into the bottoms of each of three Petri dishes and let solidify. Assemble the filter funnel on the flask of filtration unit (Fig. 20.18).

1. Remove the funnel top and place, with sterile forcep, a sterile membrane filter in place with the grid side up and centre it. Place the funnel top and fix the clamp to hold the assembly together.
2. Pour about 50 ml of sterile distilled water into the funnel, and then with pipette put 1ml of the water sample in the funnel.
3. Apply the vacuum gently. Rinse the sides with a little sterile distilled water just as the liquid level approaches the filter, and let the vacuum draw all the water through the filter.

FIG. 20.18
Filtration apparatus

4. Unclamp the funnel top (with vacuum still on). Remove the filter with sterile forcep and roll it carefully onto the pad of mEndo broth-MF. No bubbles of air. Label the plate **1 ml**.
5. Repeat steps 1, 2, 3 and 4, finally transferring the filter to one of the mFC plates labelled as **1 ml**.
6. Repeat steps 1, 2, 3 and 4, this time using 10 ml of water sample instead of 1 ml. Label this plate as **10 ml**.
7. Repeat steps 1, 2, 3 and 4 once more for the mFC plate. Label this as **10 ml**.
8. Place a filter on the funnel block and clamp the funnel down gently. Pour the water sample into the funnel (without vacuum), until the meniscus reads at 100 ml mark.
9. Apply the vacuum and rinse the sides with sterile distilled water until the water is drawn through.
10. Remove the funnel and transfer the filter to the last mEndo broth-MF plate and label it **100 ml**.
11. Repeat steps 8, 9 and 10, transferring the filter to the last mFC plate, labelled as **100 ml**.
12. Incubate the mEndo broth-MF plates at 35°C with the cover up for 24 hour. Insert the mFC plates inverted (bottom up) into a water tight, resealing plastic bag and incubate at 44.5°C for 24 hour immersed in water bath.

Observations and calculations. Examine the mEndo broth-MF plates with 10-15 X power of dissection microscope for red or pink with bright green metalbic sheen colonies of coliforms (those without green sheen are non-coliforms). Count the coliform colonies. Also examine the mFC plates in the same way for blue (any shade) colonies of fecal coliforms (all other are non-coliforms). Count the number of colonies. The coliform results are expressed as per **100 ml** rather than per ml.

The bacterial count per 100 ml is calculated as follows:

$$\frac{\text{Count}}{100\text{ ml}} = \frac{\text{Count on filter}}{\text{ml filtered}} \times 100$$

EXERCISE 15

Objective: ***To determine the concentration of bacteria on contaminated surfaces by Petrifilm plate method.***

Theory. Monitoring bacterial population on surfaces is important in the evaluation of sanitation programmes in food and medical industries. It is thus necessary to ensure the taste and shelf life of the product as well as the health of the consumer through such microbiological examination at regular intervals. One standard environmental monitoring procedure for detection of coliforms and other bacteria on surface is the Petrifilm plate method. Petrifilm^{TM} is a dry media (hydrated) bound to a polyethylene coated paper printed with a grid which can be used for monitoring the microbial quality of surfaces. Besides surfaces these can also be used for sampling milk, water, or other fluids. Petrifilm plates eliminate the need for media praparation and sterilization and can be stored for prolonged periods before use.

Materials. Petrifilm^{TM} coliform count plates; Pipettes (1 ml); Sterile nutrient broth or 0.1% (w/v) peptone water; Food, milk or water sample prepared in the appropriate manner; Incubator set at 35 ± 0.5°C.

Procedure. 1. Preparation (Hydration) of the Petrifilm plate. Place a Petrifilm plate on a flat surface. Lift the top film and dispense 1ml of sterile nutrient broth or 0.1% peptone onto the centre of the bottom film (Fig. 20.19a). Join the surfaces of the film, beginning at the hinged end using a rolling motion and avoiding entrapment of air bubbles (Fig. 20.19b,c). Distribute the broth/peptone with a down-

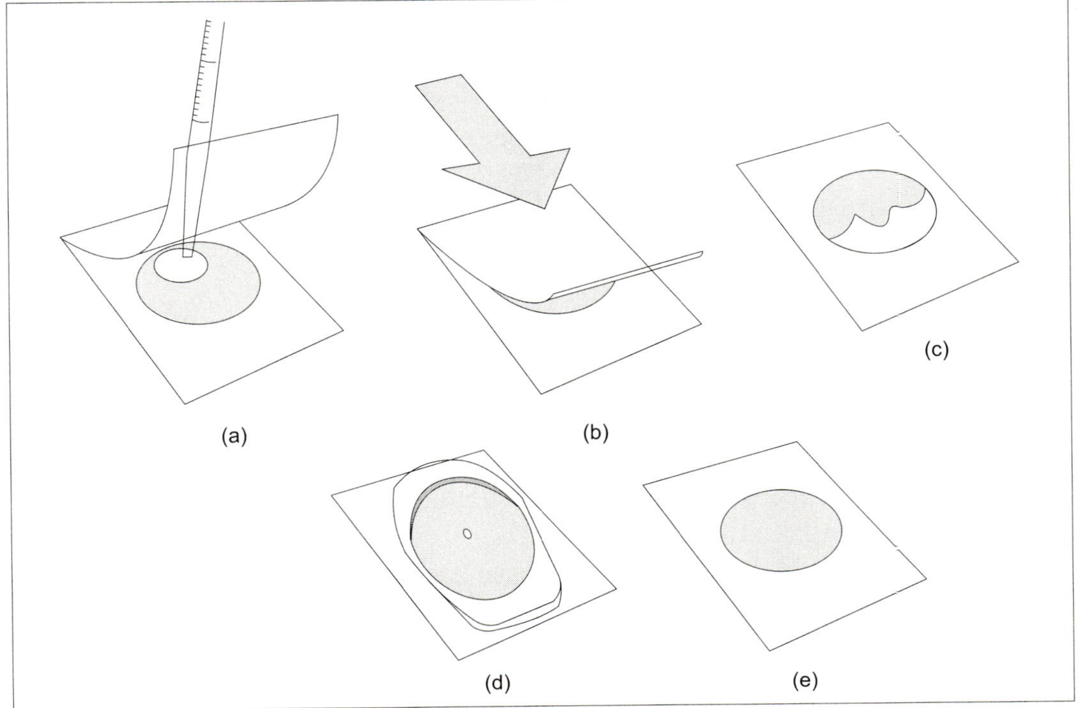

FIG. 20.19

Successive steps of the procedure for hydrating and applying a water sample to a Petrifilm^{TM} plate

ward pressure on the centre of the plastic spreader (recessed side down). Do not slide the spreader across the film (Fig. 20.19d). Remove the spreader using a vertical motion (Fig. 20.19e) and leave the plates undisturbed for one minute to allow the gel solidification. Allow the gel to completely solidify for 30 minutes before using for surface sampling.

2. Surface sampling. Lift the top film of the Prehydrated Petrifilm™ plate (gel will adhere to the top film). Place the gel and top film in contact with the surface to be sampled. Firmly rub your fingers over the entire film side of the gelled area to ensure good contact with the surface.

Lift the film from the surface and rejoin the top and bottom sheets of the Petrifilm. Incubate the plates in horizontal position with the clear side up at 35 ± 0.5°C for 24 hour.

3. Water, food or milk sampling. Place Petrifilm on a flat surface. Lift the top film, Dispense 1ml of the sample onto centre of the bottom film, keeping pipette perpendicular to Petrifilm. (Fig. 20.19a) Lower the top film using a rolling motion and avoiding entrapment of air bubbles. Do not let the top film drop (Fig. 20.19b, c). With the flat side down, place the spreader on the top film over inoculum ie. sample (on bottom film) (Fig. 20.19d). Apply gentle pressure on the spreader to distribute the inoculum over the circular area defined by the spreader. Do not twist or slide the spreader. Lift spreader using a vertical motion, and wait for 1 min to solidify the gel (Fig. 20.19 e). Incubate at 35 ± 0.5°C for 24 hour.

Observations and calculations. Remove the incubated plates from the incubator and record the number of coliforms and *E. coli*. The indicator present in the Petrifilm reacts with β-glucornidase produced by *E. coli* to form blue precipitate in the medium. Hence all blue colonies (regardless of gas production) are counted as *E. coli*. Coliforms ferment the lactose in the medium to produce gas which is trapped between the films of the plate. All colonies associated with gas bubbles (bubbles are less than one colony diameter from the colony) are counted as coliforms. Red colonies without gas bubbles are other Gram-negative organisms. Coliforms and *E. coli* bacteria are routinely found on surfaces in public restrooms, such as the bottom of toilet seats, sinks, tops, and floor.

For water and food sampling the incubated plates are removed and results are recorded as above. The number of coliforms or *E. coli* is the amount present in 1 ml of the original sample.

EXERCISE 16

Objective: *To detect coliphages in wastewater/sewage*

Theory. Enteric viruses infecting the intestinal tract of humans and animals are excreted in feces and can be isolated from domestic wastewater/sewage. There are many applications of coliphages as environmental indicators. These include their use as indicators of sewage contamination, efficiency of water and wastewater treatment, and survival of enteric viruses and bacteria in the environment. Coliphages are thus good indicators of fecal coliforms (bacterial hosts) in sewage.

Materials. Sewage or water sample containing coliphages, 1.0 ml; Nutrient broth culture (3-4 h) of **E. coli;* Tubes containing 9 ml of Tris-buffered or other buffered saline; Tubes containing 3 ml of soft (top) agar (0.7% of NA); Petri dishes with 10-12 ml of bottom agar (NA); Pipettes (1 ml); Water bath (45 – 48°C); Incubator (37°C).

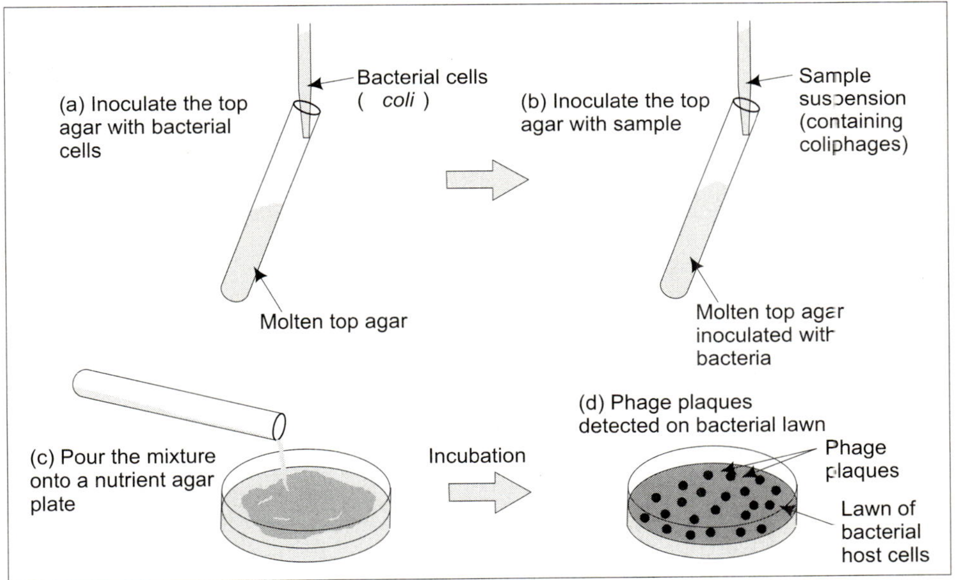

FIG. 20.20

Steps of the procedure for the preparation of a bacterial lawn in the top layer of agar to detect coliphage plaques

---

**E. coli* strain ATCC 15597 usually produces the greatest number of plaques from sewage sample. A colony of this strain is inoculated into 3 ml of trypticase soy broth and incubated overnight at 35°C. Inoculate 1 ml of this culture into a fresh flask containing 100 ml of broth, before three hours of phage assay, and place in a shaking water bath at 35°C to 37°C. Incubate for 3 hours. This will the bacterial growth in log phase.

Procedure. Dilute the sample (sewage or water containing coliphages) 1:10 and 1:100 with Tris buffer. Melt four tubes of soft agar (3 ml of 0.7% agar/tube) by placing in a steam bath or autoclaving. Place the agar tubes in a water bath at 45 – 48°C and allow the temperature of the agar to adjust at 45°C.

Add 1.0 ml of a log phase broth culture of *E. coli* and 1.0 ml of undiluted sample to the first tube (Fig. 20.20 a,b). Remove the tube from the water bath and gently rock between your hands to mix the suspension for 2-3 seconds. Wipe any water from the tube and pour the agar over the Petri dish containing bottom agar (Fig. 20.20c). Quickly rotate the plate to spread the top agar over entire surface. Repeat the mixing of bacterial cells and sample, and subsequent plating of agar over Petri dish (as described above), using 1 ml of bacteria and 1 ml of each sample dilution (Fig. 20.20 a–c).

After the agar has solidified, invert the Petri dishes and incubate at 37°C for 24 hour. Knock any moisture off the lid of dish.

Count the number of plaques on each dilution (Figs. 20.20d; 20.21) and calculate the phage concentration in the original sample.

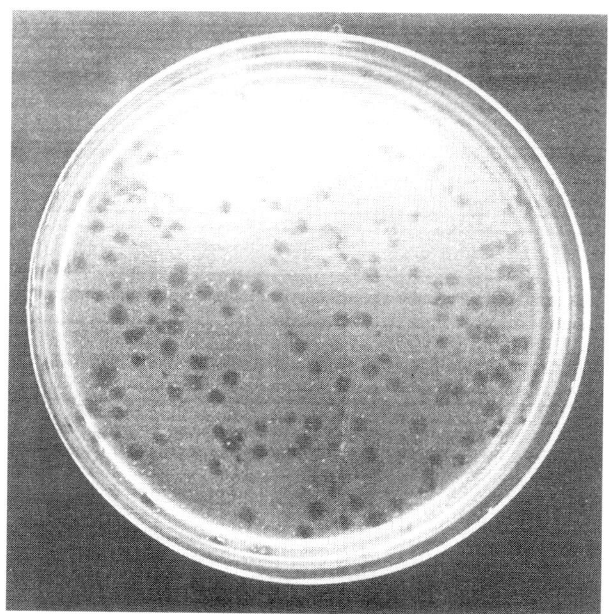

FIG. 20.21

A Petri dish showing the phage plaques on a bacterial lawn

EXERCISE 17

Objective: ***Sampling of bioaerosols by impingement and filtration methods for analysis of airborne microorganisms***

Theory. A bioaerosol is a collection of airborne biological particles. They are generated by a variety of processes (natural and man-made). Bioaerosols transmit important pathogenic microbes; deteriorate building materials; and contaminate indoor environments and thereby produce adverse human health effects. Microbial flora of bioaerosols is quantified by three basic methods: **impingement**, the trapping of airborne particles in a liquid matrix; **impaction**, the forced deposition of airborne particles on a solid surface; and **filtration**, the trapping of airborne particles by size exclusion.

Materials. All glass liquid impiniger; 20 ml of 0.1% peptone solution; Erlenmeyer flask (500 ml or 1 lit); Rubber/plastic tubings to connect impinger and cassette to vacuum source; Vacuum pump/vacuum source; Graduated cylinder (100 ml); Dilution blank with 0.1% peptone or phosphate buffered saline; Pipettes (1 ml and 10 ml); Sterile 0.45 μm pore size, 47-mm diam. membrane filters; Filter unit (as in Exercise 14); Sterile 0.45 μm pore size, 37-mm diam filters; Vortex mixer; Air monitoring cassette (37-mm); Forceps; Gas burner; Nutrient agar (NA) plates; Sabouraud dextrose agar (SDA) plates.

Procedure. 1. Impingement method. Set up the all glass impinger (Fig. 20.22). The commonly used liquid impinger is the AG1-30 (ACE Glass, Vineland, NJ).

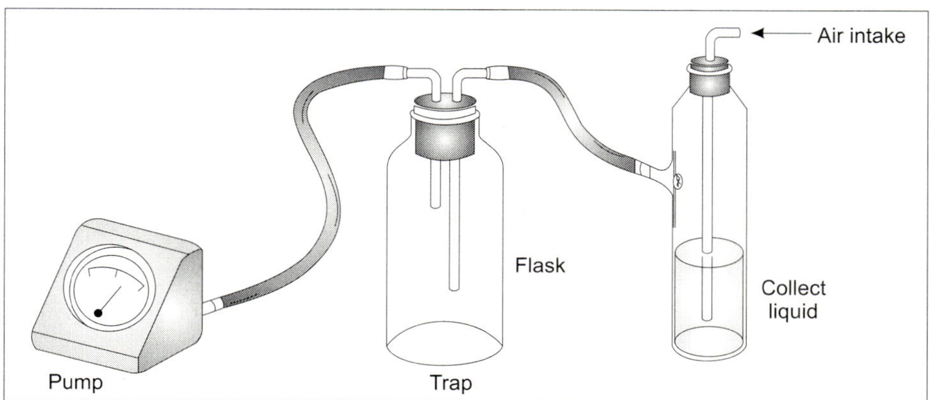

FIG. 20.22

A glass liquid impinger for collecting bioaerosols

Add 20 ml of 0.1% peptone to the reservoir followed by 0.1 ml of anti-foam (Anit-foam B, Sigma Chem.. Co.).

Turn the vacuum source (pump) on for 10 minutes. Remove 0.5 ml of fluid from the reservoir with a 1-ml pipette and place 0.1 ml on one agar plate of NA and spread plate the sample by routine isolation method of **bacteria**. Place another 0.1 ml on a plate of SDA for detection of **fungi**.

2. Filtration mehtod. Remove 6 ml of liquid from the reservoir of impinger with a 10 ml pipette and pass 5 ml through a 0.45 μm membrane filter as described in Exercise 14. Place the filter on NA plate for bacteria. Repeat the procedure, but place the membrane filter on SDA for fungi. Incubate the NA plates at 37°C for 24-48 hours and SDA plates at 25 ± 1°C for 2 to 7 days, and examine for detection of bacteria and fungi respectively.

Connect the 37 mm air monitoring cassette (disposable plastic filter cassette holding 0.45 μm membrane filters) (Fig. 20.23) to the vacuum source. Put known volume of water sample over the filter.

FIG. 20.23

A 37 mm polystyrene 3-piece monitoring cassette

Turn the vacuum source on for 10 minutes. Remove the membrane filter from the cassette with a pair of flamed forceps and place on a NA plate for **bacteria**.

Repeat the same procedure placing the membrane filter on SDA plate for **fungi**. Incubate the NA and SDA plates as described under impingement method.

Observations and calculations. Examine the incubated plates and count the number of bacterial (NA) and fungal (SDA) colonies.

Calculate the number of bacteria and fungi per cubic meter of air. The AG1-30 limiting orifice at the end of glass tube (submerged into the collection liquid), limits the amount of air passing through the liquid to 12.5 litres min^{-1}.

Micorbial concentration (as number per cubic meter of air) is calculated as follows:

$$\text{Air volume (l)} = \text{Sampling time (min.)} \times 12.5 \text{ l/min.}$$

$$\begin{array}{c}\text{No. of microbes} \\ \text{collected by impinger}\end{array} = \begin{array}{l}\text{No. of microbes} \\ \text{in volume assayed} \\ \text{(CFU/ml)}\end{array} \times \begin{array}{l}\text{ml remaining in} \\ \text{impinger after operation*}\end{array}$$

$$\begin{array}{l}\text{No. of microbes per} \\ \text{volume (l) of air} \\ \text{(CFU)}\end{array} = \frac{\text{No. of microbes collected by impinger}}{\text{Volume of air}}$$

*Volume of the liquid in the impinger may decrease during operation due to evaporation.

GLOSSARY

Abscess. A circumscribed pus-filled lesion characteristic of staphylococcal skin disease.

Acidophiles. Microorganisms that show a preference for growth at low p^H, as some bacteria grow at very low p^H values, ca 2.0.

Actinomycetes. Bacteria in which species are characterized by the formation of branching and/or true filaments.

Acyclovir (Zovirax). A drug used as a topical ointment for herpes simplex and injected for herpes encephalitis.

Adjuvant. A substance such as aluminium sulphate that increases the efficiency of a vaccine.

Aerobic. Using oxygen as a terminal electron acceptor.

Aerosol. A fine suspension of particles or liquid droplets sprayed into the air.

Agar. A derivative of marine seaweed used as a solidifying agent in many microbiological media.

Agglutination. A type of antigen-antibody reaction that results in visible clumps of organisms or other material.

Agglutinins. Antibodies that participate in agglutination reactions.

Agranulocytes. The destruction of neutrophils (granulocytes) resulting from the reaction of antibodies with antigens on the neutrophil surface, a form of type II hypersensitivity.

Alginate. A carbohydrate thickening agent used in ice cream, soups and other foods, industrially produced by microorganisms.

Allergen. An antigenic substance that stimulates an allergic reaction in the body.

Alloantigens. Antigens that exist in certain but not all members of a given species, examples are the A,B and Rh factors in humans.

Allochthonous. An organism or substance foreign to a given system.

Allograft. A tissue graft between two members of the same species, such as between two humans.

Anaphylaxis. A life-threatening allergic reaction in which series of mediators cause contractions of smooth muscle throughout the body.

Antibody. A highly specific protein molecule produced by plasma cells in the immune system, antibodies function in humoral immunity.

Antigen. Any chemical substance that elicits a response by the body's immune system.

Antiglobulin antibody. Any antibody that reacts with human antibodies.

Antitoxins. Antibodies that circulate in the blood stream and provide protection against toxins by neutralizing them.

Autoantigens. A person's own proteins and other organic compounds that elicit a specific response in the body.

Autochthonous. From within. Indigenous organism or substance.

Autograft. Tissue taken from one part of the body and grafted to another.

Autotroph. An organism that can synthesise all of its needed energy from incrganic sources.

Bacteriocins. A group of bacterial proteins toxic to other bacteria.

Bioaccumulation. General term used for a process by which chemicals are taken up by organisms from ambient medium directly or through consumption of food containing the chemicals i.e via. Food chain.

Bioconcentration. A process by which there is a net accumulation of a chemical directly from the ambient medium into organisms resulting from simultaneous uptake and elimination.

Bioconversion. A biocatalyst-mediated conversion of one substance to another.

Biodegradation. The process of chemical breakdown of a substance to smaller products caused by microorganisms or their enzymes.

Biodeterioration. The chemical or physical alteration of a product that decreases the usefulness of that product for its intended purpose, caused by microorganisms or their enzymes.

Biological oxygen demand (BOD). Amount of oxygen in any medium (water) required to oxidise organic residues.

Biomagnification. An increase in the concentration of a chemical, such as pesticide, as the substance is passed to higher members of a food chain. See also Bioaccumulation.

Biotransformation. Enzyme-catalysed conversion of one chemical, other than the normal body constituents of live organisms, into another. Normal metabolism refers to such conversions restricted to carbohydrates, fats, proteins etc. taking place inside the body.

Blanching. A process in which food is subjected to steam for two to five minutes in order to destroy cellular enzyme, and enhance preservation.

Cancer. A condition characterized by the radiating spread of cells that reproduce at an uncontrolled rate.

Corcinogens. Cancer-causing substances.

Cellular immunity. Immunity arising form the activity of T-lymphocytes on or near the body cells, also called tissue immunity and cell-mediated immunity.

Chancre. A circular purplish hard ulcer with a raised margin that occurs during primary syphilis.

Chimera. A plasmid engineered to contain a fragment of foreign DNA.

Clone. A collection or colony of identical cells arising from a single cell.

Coliforms. Gram-negaive, lactose-fermenting, enteric rods as *E. coli.*

Colostrum. The first milk secreted from the mammary gland of animals or humans.

Complement. A group of proteins that functions in a cascading series of reactions during the response by the body to certain antigens, the complement cascade is stimulated by antigen-antibody activity.

Cosmid. Phage-plasmid artificial hybrids.

Disease. Any change from the general state of good health.

Disinfectant. A chemical used to kill pathogenic microorganisms on a lifeless object such as a table top.

Droplets. Airborne particles of mucus and sputum from the respiratory tract that contain disease organisms.

Eaton agent. An alternative name for *Mycoplasma pneumaniae.*

Effluent. The liquid discharge from sewage treatment and industrial plants.

Edema. A swelling of the tissues brought about by an accumulation of fluid.

Endotoxin. A metabolic poison produced chiefly by Gram-negative bacteria, endotoxins are part of the bacterial cell wall and consequently, are released on cell disintegration, they are composed of lipid polysaccharide-peptide complexes.

Enterotoxin. A toxin that is active in the gastrointestinal tract of the host.

Enterovirus. A virus that infects intestinal cells.

Episome. A plasmid attached to the chromosome of a bacterium.

Erythema. A zone of redness in the skin due to accumulation of blood.

Exotoxin. A metabolic poison produced chiefly by Gram-positive bacteria, exotoxins are released to the environment; they are composed of protein and affect various organs and systems of the body.

Fab fragment. The portion of the antibody molecule that combines with determinant sites of the antigen.

Fe fragment. The portion of the antibody molecule that combines with phagocytes, viral receptor sites and complement.

Fermentation. Anaerobic respiration in which intermediaries in the process are used as electron acceptors, also refers to the industrial use of microorganisms.

Fimbriae. Short, hairlike structures used by bacteria for attachment, sometimes used as an alternative expression for pili.

Fomites. Inanimate objects such as clothing or utensils that carry disease organisms.

Fungemia. Dissemination of fungi through the circulatory system.

Gasohol. A mixture of gasoline and ethanol used as a fuel.

Genome. The nucleic acid core of the virus.

Gonococcus. A colloquial expression for *Neisseria gonorrhoeae.*

Halophiles. Organisms requiring NaCI for growth; extreme halophiles grow in concentrated brines.

Hansen's disease. An alternative name for leprosy.

Hemagglutination. The agglutination of red blood cells.

Hemagglutinin. An enzyme on the surface spikes of certain influenza viruses that allows the virus to bind to red blood cells.

Hemolysins. Enzymes that dissolve red blood cells, produced by streptococci, staphylococci, gas gangrene bacilli, and other microorganisms.

Humoral immunity. Immunity arising from the activity of antibodies directed against antigens in the blood stream.

Humus. The organic portion of the soil remaining after microbial decomposition.

Hybridoma. A mass of cells produced by the fusion of myeloma cells with antigen-stimulated plasma cells, produces monoclonal antibodies.

Icosahedron. A symmetrical figure composed of 20 triangular faces and 12 points, one of the major shapes taken by the virus.

Imidazoles. A group of antifungal drugs that interfere with sterol synthesis in fungal cell membranes, includes miconazoles and ketoconazoles.

Immobilised enzyme. An enzyme bound to a solid support.

Immunoglobulin. An alternative term for antibody.

Inducer. A substance that may activate the operon of the cell by combining with and negating the repressor protein.

Inflammation. A non-specific defensive response to injury, usually characterized by red color from blood accumulation, warmth from the heat of blood, swelling from fluid accumulation, and pain from injury to local nerves.

Interferon. An antiviral protein produced by body cells on exposure to viruses; interferon triggers production of a second protein that binds to mRNA coded by the virus and thereby inhibits viral replication.

Interleukins. Lymphokines produced by white blood cells that act on other white blood cell, important in cellular immunity.

Iodophores. Complexes of iodine and detergents that release iodine.

Isograft. Tissue taken from an identical twin and grafted to the other twin.

Job's syndrome. An immune disorder characterized by defective chemctaxis between phagocyte and microorganism.

Killer T-lymphocyte. A type of T-lymphocyte that attracts and destroys cells altered by the presence of antigens; important in the destruction of cancer cells also called a killer cell.

Klebs-Loffler bacillus. A common term for *Corynebacterium diphtheriae.*

Koch-Weeks bacillus. An alternative name for *Haemophilus aegypticus*

Leukemia. A cancer of the white blood cells.

Leukocidin. An enzyme that destroys phagocytes thereby preventing phagocytosis of the parasite.

Lymphoblast. The young cell to which the T-lymphocyte reverts, lymphoblasts secrete lymphokines.

Lymphocyte. A type of leukocyte that functions in the immune system.

Lymphokines. Proteins that increase the efficiency of phagocytosis at the antigen sites in cellular immunity.

Lymphopoietic cells. Primitive cells that arise from stem cells and are modified to form B-lymphocytes or T-lymphocytes.

Lysogeny. The phenomenon in which a virus remains in the cell cytoplasm as a fragment of DNA or attaches to the chromosome, but fails to replicate in or destroy the cell.

Macrophages. Large cells derived from monocytes and found within the tissues, macrophages actively engulf foreign bodies and comprise the reticuloendothelial system (RES).

Magnetosome. A cytoplasmic body in certain bacteria that assists orientation to the environment by aligning with the magnetic field.

Malt. Digested barley grain used in beer fermentations.

Mast cells. Connective tissue cells to which IgE fixes in type I hypersensitivity reactions, the cells degranulate and release histamine during allergic attacks.

Memory cells. Cells derived from B-lymphocytes or T-lymphocytes that react rapidly upon the future recurrence of antigens in the tissues.

Meningitis. A general term for infection of the meninges due to any of several bacteria, fungi, viruses or protozoa.

Mesophiles. Organisms that grow at the temperature range of 20-40°C.

Microorganism. A microscopic form of life including bacteria, viruses, fungi, protozoa and some multicellular parasites.

Mole. The quantity of a substance whose weight in grams is numerically equivalent to the molecular weight of that substance.

Monocyte. A leukocyte with a large bean-shaped nucleus, functions in phagocytosis.

Mordant. A substance that enhances the binding and stability of a dye to stained material.

Mycoplasmas. A group of tiny bacteria that lack cell walls and are seen only with the electron microscope.

Myeloma. A mass of cancerous cells.

Nanometer. A unit of measurement equivalent to one billionth of a meter, the unit is designated as nm and is often used in measuring viruses and the wavelength of energy.

Neoplasm. An uncontrolled growth of cells, often called a tumour.

Neurotoxin. A toxin that is active in the nervous system of the host.

Neutralization. A type of antigen-antibody reaction in which the activity taking place between reactants is not visible.

Night soil. Human feces sometimes used as an agricultural fertilizer.

Okazaki fragments. Segments of DNA that combine with one another to form a DNA molecule during chromosomal duplication.

Oncogene. A region of DNA in human cells thought to induce uncontrolled growth of the cell if permitted to function.

Oncology. The study of tumors and cancers.

Operon. The unit of gene activity that expresses a particular trait, also the unit that controls protein synthesis.

Opsonins. Antibodies or complement components that encourage phagocytosis.

Opsonization. Enhanced phagocytosis due to the activity of antibodies or complement.

Osmophiles. Oraganisms that grow best or only in or on media of relatively high osmotic pressure.

Papilloma. A tumour of the skin tissue.

Pasteurization. A heating process that destroys pathogenic bacteria in fluid such as milk and lowers the overall number of bacteria in the fluid.

pH. An abbreviation for the negative logarithm of the amount of hydrogen ion concentration in 1litre of solution, the pH scale extends from 1 to 14 and indicates the degree of acidity or alkalinity of a solution.

Phagocyte. A cell that practices phagocytosis.

Phagocytosis. A process in which solid particles are taken into the cell, important in nutritional processes and in defence against disease.

Phagosome. A vesicle that contains particles of phagocytized material.

Picornavirus. A small virus containing RNA in its genome.

Pili. Short, hairlike appendages of bacteria that anchor the cell to a surface, pili are also involved in conjugation between bacteria.

Pinocytosis. A type of phagocytosis in which materials dissolved in fluid are taken into the cell.

Plaque. A clear area on a lawn of bacteria where viruses have destroyed that bacteria, also the gummy layer of gelatinous material consisting of bacteria and organic matter on the teeth.

Plasmid. A small, closed loop molecule of DNA apart from the chromosome, plasmids carry genes for drug resistance and pilus formatron and are used in genetic engineering experiments.

Potable water. Water fit to drink.

Precipitins. Antibodies that participate in precipitation reactions.

Prions. Infectious particles of protein, possibly involved in human diseases of brain.

Properdin. A protein that functions in the alternative pathway of complement activation.

Prophage. The DNA segment of a temperate phage.

R factors. Plasmids that occur frequently in Gram-negative bacteria and carry genes for drug resistance.

Reagin. An alternative name for the IgE that stimulates anaphylaxis in the body.

Recalcitrant. Resistance to degradation and metabolism by microbes.

Repressor protein. A protein that inhibits the activity of certain genes, lysogeny is established when repressor protein is produced under direction of a virus.

Resolving power. The numerical value of a lens system that indicates the size of the smallest object that can be seen clearly when using that system.

Retrovirus. The RNA virus that uses reverse transcriptase to synthesise DNA from RNA.

Reverse transcriptase. An enzyme that synthesizes a DNA molecule from the code supplied by a RNA molecule.

Ropy bread. Bread that has become soft and stringy due to capsular material deposited by a bacterium such as *Bacillus subtilis.*

Rum. Distilled spirits produced by fermentation of molasses.

Sabin vaccine. A type of polio vaccine prepared with attenuated viruses, the vaccine is taken orally.

Sake. A type of rice beer produced primarily in the Orient.

Sarcoma. A tumour of the connective tissues.

Septicemia. A generalized bacterial infection of the blood stream due to any of several organisms including streptococci and staphylococci, once known as blood poisoning.

Serology. The branch of immunology that studies serological reactions.

Serotype. A rank of classification below the species level based on an organism's reaction with antibodies in serum, used for several bacteria, especially *Salmonella.*

Sewage. The refuse liquids or waste matter carried by sewers.

Sexduction. A process of recombination in which chromosomal genes pass from a donor cell to a recipient cell while attached to the F factor.

Siderophores. Iron chelators, produced by some microbes when grow or grown in iron-deficient media, that solubilise ferric hydroxide making soluble iron available.

Silage. A type of animal feed produced by fermenting grains and other plants in silos, the huge cylindrical structures that often stand next to barns.

Sludge. The solid fraction of sewage.

Starter culture. A quantity of bacteria added to milk in the industrial production of dairy products.

Stem cell. A primordial cell of the bone marrow from which hemapoietic and lymphopoietic cells develop.

Sterilization. The removal of all the life forms, especially bacterial spores.

Stormy fermentation. Fermentation and curdling of milk accompanied by gas accumulation that forces the curd apart.

Suppressor T-lymphocyte A T-lymphocyte that interferes with the activity of B-lymphocyte.

Synthetic vaccine. A vaccine that contains chemically synthesized parts of microorganisms, such as proteins normally found in viral capsids.

Target cells. Cells to which IgG fixes during type II hypersensitivity.

Thermal death point. The temperature required to kill an organism in a given length of time.

Thermophiles. Organisms that grow at a high temperature ranges of 40° to 90°C.

Tissue typing. An immunological procedure used to locate compatible tissue types for transplantation.

Titer. The most dilute concentration of antibody that will yield a positive reaction with specific antigen, a method of expressing the amount of antibody in a sample of serum.

T-lymphocyte. A lymphocyte that is modified in the thymus gland and is associated with the system of cellular immunity, also called a T-cell.

Toxin. A poisonous substance produced by a species of microorganism, bacterial toxins are classified as exotoxins or endotoxins.

Toxoid. An immunizing agent produced from an exotoxin that elicites antitoxin production by the body.

Transduction. A type of bacterial recombination in which a virus transports fragments of DNA from a donor cell to a recipient cell.

Transposon. A segment of DNA that moves from one site on a DNA molecule to another site, transposons carry information for protein synthesis, also known as jumping genes.

Ultrapasteurization. A pasteurization process in which milk is heated at 82°C for 3 seconds.

Vaccine. Any antigenic preparation administered to stimulate the recipient's immune defense mechanisms with respect to given pathogens or toxic agents.

Vaccinia. The alternative name for cow-pox.

Vaginitis. A general term for disease of the vagina.

Varicella. An alternative name for chicken pox, means "little vessel", a reference to small chickenpox lesions.

Varicella-Zoster immune globulin (VZIG). A preparation of purified antibodies from blood donors that give some protection to chicken pox.

Vector. A living organism that transmits the agents of disease.

Virion. A completely assembled virus outside its host cell.

Viroids. Tiny fragments of nucleic acid associated with certain plant diseases, possibly associated with animal disease.

V-Z virus. The name given to the virus that causes varicella (chicken pox) and herpes zoster (shingles).

Wandering cells. Cells of the reticulo-endothelial system that move about actively within the tissues.

Whisky. Distilled spirits produced by fermentation of malted cereal grains.

Wort. The fluid portion of mashed barley grain used in beer production.

Xenobiotic. A synthetic product not formed by natural biosynthetic processes; a foreign substance or poison.

Xenograft. A tissue graft between members of different species such as between an animal and a human.

Yeast. A type of fungus that is unicellular and resembles bacteria in culture.

Zoonosis. An animal disease that may be transmitted to humans.

Zymogenous. Term used to describe soil microbes that grow rapidly on exogenous substrates.

RECOMMENDED READINGS

APHA 1998. *Standard Methods for the Examination of Water and Wastewater.* 20th edn. American Public Health Association, Washington, DC.

Atlas, R.M. and Bartha, R. 1993. *Microbial Ecology.* 3rd edn. Benjamin Cummings, Redwood City.

Berg, G. 1987. *Methods for Recovering Viruses from the Environment.* CRC Press, Boca Raton.

Bitton, G. 1999. *Wastewater Microbiology.* 2nd edn. Wiley-Liss, New York.

Burge, H.A. 1995. *Bioaerosols.* Lewis Publishers, Boca Raton.

Burlage, R.S., Atlas, R., Stahl., D., Gessey, G. and Sayler, G. (eds.). 1998. *Techniques in Microbial Ecology.* Oxford University Press, Oxford.

Burris, R.H. and Evans, H.U. (eds). 1992. *Biological Nitrogen Fixation.* Chapman & Hall, London.

Chapelle, F.H. 1993. *Groundwater Microbiology and Geochemistry.* John Wiley & Sons, New York.

Cockerham, L.G. and Shane, B.S. 1994. *Basic Environmental Toxicology.* CRC Prees, Boca Raton.

Curl, E.R. and Truelove, B. 1986. *The Rhizosphere.* Springer-Verlag, Berlin.

Edwards, C. (ed). 1990. *Microbiology of Extreme Environments.* Open University Press, Milton Keynes.

Ehrlich, H.L. 1996. *Geomicrobiology.* 3rd edn. Marcel Dekker, New York.

Ford, T.E. (ed). 1993. *Aquatic Microbiology; An Ecological Approach.* Blackwell Scientific Publications, Oxford.

Gregory, P.H. 1973. *The Microbiology of the Atmosphere.* Leonard Hill Books, Aylesbury, Bucks.

Haas, C.N., Rose, J.B. and Gerba, C.P. 1999. *Quantitative Microbial Risk Assessment.* John Wiley & Sons, New york.

Heitz, E., Flemming, H.C. and Sand, W. (eds). 1996. *Microbially Influenced Corrosion of Materials.* Springer-Verlag, Berlin.

Hughes, M.N. and Poole, R.K. 1989. *Metals and Microorganisms.* Chapman & Hall, London.

Hurst, C.D. (ed.). 1996. *Modelling Disease Transmission and its Prevention by Disinfection.* Cambridge University Press, Cambridge.

Hurst, C.J., Crawford, R.L., Knudsen, G.R., McInerney, M.J. and Stetzenbach, L.D.(eds). 2003. *Manual of Environmental Microbiology.* 2nd edn. ASM Press, Washington, DC.

Hurst, C.J., Knudsen, G.R., McInerney, M.J., Stetzenbach, L.D. and Walter, M.V. (eds). 1997. *Manual of Environmental Microbiology.* ASM Press, Washington, D.C.

Jay, J.M. 1996. *Modern Food Microbiology.* 5th. edn. Chapman & Hall, London.

Killham, K. 1994. *Soil Ecology.* Cambridge University Press, Cambridge.

Lappin-Scott, H.M. and Costerton, J.W. (eds). 1995. *Microbial Biofilms.* Cambridge University Press, Cambridge.

Lovelock, J. 1995. The *Ages of Gaia.* W.W. Norton, New York.

Maier, R.M., Pepper, I.L. and Gerba, C.P. (eds). 2000. *Environmental Microbiology.* Academic Press, London.

Marshall, K.C. 1992. *American Society Microbiology News.* **58**: 502-507.

McFeters, G.(ed). 1990. *Drinking Water Microbiology.* Springer-Verlag, Berlin.

Means, J.L. and Hinchee, R.E. (eds). 1994. *Emerging Technology for Bioremediation of Metals.* Lewis Publishers, London.

Mitchell, R. (ed). 1992. *Environmental Microbiology.* Wiley-Liss Inc., New York.

Muilenberg, M. and Burge, H. 1996. *Aerobiology.* Lewis Publishers, New York.

Norris, R.D. 1994. *Handbook of Bioremediation.* Lewis Publishers, Boca Raton.

Pepper, I.L. and Gerba, C.P. 2004. *Environmental Microbiology: A Laboratory Manual,* 2nd. edn. Elsevier Academic Press, London.

Pepper, I.L., Gerba, C.P. and Brusseau, M.L. 1996. *Pollution Science.* Academic Press, London.

Rao, V.C. and Melnick J.L. 1986. *Environmental Virology.* ASM Press, Washington, DC.

Rheinheimer, G. 1985. *Aquatic Microbiology.* 3rd. edn. John Wiley & Sons, New York.

Rochelle, P.A. (ed). 2001. *Environmental Molecular Microbiology: Protocols and Applications.* Horizon Press, Norfolk, England.

Sayler, G. (ed). 1998. *Techniques in Microbial Ecology.* Oxford University Press, Oxford.

Silver, S. and Phung, L.T. 1996. *Annual Review of Microbiology.* **50**: 753-789.

Skipper, H.D. and Turco, R.F. (eds). 1995. *Bioremediation: Science and Applications.* SSSA, Madison.

Young, L.Y. and Cerniglia, C.E. (eds). 1995. *Microbial Transformation and Degradation of Toxic Chemicals.* Wiley & Sons, New York.

Wetzel, R.G. and Likens, G.E. 1991. *Limnological Analysis.* 2nd edn. Springer-Verlag, Berlin.

INDEX